AF598456

# Gmelin Handbook of Inorganic Chemistry

8th Edition

# Gmelin Handbook of Inorganic Chemistry

**8th Edition**

**Gmelin Handbuch der Anorganischen Chemie**

Achte, völlig neu bearbeitete Auflage

| | |
|---|---|
| Prepared and issued by | Gmelin-Institut für Anorganische Chemie der Max-Planck-Gesellschaft zur Förderung der Wissenschaften<br>Director: Ekkehard Fluck |

| | |
|---|---|
| Founded by | Leopold Gmelin |
| 8th Edition | 8th Edition begun under the auspices of the Deutsche Chemische Gesellschaft by R. J. Meyer |
| Continued by | E. H. E. Pietsch and A. Kotowski, and by Margot Becke-Goehring |

Springer-Verlag Berlin · Heidelberg · New York · Tokyo 1987

STAFF GMELIN HANDBOOK

Dr. G. Bär, D. Barthel, Dr. N. Baumann, Dr. W. Behrendt, Dr. L. Berg, Dipl.-Chem. E. Best, K. D. Bonn, P. Born-Heck, Dipl.-Ing. V. A. Chavizon, E. Cloos, Dipl.-Phys. G. Czack, I. Deim, Dipl.-Chem. H. Demmer, R. Dowideit, Dipl.-Chem. M. Drößmar, U. Duggen, Dr. P. Eigen, M. Engels, Dr. H.-J. Fachmann, Dr. J. Faust, Dr. Chr. Fianda, Dr. W.-D. Fleischmann, V. Frick, G. Funk, Dipl.-Ing. N. Gagel, Dr. U. W. Gerwarth, Dipl.-Phys. D. Gras, Dipl.-Bibl. W. Grieser, Dr. R. Haubold, Dipl.-Min. H. Hein, G. Heinrich-Sterzel, A. Heise-Schuster, H.-P. Hente, H. W. Herold, U. Hettwer, Dr. W. Hoffmann, Dr. W. Huisl, Dr. S. Jäger, Dr. R. Jotter, Dr. J. von Jouanne, Dr. B. Kalbskopf, Dipl.-Chem. W. Karl, H.-G. Karrenberg, Dipl.-Phys. H. Keller-Rudek, B. Kirchner, A. Klusch, Dipl.-Phys. E. Koch, Dipl.-Chem. K. Koeber, Dipl.-Chem. H. Köttelwesch, R. Kolb, E. Kranz, Dipl.-Chem. I. Kreuzbichler, Dr. A. Kubny, Dr. M. Kunz, Dr. W. Kurtz, M. Langer, Dr. U. Lanzendörfer, Dr. B. Ledüc, Dr. A. Leonard, Dipl.-Chem. H. List, H. Mathis, E. Meinhard, Chr. Metz, K. Meyer, Dr. M. Mirbach, Dipl.-Chem. B. Mohsin, Dr. U. Neu-Becker, V. Neumann, K. Nöring, Dipl.-Chem. R. Nohl, Dipl.-Min. U. Nohl, Dr. W. Petz, I. Rangnow, Dipl.-Phys. H.-J. Richter-Ditten, Dipl.-Chem. H. Rieger, B. Riegert, E. Rieth, A. Rosenberger, E. Rudolph, G. Rudolph, Dipl.-Chem. S. Ruprecht, Dr. B. Sarbas, V. Schlicht, Dipl.-Chem. D. Schneider, Dipl.-Min. P. Schubert, Dr. B. Schwager, A. Schwärzel, Dipl.-Ing. H. M. Somer, Dr. C. Strametz, Dr. U. Taubald, M. Teichmann, Dr. D. Tille, Dipl.-Ing. H. Vanecek, Dipl.-Chem. P. Velić, Dipl.-Ing. U. Vetter, H.-M. Wagner, Dipl.-Phys. J. Wagner, R. Wagner, Dr. E. Warkentin, Dr. C. Weber, Dipl.-Chem. A. Wietelmann, Dr. M. Winter, Dr. B. Wöbke, K. Wolff

STAFF GMELIN ONLINE DATA SYSTEM

Dr. R. Froböse, Dr. P. Kuhn, Dr. G. Olbrich, Dr. B. Roth

Dr. I. Barthelmess, Dipl.-Min. M.-B. Cinarz, Dr. J. Fippinger, Dr. G. Friedrich, Dr. B. Fröhlich, M. Klöffler, R. Lohmeyer, Dr. M. Pauluth, Dr. K. Schücke

## Volumes published on "Sulfur" (Syst. No. 9)

**Sulfur A 1**
History (in German) – 1942

**Sulfur A 2**
Occurrence. Technology of Sulfur and Its Compounds. Colloidal Sulfur. Toxicity (in German) – 1953

**Sulfur A 3**
The Element. Preparation in Pure State. Properties (in German) – 1953

**Sulfur B 1**
Hydrides and Oxides of Sulfur (in German) – 1953

**Sulfur B 2**
Sulfur-Oxygen Acids (in German) – 1960

**Sulfur B 3**
Compounds (concluded) (in German) – 1963

**Sulfur-Nitrogen Compounds 1**
Compounds with Sulfur of Oxidation Number VI (in German) – 1977

**Sulfur Suppl. Vol. 1**
Thionyl Halides (in German) – 1978

**Sulfur Suppl. Vol. 2**
Sulfur Halides (in German) – 1978

**Sulfur Suppl. Vol. 3**
Sulfur Oxides (in German) – 1980

**Sulfur Suppl. Vol. 4 a/b**
Sulfanes – 1983

**Sulfur-Nitrogen Compounds 2**
Compounds with Sulfur of Oxidation Number IV – 1985

**Sulfur-Nitrogen Compounds 3**
Compounds with Sulfur of Oxidation Number IV – 1987

**Sulfur-Nitrogen Compounds 4**
Compounds with Sulfur of Oxidation Number IV – 1987 (present volume)

# Gmelin Handbook of Inorganic Chemistry

8th Edition

---

# S

# Sulfur-Nitrogen Compounds

Part 4

Compounds with Sulfur of Oxidation Number IV

With 36 illustrations


AUTHORS Norbert Baumann, Hans-Jürgen Fachmann, Brigitte Heibel, Hannelore Keller-Rudek, Alfons Kubny

EDITORS Norbert Baumann, Brigitte Heibel, Alfons Kubny

INDEX AND NOMENCLATURE Ursula Hettwer

CHIEF EDITOR Brigitte Heibel


**System Number 9**

Springer-Verlag Berlin · Heidelberg · New York · Tokyo 1987

LITERATURE CLOSING DATE: 1984
IN MOST CASES MORE RECENT DATA HAVE BEEN CONSIDERED

Library of Congress Catalog Card Number: Agr 25-1383

ISBN 3-540-93553-3 Springer-Verlag, Berlin · Heidelberg · New York · Tokyo
ISBN 0-387-93553-3 Springer-Verlag, New York · Heidelberg · Berlin · Tokyo



Typesetting, printing, and bookbinding: LN-Druck Lübeck

## Preface

The present volume describes cyclic sulfur-nitrogen compounds containing C atoms, C and O atoms, and C and P atoms in six-, seven-, eight-, and nine-membered S–N ring systems. At least one S atom can be regarded as having an oxidation number IV in one resonance structure. That means the sulfur in fact has an oxidation number higher than II and lower than VI. Compounds with an $S^{IV}$-C bond in the ring are excluded. The volume is a continuation of "Sulfur-Nitrogen Compounds" Part 3, and closes the description of cyclic $S^{IV}$–N compounds.

The compounds are arranged according to the number of atoms in the S–N ring system, and for a given number of ring atoms, in order of increasing number of heteroatoms and decreasing S:N ratio. Compounds containing two types of heteroatoms are described after those with one type. Neutral compounds are described before ions, saturated rings before unsaturated ones, and compounds not substituted at sulfur before S-substituted ones.

Especially interesting are the structure of the rings and their aromatic character, ring closure, and ring cleavage. In most cases only substituted rings are known; for the parent compound calculations were sometimes made. With respect to the „inorganic" character of Gmelin Handbook, the described compounds often must be restricted to the „simplest" examples of a ring system.

The nomenclature follows IUPAC recommendations. In contrast to the text where complicated formulas for cyclic species are illustrated, it has been necessary to write these formulas in a linear form in the formula index. These linear formulas are always added below their structural formulas in the text. Monocyclic rings are disjoined and set in brackets, e.g., [=S=N–$CH_2$–$CH_2$–$CH_2$–N=], 4,5-dihydro-3H-1,2,6-thia($S^{IV}$)diazine. Aromatic bonds are indicated by dots, e.g., [(∸S∸N∸CH∸N∸$)_2$], 1,5,2,4,6,8-dithiatetrazocine. If a condensed ring system is described, the S–N-containing ring is disjoined, and the second ring is taken as a unit in the linear formula, e.g., [∸S∸N∸{1,2-$C_6H_4$}∸S∸N∸], 1,3,2,4-benzo-dithiazine. In other cases, it also proved necessary to disjoin the second ring, e.g., [∸S∸N∸S∸N∸{C∸S∸N∸S∸C}∸N∸], 1,3,2-dithiazolo[4,5-e][1,3,2,4,7]dithiatriazepine. In many cases in the text a reduced formula is used, e.g., $S_4N_4C_2$ for the last cited compound.

In addition to the formula index for part 3 and part 4 of sulfur-nitrogen compounds, a ring index for the two volumes is given. It not only facilitates finding certain compounds, but also shows the puzzling manifold of heterocycles containing the $S^{IV}$–N groups.

Frankfurt am Main
July 1987

Brigitte Heibel

## Remarks on Abbreviations and Standards

A number of abbreviations and short names usual in the literature for **solvents** and other **chemical substances** are used, mostly without definitions, in the text and in the tables of this volume:

| | |
|---|---|
| DBU | 1,8-diaza-bicyclo[5.4.0]undec-7-ene, $C_9H_{16}N_2$ |
| DMF | dimethylformamide, $HCON(CH_3)_2$ |
| DMSO | dimethyl sulfoxide, $(CH_3)_2SO$ |
| THF | tetrahydrofuran, $C_4H_8O$ |
| TMS | tetramethylsilane, $Si(CH_3)_4$ |

**Physical data** are cited in short form using abbreviations:

| | |
|---|---|
| $D_m$ | measured density |
| $D_x$ | density calculated from X-ray data |
| m.p. | melting point; dec.: melting with decomposition |
| b.p. | boiling point, often measured at low pressure (given in Torr behind the b.p.) |
| IR | (infrared) spectrum. The medium of measurement (e.g., KBr or a solvent) is given in parentheses. For the absorption maxima (wave numbers in $cm^{-1}$), intensity and shape are abbreviated as usual: s (strong), vs (very strong), w (weak), vw (very weak), vvw (very very weak), m (medium), br (broad), sh (shoulder) |
| Ra | (Raman) spectrum |
| UV | (ultraviolet) spectrum includes the visible spectrum. The medium of measurement is set in parentheses. In addition to $\lambda_{max}$ (wavelength of absorption maximum in nm) in many cases ε (extinction coefficient in $L \cdot mol^{-1} \cdot cm^{-1}$) is given |
| NMR | (nuclear magnetic resonance) spectra. Solvent and standard are given in parentheses. For δ (chemical shift in ppm) downfield shift is indicated by a positive sign as recommended by JUPAC. Standard substances (if not otherwise cited) are TMS ($=Si(CH_3)_4$) as internal standard for $^1H$ and $^{13}C$ NMR, $CFCl_3$ as internal standard for $^{19}F$ NMR, and 85% $H_3PO_4$ as external standard for $^{31}P$ NMR. The multiplicity of the signal is given in parentheses: s (singlet), d (doublet), dd (double doublet), t (triplet), q (quartet), quint (quintet), sept (septet), m (multiplet). The assignment is given behind the multiplicity; if necessary for clarity the assigned atoms are underlined<br>J (coupling constant in Hz) is fixed by the number of bonds involved and the coupling nuclei (e.g., $^1J(^{13}C, H)$), or by listing the coupling groups (e.g., $J(CH_2, CH_3)$) |
| MS | (mass spectrum). m/e = mass/charge; $M^+$ = molecular ion (1+) |
| PE | (photoelectron) spectrum. Ionization energy $E_i$ in eV |
| ESR | (electron spin resonance) spectrum; g value; hyperfine coupling constant a in G |

# Table of Contents

# 9 Six Atom S–N–C, S–N–C–O, and S–N–C–P Ring Systems

## 9.1 $S_2N_3C$ Ring

### 9.1.1 1,3,2,4,6-Dithiatriazine and Derivatives

R = H, F, Cl, $NH_2$, $CH_3$, $CF_3$, $C_6H_5$

**[∸(S∸N)$_2$∸CR∸N∸]** (Reduced formula in the text $S_2N_3CR$)

Most of the monomeric ring compounds were treated only theoretically. Experimental data exist only for $S_2N_3C(CF_3)$ and for norbornadiene adducts, e.g. $S_2N_3C(C_6H_5) \cdot C_7H_8$ (see Fig. 3, p. 5).

#### 9.1.1.1 $S_2N_3CH$

The parent compound, $S_2N_3CH$, of this 8$\pi$-electron heterocyclic system has not been synthesized until now, but a series of MNDO (modified neglect of diatomic overlap) calculations was carried out using a $C_{2v}$ symmetry restriction. Calculations yield heats of formation that favor the triplet state ($b_1^2$, $a_2^2$, $b_1^2$, $a_2^1$, $b_1^1$) over the singlet state ($b_1^2$, $b_1^2$, $a_2^2$, $a_2^2$) by 15.3 kcal/mol. $\pi$-orbital energies and distributions for these two states of $S_2N_3CH$ are given in the paper [1].

On the basis of a quantum chemical HFS–SCF MO analysis, it was not possible to obtain a singlet ground state for a planar structure [3, 4].

The coupling of two molecules with triplet configuration results in the formation of a singlet ground state dimer (see p. 2) [1].

#### 9.1.1.2 $S_2N_3C(CF_3)$

The heterocycle forms in a small yield from the dechlorinating reduction of $S_2N_3CCl_2(CF_3)$, see p. 10, with equimolar amounts of Zn in liquid $SO_2$ at t $\leqq$ 0°C. The mass spectrum (electron impact, 70 eV) shows the molecular peak at m/e = 187 $M^+$ (100% relative abundance) and the fragments m/e = 173 $S_2N_2CCF_3^+$ (17%), 127 $SNCCF_3^+$ (16%), 92 $S_2N_2^+$ (79%), 78 $S_2N^+$ (44%), 69 $CF_3^+$ (91%), 64 $S_2^+$ (24%), and 50 $CF_2^+$ (42%) [2].

#### 9.1.1.3 $S_2N_3CR$ (R = F, Cl, $NH_2$, $CH_3$, $C_6H_5$)

The compounds have not been synthesized until now. Attempts to prepare $S_2N_3C(C_6H_5)$ yielded the dimer $(S_2N_3C(C_6H_5))_2$ (see p. 2).

MNDO calculations using a $C_{2v}$ symmetry restriction yield heats of formation for $S_2N_3CR$ with R = F, $NH_2$, $C_6H_5$ that favor the triplet state ($b_1^2$, $a_2^2$, $b_1^2$, $a_2^1$, $b_1^1$) over the singlet state ($b_1^2$, $b_1^2$, $a_2^2$, $a_2^2$) by 12 to 15 kcal/mol [1]. From a quantum chemical HFS–SCF MO analysis it was not possible to obtain a singlet ground state for a planar structure of $S_2N_3CR$ with R = Cl and $CH_3$ [3, 4], whereas for $S_2N_3CR$ with R = $NH_2$ a singlet ground state was obtained [4].

**References:**

[1] Boeré, R. T., French, C. L., Oakley, R. T., Cordes, A. W., Privett, J. A. J., Craig, S. L., Graham, J. B. (J. Am. Chem. Soc. **107** [1985] 7710/7).

[2] Höfs, H.-U., Hartmann, G., Mews, R., Sheldrick, G. M. (Z. Naturforsch. **39b** [1984] 1389/92).

[3] Treu Jr., O., Trsic, M. (J. Mol. Struct. **133** [1985] 1/9).

[4] Chivers, T., Edelmann, F., Richardson, J. F., Smith, N. R. M., Treu Jr., O., Trsic, M. (Inorg. Chem. **25** [1986] 2119/25).

## 9.1.2 Dimeric 1,3,2,4,6-Dithiatriazine and Derivatives

### 9.1.2.1 $(S_2N_3CH)_2$

The parent dimer has not been synthesized until now. MNDO calculations on a model $(S_2N_3CH)_2$ dimer have been carried out using a pair of coplanar $S_2N_3CH$ (see p. 1) units in the triplet ground state. The net "bond" between the two halves arises from the in-phase overlap of the $3b_1$ and $2a_2$ orbitals on each ring, see **Fig. 1** [1].

### 9.1.2.2 $(S_2N_3C(C_6H_5))_2$

The title compound was formed when a solution of $Sb(C_6H_5)_3$ in $CHCl_3$ was added dropwise to a stirred solution of $S_2N_3CCl_2(C_6H_5)$ (see p. 11) (1:1 mole ratio) in $CHCl_3$. A buff-colored, microcrystalline precipitate of $(S_2N_3C(C_6H_5))_2$ was immediately produced in a 78% yield [1, 2].

The crystal structure was obtained by X-ray diffraction. The small, extremely thin platelets are triclinic, space group $P\bar{1}$-$C_i^1$ (No. 2) with a = 6.093(2), b = 10.874(4), c = 12.458(5) Å, $\alpha$ = 109.13(3)°, $\beta$ = 100.59(3)°, $\gamma$ = 92.52(3)°; Z = 2. V = 762(1) $Å^3$. R = 0.051 for 1327 reflections. Fractional atomic coordinates and isotropic thermal parameters are given in the paper [1].

The crystals consist of dimeric $(S_2N_3C(C_6H_5))_2$ units stacked in a head-to-tail pattern, with the stacks lying approximately along the yz diagonal. Adjacent dimer units are related by a center of symmetry and are equally spaced at a phenyl-ring-center to $S_2N_3C$-ring-center distance of 3.65 Å, a distance which is only slightly longer than the sum of the van der Waals radii of sulfur and a phenyl ring (3.55 Å).

The molecule (shown in **Fig. 2**, p. 4) consists of a cofacial arrangement of two $S_2N_3C$ rings, each of which adopts a half-chair conformation, the unique nitrogens N(5) and N(6) being tipped out of the plane of the remaining five endocyclic atoms, which are both coplanar to within 0.02 Å, to produce dihedral angles of 16.0° and 16.2°. The displacements of nitrogen atoms N(5) and N(6) are in opposite directions. The two five-atom SNCNS planes are mutually inclined with a dihedral angle of 12.5°, so that while the S–S contacts are 2.524(3) and 2.530(3) Å, the phenyl ring centers are separated by 3.73 Å, and the remote C···C distance is 4.046 Å.

Selected bond lengths and bond angles are given in Table 1.

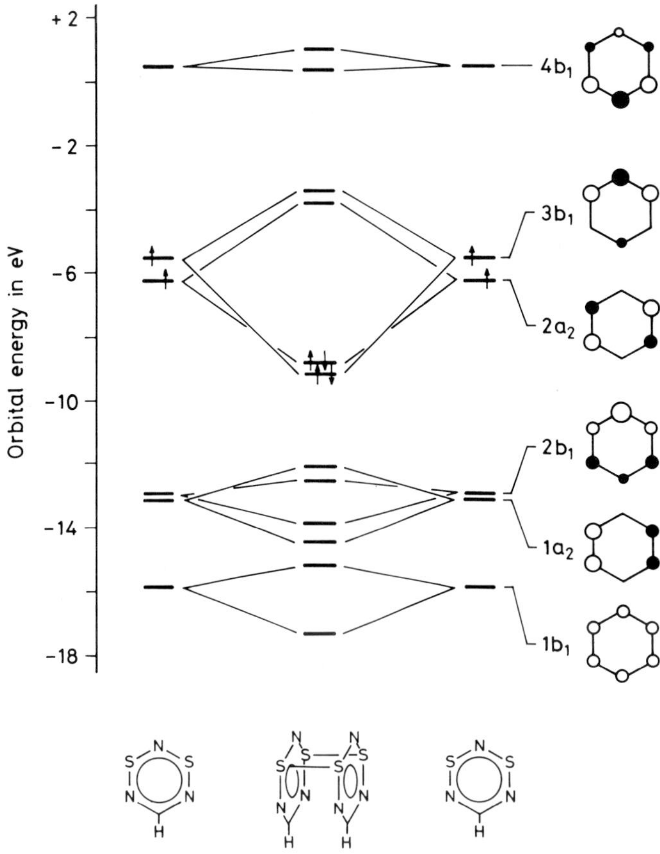


Fig. 1. Formation of singlet ground state $(S_2N_3CH)_2$ dimer by the mixing of orbitals of triplet ground state $S_2N_3CH$ monomers.

Table 1
Selected Bond Lengths and Bond Angles of $(S_2N_3C(C_6H_5))_2$.

| bond length | in Å | bond angle | in ° |
|---|---|---|---|
| S(1)–S(3) | 2.526(3) | N(1)–S(1)–N(5) | 114.6(3) |
| S(2)–S(4) | 2.532(3) | N(2)–S(2)–N(5) | 113.5(3) |
| S(1)–N(1) | 1.590(5) | N(3)–S(3)–N(6) | 113.8(3) |
| S(2)–N(2) | 1.599(5) | N(4)–S(4)–N(6) | 114.1(3) |
| S(3)–N(3) | 1.581(5) | S(2)–S(1)–S(3) | 90.78(8) |
| S(4)–N(4) | 1.600(5) | S(1)–S(2)–S(4) | 89.33(8) |
| S(1)–N(5) | 1.622(5) | S(1)–N(5)–S(2) | 116.6(3) |
| S(2)–N(5) | 1.624(5) | S(3)–N(6)–S(4) | 116.3(3) |
| S(3)–N(6) | 1.635(5) | S(1)–N(1)–C(1) | 122.5(4) |
| S(4)–N(6) | 1.620(6) | S(2)–N(2)–C(1) | 123.0(4) |

Table 1 (continued)

| bond length | in Å |
|---|---|
| N(1)–C(1) | 1.339(7) |
| N(2)–C(1) | 1.350(7) |
| N(3)–C(8) | 1.334(8) |
| N(4)–C(8) | 1.347(7) |
| C(1)–C(2) | 1.471(2) |
| C(8)–C(9) | 1.470(8) |

| bond angle | in ° |
|---|---|
| S(3)–N(3)–C(8) | 123.3(5) |
| S(4)–N(4)–C(8) | 122.4(5) |
| N(1)–C(1)–N(2) | 127.2(5) |
| N(3)–C(8)–N(4) | 127.4(6) |
| N(1)–C(1)–C(2) | 116.2(5) |
| N(2)–C(1)–C(2) | 116.6(5) |
| N(3)–C(8)–C(9) | 117.0(6) |
| N(4)–C(8)–C(9) | 115.5(6) |

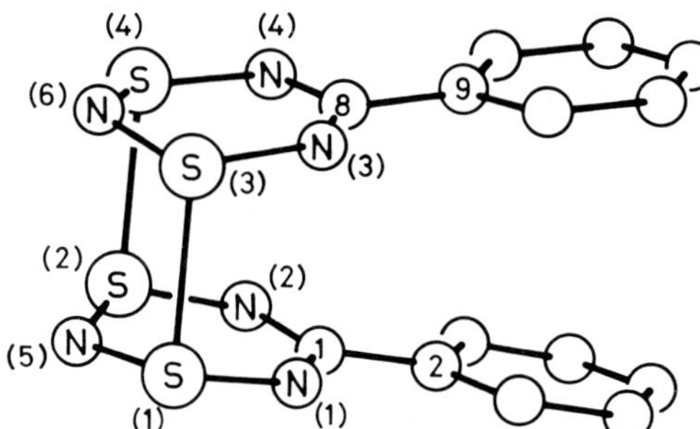


Fig. 2. Molecular structure of $(S_2N_3C(C_6H_5))_2$.

For the IR spectrum (1600 to 250 $cm^{-1}$region), 26 absorption bands are listed in the paper [1]. Mass spectrum (70 eV): m/e = 195 $S_2N_3CC_6H_5^+$ (19%), 181 $S_2N_2CC_6H_5^+$(6%), 149 $SN_2CC_6H_5^+$ (21%), 135 $SNCC_6H_5^+$ (3%), 103 $NCC_6H_5^+$ (25%), 92 $S_2N_2^+$ (12%), 78 $S_2N^+$ (11%), 76 (12%), 64 $S_2^+$ (9%), 46 $SN^+$ (100%) [1, 2].

The microcrystalline yellow solid melts at 115°C [1, 2] and decomposes on attempted sublimation at 80°C/0.01 Torr to yield, inter alia, $S_4N_4$ [1]. It is remarkably insoluble in most common organic solvents. The compound is perfectly air-stable in the solid state but appears to be extremely sensitive to oxidation in solution. There is no evidence for the dissociation of the dimer into its constituent halves. From ESR spectroscopy it was shown that dilute solutions of the compound in $CH_2Cl_2$ exhibit a five-line signal (with $a_N = 0.50$ mT) characteristic of the $NSN^-$ radical anion [1].

A suspension of the compound in $CCl_4$ reacts with $Cl_2$ during 3 h to give $S_2N_3CCl_2(C_6H_5)$, see p. 11. A slurry of $(S_2N_3C(C_6H_5))_2$ in $CH_3CN$ reacts with excess norbornadiene, $C_7H_8$, for 16 h to give the adduct $S_2N_3C(C_6H_5) \cdot C_7H_8$ in 63% yield [1, 2]. Its molecular structure is shown in **Fig. 3** [3].

**References:**

[1] Boeré, R. T., French, C. L., Oakley, R. T., Cordes, A. W., Privett, J. A. J., Craig, S. L., Graham, J. B. (J. Am. Chem. Soc. **107** [1985] 7710/7).

[2] Boeré, R. T., Cordes, A. W., Oakley, R. T. (J. Chem. Soc. Chem. Commun. **1985** 929/30).

[3] Cordes, A. W., Craig, S. L., Privett, J. A. J., Oakley, R. T., Boeré, R. T. (Acta Cryst. C **42** [1986] 508/9).

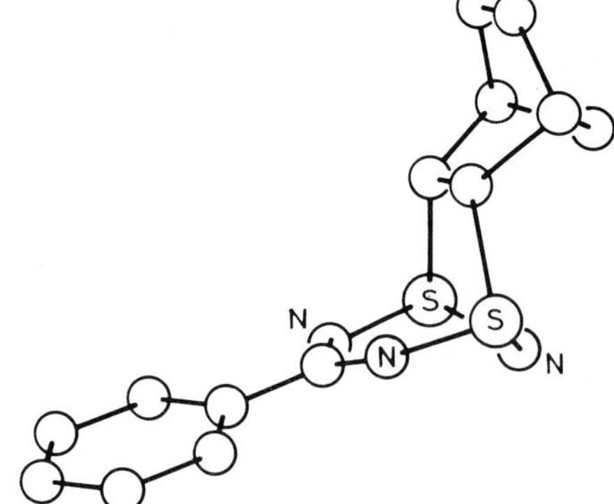


Fig. 3. Molecular structure of $S_2N_3C(C_6H_5) \cdot C_7H_8$.

## 9.1.3 Polymeric 1,3,2,4,6-Dithiatriazines

### 9.1.3.1 $(S_2N_3C(N(C_2H_5)_2))_n$

The title compound forms in 38% yield by adding a solution of $((CH_3)_3Si)_2Hg$ in $(C_2H_5)_2O$ to a vigorously stirred solution of $S_2N_3CCl_2(N(C_2H_5)_2)$ (see p. 9) (1:1 mole ratio) in $(C_2H_5)_2O$ at −78°C.

The IR spectrum (Nujol) of the purple microcrystals shows 18 bands in the range 1500 to 360 $cm^{-1}$. Mass spectrum (70 eV): m/e = 190 $S_2N_3CN(C_2H_5)_2^+$ (72%), 176 $S_2N_2CN(C_2H_5)_2^+$ (31%), 162 $S_2N_3CNHC_2H_5^+$ (16%), 144 $SN_2CN(C_2H_5)_2^+$ (17%), 119 $S_2N_2CNH^+$ (13%), 112 $N_2CN(C_2H_5)_2^+$ (12%), 98 $NCN(C_2H_5)_2^+$ (46%), 83 $NCN(CH_2)C_2H_5^+$ (72%), 78 $S_2N^+$ (61%), 72 $N(C_2H_5)_2^+$ (71%), 55 (93%), 46 $SN^+$ (100%), 42 $NCNH_2^+$ (50%).

The reaction with excess norbornadiene in $(C_2H_5)_2O$ at −20°C yields the adduct $S_2N_3C(N(C_2H_5)_2) \cdot C_7H_8$.

**Reference:**

Chivers, T., Edelmann, F., Richardson, J. F., Smith, N. R. M., Treu Jr., O., Trsic, M. (Inorg. Chem. **25** [1986] 2119/25).

### 9.1.3.2 $(S_2N_3C(N(i\text{-}C_3H_7)_2))_n$

The polymeric compound was obtained in 67% yield by adding a solution of $((CH_3)_3Si)_2Hg$ in $(C_2H_5)_2O$ to a vigorously stirred solution of $S_2N_3CCl_2(N(i\text{-}C_3H_7)_2)$ in $(C_2H_5)_2O$ (1.1:1.0 mole ratio) at −78°C and warming the reaction mixture to 0°C.

The IR spectrum (Nujol) of the purple-black microcrystals shows 16 bands in the range 1473 to 350 $cm^{-1}$. Mass spectrum (70 eV): m/e = 218 $S_2N_3CN(C_3H_7)_2^+$ (17%), 204 $S_2N_2CN(C_3H_7)_2^+$ (11%), 172 $SN_2CN(C_3H_7)_2^+$ (7%), 147 $S_2N_2CNC_2H_5^+$ (17%), 134 (14%), 130 (12%), 126 $NCN(C_3H_7)_2^+$ (26%), 111 $(C_3H_7)(C_3H_6)NC^+$ (22%), 100 $(C_3H_7)_2N^+$ (15%), 92 $S_2N_2^+$ (12%), 88 (45%), 84 $C_3H_7NCNH^+$ (42%), 78 $S_2N^+$ (22%), 69 $C_3H_7NC^+$ (82%), 64 $S_2^+$ (69%), 58 $C_3H_7NH^+$ (22%), 46 $SN^+$ (72%), 43 $H_2NCNH^+$ (100%).

Addition of excess norbornadiene to a solution of the polymeric compound in $(C_2H_5)_2O$ at −20°C and warming at 23°C for 2 h gives the adduct $S_2N_3C(N(i\text{-}C_3H_7)_2) \cdot C_7H_8$ in 57% yield.

**Reference:**

Chivers, T., Edelmann, F., Richardson, J. F., Smith, N. R. M., Treu Jr., O., Trsic, M. (Inorg. Chem. **25** [1986] 2119/25).

### 9.1.4 Salts of 5-Substituted 1-Chloro-1H,3H-1,3,2,4,6-dithiatriazine Ion (1+)

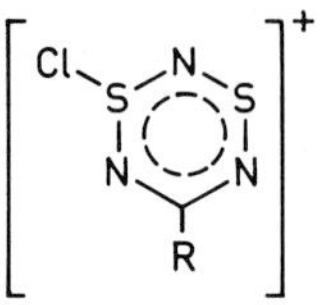

$R = N(CH_3)_2$; $X^- = SbCl_6^-$
$R = N(C_2H_5)_2$; $X^- = BCl_4^-$, $SbCl_6^-$, $\frac{1}{2}\,SnCl_6^{2-}$
$R = N(i\text{-}C_3H_7)_2$; $X^- = \frac{1}{2}\,SnCl_6^{2-}$

**[∸SCl∸N∸S∸N∸CR∸N∸]$^+$X$^-$** (Reduced formula in the text $S_2N_3CClR^+X^-$)

The salts given in the following table are prepared by dropwise addition of 2 molar equivalents of $SbCl_5$, $BCl_3$ in n-$C_6H_{14}$, and $SnCl_4$ to a solution of $S_2N_3CCl_2$ ($NR'_2$) with $R' = CH_3$, $C_2H_5$, i-$C_3H_7$ in $CCl_4$ or n-$C_6H_{14}$ at 23°C.

| salt | color | IR spectrum (Nujol), ν in cm$^{-1}$ | |
|---|---|---|---|
| | | ν($CN_2$) | ν(S–Cl) |
| $S_2N_3CCl(N(CH_3)_2)^+SbCl_6^-$ | orange | 1614 | 450 |
| $S_2N_3CCl(N(C_2H_5)_2)^+BCl_4^-$ | pale cream | 1596 | 455 |
| $S_2N_3CCl(N(C_2H_5)_2)^+SbCl_6^-$ | orange | 1589 | 449 |
| $(S_2N_3CCl(N(C_2H_5)_2)^+)_2SnCl_6^{2-}$ | orange | 1595 | 446 |
| $(S_2N_3CCl(N(i\text{-}C_3H_7)_2)^+)_2SnCl_6^{2-}$ | golden yellow | 1575 | 444 |

**Reference:**

Chivers, T., Edelmann, F., Richardson, J. F., Smith, N. R. M., Treu Jr., O., Trsic, M. (Inorg. Chem. **25** [1986] 2119/25).

### 9.1.5 5-Substituted 1,3-Dichloro-1H,3H-1,3,2,4,6-dithiatriazines

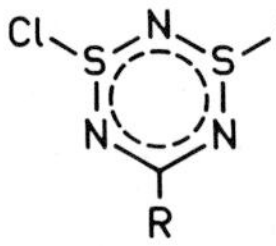

$R = N(CH_3)_2$, $N(C_2H_5)_2$, $N(i\text{-}C_3H_7)_2$, $CF_3$, $C_6H_5$

**[∸(SCl∸N)$_2$∸CR∸N∸]** (Reduced formula in the text $S_2N_3CCl_2R$)

The 5-substituted six-membered heterocycles are formed from other ring compounds: from the five-membered $S_3N_2Cl^+Cl^-$ (when reacted with $(CH_3)_2NC(=NH)NH_2 \cdot HCl$, $R = N(CH_3)_2$), from the six-membered $S_3N_3Cl_3$ (with RCN, $R = N(CH_3)_2$, $N(C_2H_5)_2$, $N(i\text{-}C_3H_7)_2$, $CF_3$) or from the nine-membered bicyclic $S_3N_5CR$ (with $Cl_2$ or $S_3N_3Cl_3$, $R = N(CH_3)_2$, $C_6H_5$).

From X-ray crystallography it was shown that the ring skeleton of the compounds with $R = N(CH_3)_2$, $N(C_2H_5)_2$, and $C_6H_5$ are nearly planar; the heterocycle with $R = CF_3$, however, has an approximate boat conformation.

### 9.1.5.1 $S_2N_3CCl_2(N(CH_3)_2)$

The title compound was prepared in 7% yield by reacting equimolar amounts of $S_3N_2Cl^+Cl^-$ and N, N-dimethylguanidine hydrochloride, $(CH_3)_2NC(=NH)NH_2 \cdot HCl$, in $CH_2Cl_2$ at room temperature for 3 h and at reflux temperature for 20 h [1]. Alternatively, the compound has been obtained in 93% yield by slow addition of a solution of dimethylcyanamide, $(CH_3)_2NCN$, in $CCl_4$ with vigorous stirring to a solution of excess $S_3N_3Cl_3$ in $CCl_4$ at ca. 65°C. After addition was complete the reaction mixture was heated at reflux for ca. 3 h, until the solution had become orange, and then cooled. The crude product was recrystallized from $CCl_4$-n-$C_5H_{12}$ (1:1) [2, 3]. The title compound is also produced by a quantitative ring contraction reaction of $S_2N_4C_2Cl_2(N(CH_3)_2)_2$ [5, 7C] (see p. 133) in $CDCl_3$ or $CH_3CN$ at 23°C [2, 3], or from the reaction of the bicyclic $S_3N_5C(N(CH_3)_2)$ (see p. 147) with equimolar amounts of $S_3N_3Cl_3$ in $CCl_4$ at 65°C [4].

Crystal data from an X-ray diffraction study are given in Table 2.

Table 2
Crystal Data of 5-Substituted 1,3-Dichloro-1H,3H-1,3,2,4,6-dithiatriazines, $S_2N_3CCl_2R$.

| R | $N(CH_3)_2$ | $N(C_2H_5)_2$ | $CF_3$ | $C_6H_5$ |
|---|---|---|---|---|
| symmetry | monoclinic | monoclinic | orthorhombic | triclinic |
| space group | $P2_1/c\text{-}C^5_{2h}$ (No.14) | $P2_1/n\text{-}C^5_{2h}$ (No.14) | $P2_12_12_1\text{-}D^4_2$ (No. 19) | $P\bar{1}\text{-}C^1_i$ (No. 2) |
| a in Å | 6.603(6) | 9.464(3) | 4.749(4) | 6.028(1) |
| b in Å | 8.345(7) | 8.921(1) | 11.015(6) | 9.985(2) |
| c in Å | 17.747(15) | 13.421(4) | 15.361(6) | 10.157(3) |
| β in ° | 112.83(6) | 108.29(1) | — | 106.73(2)* |
| V in Å³ | 901.29 | 1075.9(5) | 803.5 | 513.3(4) |
| Z | 4 | 4 | 4 | 2 |
| $D_x$ in g/cm³ | 1.718 | 1.612 | 2.134 | 1.72 |
| $R(R_w)$ | 0.062 | 0.043(0.044) | (0.056) | 0.032 |
| reflections | 1168 | 1615 | 1335 | 1547 |
| Ref. | [1] | [2] | [5] | [9] |

* $\alpha = 117.33(2)°$, $\gamma = 90.31(2)°$

The molecular structure of the compound, shown in **Fig. 4**, p. 8, consists of a nearly planar six-membered ring with an average deviation of 0.08 Å from the ring plane. The molecule has a noncrystallographic mirror plane through N(2), C(1), and N(4) perpendicular to the ring. The exocyclic N atom N(4) lies 0.12 Å below the ring plane and the two methyl C atoms C(2) and C(3) approximately 0.26 Å below the ring plane. The neighborhood of N(4) is effectively planar, indicating an $sp^2$ hybridization of the N atom and consistent with the short N(4)–C(1) distance (1.324 Å). The two Cl atoms lie on the same side of the ring. Atomic coordinates and anisotropic thermic parameters are listed in the paper [1], principal bond lengths and bond distances are given in Table 3. The yellow crystals, recrystallized from $CCl_4$-n-hexane, melt at 93°C [1]. $^1H$ NMR ($CDCl_3$/TMS): δ (in ppm) = 3.2 [1], 3.24 [2, 3], 3.25 [4] (s, $CH_3$). $^{13}C$ NMR ($CDCl_3$/TMS): δ (in ppm) = 36.7 [1, 3], 36.65 [2, 4] ($CH_3$), and 152.5 [1], 152.25 [2, 4] (C=N). For the IR spectrum (Nujol), 12 absorption bands between 1660 and 435 $cm^{-1}$ are listed in the paper [1]. The strong bands at 1572 and 438 $cm^{-1}$ have been attributed to a ring vibration $\nu(CN_2)$ and to ν(S–Cl), respectively [4]. A published mass spectrum [1] was found to be incorrect [2, 3].

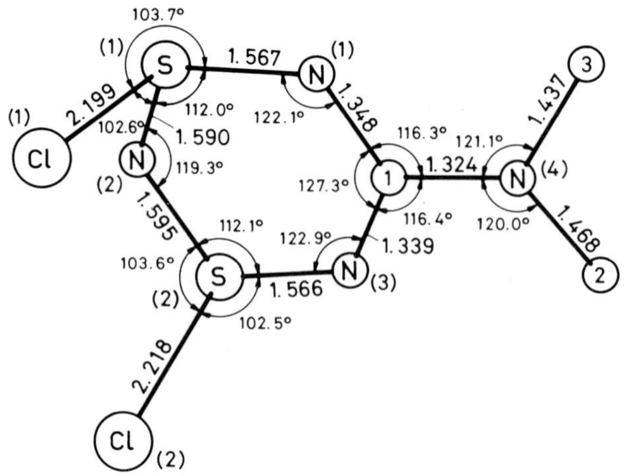


Fig. 4. Molecular structure of $S_2N_3CCl_2(N(CH_3)_2)$.
Bond lengths in Å, bond angles in °.

Table 3

Selected Bond Lengths (in Å) and Bond Angles (in °) in 5-Substituted 1,3-Dichloro-1H,3H-1,3,2,4,6-dithiatriazines, $S_2N_3CCl_2R$. The numeration of the atoms is given in Fig. 4.

| R | $N(CH_3)_2$ | $N(C_2H_5)_2$ | $CF_3$ | $C_6H_5$ |
|---|---|---|---|---|
| bond length* | | | | |
| S(1)–N(2) | 1.590(6) | 1.600(3) | 1.604(4) | 1.600(1) |
| N(2)–S(2) | 1.595(8) | 1.591(3) | 1.603(4) | 1.615(2) |
| S(2)–N(3) | 1.566(7) | 1.566(3) | 1.613(4) | 1.581(1) |
| N(3)–C(1) | 1.339(9) | 1.345(4) | 1.331(5) | 1.344(2) |
| C(1)–N(1) | 1.348(11) | 1.357(4) | 1.327(5) | 1.327(2) |
| N(1)–S(1) | 1.567(7) | 1.565(4) | 1.615(4) | 1.600(1) |
| S(1)–Cl(1) | 2.199(4) | 2.203(2) | 2.090(2) | 2.180(1) |
| S(2)–Cl(2) | 2.218(4) | 2.226(1) | 2.094(2) | 2.129(1) |
| C(1)–R(N or C) | 1.324(10) | 1.324(5) | 1.521(6) | 1.475(2) |
| bond angle | | | | |
| S(1)–N(2)–S(2) | 119.3(4) | 118.5(2) | 120.5(2) | 118.41(9) |
| N(2)–S(2)–N(3) | 112.1(3) | 113.2(1) | 110.2(2) | 110.84(8) |
| S(2)–N(3)–C(1) | 122.9(6) | 122.9(2) | 119.2(3) | 122.1(1) |
| N(3)–C(1)–N(1) | 127.3(7) | 126.6(4) | 133.3(4) | 128.4(2) |
| C(1)–N(1)–S(1) | 122.1(5) | 121.9(2) | 119.4(3) | 122.0(1) |
| N(1)–S(1)–N(2) | 112.0(3) | 113.2(2) | 110.1(2) | 112.16(8) |
| N(1)–S(1)–Cl(1) | 103.7(3) | 103.2(1) | 100.7(2) | 100.49(6) |
| N(2)–S(2)–Cl(2) | 103.6(3) | 102.5(1) | 107.4(2) | 103.56(6) |
| N(1)–C(1)–R(N or C) | 116.3(6) | 116.2(3) | 113.1(4) | 116.8(2) |
| N(3)–C(1)–R(N or C) | 116.4(7) | 117.1(3) | 113.5(4) | 114.8(2) |
| Ref. | [1] | [2] | [5] | [9] |

$S_2N_3CCl_2(N(CH_3)_2)$ reacts with $(CH_3)_3SiNSO$ (1:2 mole ratio) in $CCl_4$ at room temperature to give $S_3N_5C(N(CH_3)_2)$ (see p. 147), $(CH_3)_3SiCl$, $(CH_3)_3SiN{=}S{=}NSi(CH_3)_3$, and $S_3N_4CCl(N(CH_3)_2)$ (see p. 127). With equimolar amounts of $(CH_3)_3SiN{=}S{=}NSi(CH_3)_3$ in $CCl_4$ at room temperature for 15 h, the bicyclic $S_3N_5C(N(CH_3)_2)$ was the only isolated product [3, 4]. The reaction of $S_2N_3CCl_2(N(CH_3)_2)$ with sodium azide in $CH_3CN$ at 23°C produces an intense reddish purple solution ($\lambda_{max}$ = 520 nm), but no tractable products were isolated. On treatment with a slight excess of $SbCl_5$ in $CCl_4$, the dication $S_2N_3C(N(CH_3)_2)^{2+}$ readily forms in this solution [3].

A solution of $S_2N_3CCl_2(N(CH_3)_2)$ in $CCl_4$ or n-$C_6H_{14}$ reacts with $SbCl_5$ (1:2 mole ratio) at 23°C to give $S_2N_3CCl(N(CH_3)_2)^+SbCl_6^-$ (see p. 6). Addition of a solution of $((CH_3)_3Si)_2Hg$ in $(C_2H_5)_2O$ to a vigorously stirred solution of $S_2N_3CCl_2(N(CH_3)_2)$ in $(C_2H_5)_2O$ in the mole ratio ~1:1 at −78°C yields a purple-brown reaction mixture which on treatment with excess norbornadiene, $C_7H_8$, gives the adduct $S_2N_3C(N(CH_3)_2) \cdot C_7H_8$ in about 74% yield [10].

### 9.1.5.2 $S_2N_3CCl_2(N(C_2H_5)_2)$

The compound has been prepared similarly to $S_2N_3CCl_2(N(CH_3)_2)$ by the reaction of diethylcyanamide, $(C_2H_5)_2NCN$, with excess $S_3N_3Cl_3$ in $CCl_4$ at ca. 60°C for 3 h. The solution was then reduced to one-fourth volume and n-pentane was added. After 1 d at −20°C, opaque yellow needles of $S_2N_3CCl_2(N(C_2H_5)_2)$ were obtained in a 79% yield. Crystal data from an X-ray diffraction study are given in Table 2, p. 7. For selected bond lengths and bond angles, see Table 3; positional and isotropic thermal parameters are given in the paper [2]. The structural features are essentially the same as those reported for the dimethylamino derivative. $^1$H NMR spectrum ($CDCl_3$/TMS): δ (in ppm) = 1.28 (t, $CH_3$), 3.63 (q, $CH_2$); $^3J$ ($CH_3$, $CH_2$) = 7.2 to 7.3 Hz. $^{13}$C NMR spectrum ($CDCl_3$/TMS): δ (in ppm) = 13.31 ($CH_3$), 42.76 ($CH_2$), 151.49 ($NCN_2$) [2, 4]. For the IR spectrum (Nujol), 17 absorption bands between 1548 and 425 $cm^{-1}$ are listed in the paper [2]. The strong bands at 1550 and 425 $cm^{-1}$ have been attributed to a ring vibration $\nu(CN_2)$ and to $\nu$(S–Cl), respectively [4]. Mass spectrum (70 eV): m/e (relative intensity) = 225 $M^+$ − Cl (3%), 190 $M^+$ − 2Cl (89%), 144 (46%), 119 (48%), 98 (73%), 97 (81%), 83 (89%), 78 (87%), 72 (92%), 56 (74%), and 46 (100%) [2]. The compound reacts with equimolar amounts of $(CH_3)_3SiN{=}S{=}NSi(CH_3)_3$ in $CCl_4$ at 23°C, for 24 h to form the bicyclic $S_3N_5C(N(C_2H_5)_2)$ (see p. 147) [4]. Reaction with equimolar amounts of $NaN_3$ in $CH_3CN$ at 23°C for 2 h gives on vigorous

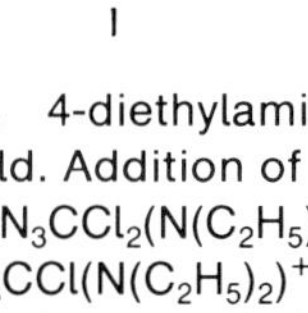

$$\left[\begin{array}{c} S{-}S \\ N \quad N \\ | \\ R \end{array}\right]^+ Cl^-$$

I

evolution of a gas, presumably $N_2$, 4-diethylamino-1,2,3,5-dithiadiazolium chloride, $S_2N_2C(N(C_2H_5)_2)^+Cl^-$ (I), in about 23% yield. Addition of the halide acceptors $BCl_3$, $SbCl_5$, and $SnCl_4$ in n-$C_6H_{14}$ or $CCl_4$ to a solution of $S_2N_3CCl_2(N(C_2H_5)_2)$ in n-$C_6H_{14}$ or $CCl_4$ (2:1 mole ratio) at 23°C yields the monocation salts $S_2N_3CCl(N(C_2H_5)_2)^+BCl_4^-$, $S_2N_3CCl(N(C_2H_5)_2)^+SbCl_6^-$, and $(S_2N_3CCl(N(C_2H_5)_2)^+)_2SnCl_6^{2-}$, respectively (see p. 6). When a solution of $((CH_3)_3Si)_2Hg$ in $(C_2H_5)_2O$ is added to a vigorously stirred solution of $S_2N_3CCl_2(N(C_2H_5)_2)$ in $(C_2H_5)_2O$ (1:1 mole ratio) at −78°C, mercury and a purple solid formulated as $(S_2N_3C(N(C_2H_5)_2))_n$ (see p. 5) form in low yield. A solution of $S_2N_3CCl_2(N(C_2H_5)_2)$ in $(C_2H_5)_2O$ reacts with a mixture of $Sb(C_6H_5)_3$ and norbornadiene, $C_7H_8$, (1:1:1.5 mole ratio) in $(C_2H_5)_2O$ at −78 to +23°C to give $(C_6H_5)_3SbCl_2$ and the adduct $S_2N_3C(N(C_2H_5)_2) \cdot C_7H_8$ in 24% yield [10].

### 9.1.5.3 $S_2N_3CCl_2(N(i\text{-}C_3H_7)_2)$

The compound was prepared analogously to $S_2N_3CCl_2(N(C_2H_5)_2)$ by the reaction of diisopropylcyanamide, $(i\text{-}C_3H_7)_2NCN$, with excess $S_3N_3Cl_3$ in $CCl_4$ at ca. 60°C for 2 h. The solution was reduced in volume before n-pentane was added. After 1 d at −20°C yellow crystals of $S_2N_3CCl_2(N(i\text{-}C_3H_7)_2)$ were isolated; yield 36% [2].

$^1H$ NMR ($CDCl_3$/TMS): δ (in ppm) = 1.39 (d, $(CH_3)_2$), 4.41 (sept, CH); $^3J(CH_3, CH) = 7.0 \pm 0.2$ Hz. $^{13}C$ NMR ($CDCl_3$/TMS): δ (in ppm) = 20.51 ($(CH_3)_2$), 48.46 (CH), 151.67 ($NCN_2$) [2, 4]. For the IR spectrum (Nujol), 21 absorption bands between 1536 and 416 $cm^{-1}$ are listed in the paper [2]. The strong bands at 1536 and 445 $cm^{-1}$ have been attributed to a ring vibration $\nu(CN_2)$ and to $\nu$(S–Cl), respectively. Mass spectrum (70 eV): m/e (relative intensity) = 253 $M^+ - Cl$ (0.5%), 218 $M^+ - 2Cl$ (41%), 172 (14%), 100 (28%), 88 (88%), 78 (20%), 69 (67%), 58 (36%), and 46 (100%).

The compound reacts with equimolar amounts of $(CH_3)_3SiN{=}S{=}NSi(CH_3)_3$ in $CCl_4$ at 23°C for 24 h to give the bicyclic $S_3N_5C(N(i\text{-}C_3H_7)_2)$ (see p. 149) [4]. Dropwise addition of $SnCl_4$ to a solution of $S_2N_3CCl_2(N(i\text{-}C_3H_7)_2)$ (2:1 mole ratio) in $CCl_4$ or n-$C_6H_{14}$ at 23°C results in the immediate formation of the salt $(S_2N_3CCl(N(i\text{-}C_3H_7)_2)^+)_2\ SnCl_6^{2-}$ (see p. 6). When a solution of excess $((CH_3)_3Si)_2Hg$ in $(C_2H_5)_2O$ is added rapidly to a vigorously stirred solution of $S_2N_3CCl_2(N(i\text{-}C_3H_7)_2)$ in $(C_2H_5)_2O$ at −78°C and the reaction mixture is allowed to warm to ~0°C, the polymer $(S_2N_3C(N(i\text{-}C_3H_7)_2))_n$ (see p. 5) is obtained in ~67% yield [10].

### 9.1.5.4 $S_2N_3CCl_2(CF_3)$

The title compound forms in a 21% yield along with 4-trifluoromethyl-1,2,3,5-dithiadiazolium chloride, $S_2N_2C(CF_3)^+Cl^-$ (see compound I, p. 9), from the reaction of $S_3N_3Cl_3$ with $CF_3CN$ (~1:3.7 mole ratio) at 50°C for 50 h in an autoclave with a teflon covering [5, 6]. Crystal data obtained from X-ray diffraction are given in Table 2, p. 7.

Selected bond lengths and bond angles are given in Table 3, p. 8. The molecular structure of $S_2N_3CCl_2(CF_3)$ is shown in **Fig. 5**. It is very similar to that of $S_3N_3Cl_3$. The two Cl atoms are in a *cis* position. The ring skeleton has an approximate boat conformation with N(2) and C(1) deviating from the best plane through S(1)S(2)N(3)N(1) by 0.37 and 0.053 Å, respectively. The S–N distances are approximately equal (1.61 Å average), and the C–N distances agree with those having been found in 1,3,5-triazines [5].

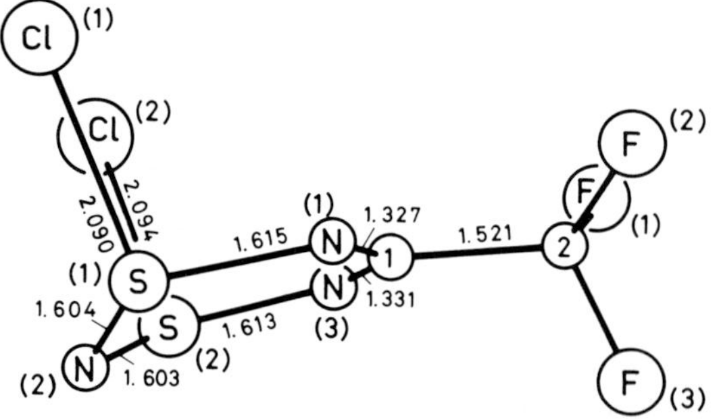


Fig. 5. Molecular structure of $S_2N_3CCl_2(CF_3)$. Bond lengths in Å.

The moisture-sensitive, light yellow solid is sublimable (25°C/$10^{-2}$ mbar). Melting point 69°C (recrystallized from $SO_2ClF$ and then resublimed), decomposition point 130°C [5].

$^{19}F$ NMR spectrum (20% in $SO_2/CFCl_3$): $\delta = -76.5$ ppm (s, $CF_3$). For the IR spectrum, 24 absorption bands between 1654 and 345 $cm^{-1}$ are listed in the paper [5]. Mass spectrum (70 eV): m/e = 257 $M^+$ (2%), 222 $M^+-Cl$ (33%), 187 $M^+-2Cl$ (100%), 173 $S_2N_2CCF_3^+$ (16%), 127 $SNCCF_3^+$ (14%), 92 $S_2N_2^+$ (78%), 78 $S_2N^+$ (40%), 69 $CF_3^+$ (83%), 64 $S_2^+$ (19%), and 50 $CF_2^+$ (34%). On heating the compound to ca. 120°C, decomposition occurs with formation of $S_2N_2C(CF_3)^+Cl^-$ (see compound I, p. 9). $S_2N_3CCl_2(CF_3)$ is dechlorinated by Zn in liquid $SO_2$ at ≦0°C with ring conservation to give $S_2N_3C(CF_3)$ (see p. 1) in a small yield or at room temperature with ring contraction to give the radical $S_2N_2C(CF_3)^\bullet$ in a 26% yield [5].

### 9.1.5.5 $S_2N_3CCl_2(C_6H_5)$

The compound was formed in a yield up to 74% when chlorine gas was passed over a slurry of $S_3N_5C(C_6H_5)$ (see p. 149) in $CCl_4$ for approximately 5 min and the resulting suspension was allowed to react for an additional 30 min. The mixture was then filtered under $N_2$ and the filtrate reduced to dryness under vacuum. The residual solid was recrystallized from $CH_2Cl_2$-hexane [7, 8, 9]. The compound can also be obtained in 85% yield by reformation of its reduction product $(S_2N_3C(C_6H_5))_2$ (see p. 2) by treatment with chlorine in $CCl_4$ [7, 8].

Crystal data obtained from X-ray diffraction are given in Table 2, p. 7. Selected bond lengths and bond angles are given in Table 3, p. 8; for fractional atomic coordinates and isotropic thermal parameters, see the paper [9]. The molecular structure is similar to that of $S_2N_3CCl_2(N(CH_3)_2)$, see Fig. 4, p. 8. The SNCNS portion of the ring is planar to within 0.060(2) Å and the third N is displaced by 0.336(2) Å from this plane on the side opposite the S-bonded Cl atoms [9]. The half-chair dihedral angle is 24.0° [8]. The S–N distances range from 1.581(1) to 1.615(2) Å and the N–C distances are 1.327(2) and 1.344(2) Å [9]. For the IR spectrum (1600 to 250 $cm^{-1}$ region), 17 absorption bands are listed in the paper [8]. Mass spectrum (70 eV): m/e = 230 $S_2N_3ClCC_6H_5^+$ (2%), 195 $S_2N_3CC_6H_5^+$ (39%), 181 $S_2N_2CC_6H_5^+$ (11%), 149 $SN_2CC_6H_5^+$ (46%), 135 $SNCC_6H_5^+$ (13%), 103 $NCC_6H_5^+$ (50%), 46 $SN^+$ (100%). The yellow, moisture-sensitive plates (from $CH_2Cl_2$-hexane) melt at 119 to 121°C.

When treated with equimolar amounts of $(CH_3)_3SiN{=}S{=}NSi(CH_3)_3$ in $CH_2Cl_2$ the compound is transformed to the bicyclic $S_3N_5C(C_6H_5)$, (see p. 149) in 78% yield. The reduction with equimolar amounts of $Sb(C_6H_5)_3$ in $CHCl_3$ yields dimeric 5-phenyl-1,3,2,4,6-dithiatriazine, $(S_2N_3C(C_6H_5))_2$ (see p. 2) [7, 8].

**References:**


[1] Roesky, H. W., Schäfer, P., Noltemeyer, M., Sheldrick, G. M. (Z. Naturforsch. **38b** [1983] 347/9).

[2] Chivers, T., Richardson, J. F., Smith, N. R. M. (Inorg. Chem. **25** [1986] 47/51).

[3] Chivers, T., Richardson, J. F., Smith, N. R. M. (Mol. Cryst. Liquid Cryst. **125** [1985] 319/27).

[4] Chivers, T., Richardson, J. F., Smith, N. R. M. (Inorg. Chem. **25** [1986] 272/5).

[5] Höfs, H.-U., Hartmann, G., Mews, R., Sheldrick, G. M. (Z. Naturforsch. **39b** [1984] 1389/92).

[6] Höfs, H.-U., Hartmann, G., Mews, R., Sheldrick, G. M. (Angew. Chem. **96** [1984] 1001/2; Angew. Chem. Intern. Ed. Engl. **23** [1984] 988/9).

[7] Boeré, R. T., Cordes, A. W., Oakley, R. T. (J. Chem. Soc. Chem. Commun. **1985** 929/30).

[8] Boeré, R. T., French, C. L., Oakley, R. T., Cordes, A. W., Privett, J. A. J., Craig, S. L., Graham, J. B. (J. Am. Chem. Soc. **107** [1985] 7710/7).

[9] Graham, J. B., Cordes, A. W., Oakley, R. T., Boeré, R. T. (Acta Cryst. C **41** [1985] 1835/6).

[10] Chivers, T., Edelmann, F., Richardson, J. F., Smith, N. R. M., Treu Jr., O., Trsic, M. (Inorg. Chem. **25** [1986] 2119/25).

## 9.2 $S_2N_2C_2$ Ring

### 9.2.1 1,3,2,4-Benzodithiadiazine

**[∸S∸N∸{1,2-$C_6H_4$}∸S∸N∸]**

The compound was formed when equimolar solutions of $C_6H_5N{=}S{=}NSi(CH_3)_3$ and $SCl_2$ in $CH_2Cl_2$ were simultaneously injected over a 7 h period into a stirred reservoir of $CH_2Cl_2$. The solvent was removed from the resulting dark green mixture in vacuum, leaving a green-black paste. Heating this residue to 50°C at 0.01 Torr yielded a dark blue sublimate which was collected on a −78°C cold finger. The sublimate was further purified by repeated sublimation in vacuum and recrystallization from pentane to give blue-black needles, m.p. 48 to 50°C, in a 33% yield [1], see also [2]. The compound can also be obtained from its norbornadiene adduct on dissolving in $CH_2Cl_2$ or $CHCl_3$ [1].

The structure of the compound has been determined by X-ray diffraction. At −115(2)°C crystals are monoclinic, space group Pc-$C_s^2$ (No. 7) with a = 5.616(1), b = 3.896(1), c = 15.434(3) Å, and β = 102.23(2)°; Z = 2; V = 330.0(1) Å$^3$, $D_x$ = 1.69 g/cm$^3$; R = 0.040, $R_w$ = 0.049 for 634 observed reflections.

The molecular structure with selected bond lengths and bond angles is shown in **Fig. 6**; for nonhydrogen atomic coordinates and isotropic thermal parameters, see the paper [1]. The molecule is planar to within 0.04 Å. The crystal structure consists of columns of parallel molecules formed by unit translations along the b axis. The uniform interplanar separation is 3.29 Å. The molecules are inclined at an angle of 57.5° to the b axis, resulting in an overlap pattern in which each ring has centered above and below it an atom from a neighboring molecule in the stack.

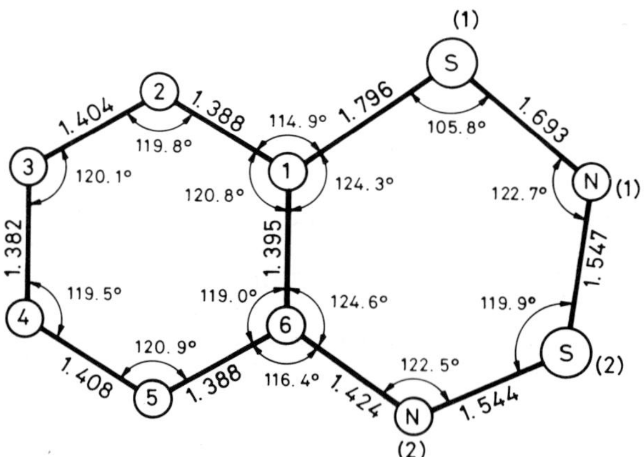


Fig. 6. Molecular structure of 1,3,2,4-benzodithiadiazine [∸S∸N∸{1,2-$C_6H_4$}∸S∸N∸]. Bond lengths in Å, bond angles in °.

An MNDO molecular orbital calculation on the 12π-electron system involving full geometry optimization within $C_s$ symmetry reveals the π-bond orders C(1)–S(1) 0.0916, C(6)–N(2) 0.0362, N(1)–S(1) 0.0224, S(2)–N(2) 0.575, and N(1)–S(2)–N(2) (av) 0.637 that provide a satisfying correlation with the experimental bond distances and a conformity with the diimide formulations A, B, and not the quinoid formulation C for the compound. The importance of these

A B C

diimide formulations can be interpreted in a molecular orbital context as reflecting a relatively weak interaction between the two "halves" of the molecule, a benzene nucleus, and an NSNS fragment. Frontier orbital (MNDO) energies (in eV) have been calculated to be E(HOMO) = −8.215, E(LUMO) = −2.286. The E(HOMO) value is consistent with the relatively low oxidation potential of the compound (see below). The LUMO energy is consistent with a remarkable susceptibility to reduction, in spite of its being an "electron-rich" $\pi$ system [1].

The $^1$H NMR spectrum (solvent?/TMS) is complicated by the absence of a twofold axis. It approximately consists of an AA'XX' pattern. The absolute distinction between the 5-, 8-, and the 6-, 7-positions remains unestablished. $\delta$ (in ppm) = 5.90 and 5.78 (H5 and H8), 6.78 and 6.62 (H6 and H7); $^3$J(H5,H6) = 8.14 Hz, $^3$J(H6,H7) = 8.2 Hz, $^3$J(H7,H8) = 7.8 Hz, $^4$J(H5,H7) = 0.9 Hz, $^4$J(H6,H8) = 1.4 Hz, $^5$J(H5,H8) = 0.2 Hz. The $\delta$ data reflect an intrinsic antiaromatic character for the compound. The relatively small ratios of the J values are consistent with a lack of bond localization in the benzene ring of the compound. $^{13}$C NMR spectrum ($CDCl_3$/TMS): $\delta$ (in ppm) = 138.8, 133.3, 130.6, 124.1, 123.8, 115.6 [1]; nearly identical values have been obtained measuring in $CDCl_3$ with $10^{-2}$ M $Cr(acac)_3$ (Hacac = acetylacetone) added [2]. For the IR spectrum (Nujol mull; 1600 to 250 $cm^{-1}$ region), 17 absorption bands are listed in the paper. UV-visible spectrum (in $CH_2Cl_2$, $\lambda_{max}$ in nm, (log $\varepsilon$)): 617 (2.7), 371 (3.0), 291 (4.3), 283 (4.3) [1, 2]; the lowest energy absorption ($\lambda_{max}$ = 617 nm) has been tentatively assigned to a $\pi_6$(HOMO) − $\pi_7$(LUMO) excitation (i.e., between orbitals localized over the NSNS fragment) [1]. Low-resolution mass spectrum (70 eV): m/e (relative intensity) = 168 $S_2N_2C_6H_4^+$ (100%), 154 $S_2NC_6H_4^+$ (12%), 136 $SN_2C_6H_4^+$ (50%), 122 $SNC_6H_4^+$ (12%), 108 $NC_6H_4^+$ (12%) [1, 2]. 1,3,2,4-Benzodithiadiazine is a volatile crystalline solid with an odor similar to that of naphthalene. It is extremely soluble in all common organic solvents and insoluble in water. It is air-stable and thermally stable and can be recovered unchanged after being heated at reflux in boiling toluene for 24 h. Electrochemical reduction and oxidation were carried out at a dropping mercury electrode and a rotating platinum electrode, respectively, in $CH_3CN$ with 0.1 M $(C_2H_5)_4N^+ClO_4^-$ as supporting electrolyte. The half-wave potentials are $E_{1/2}$(red) = −0.57 and −1.55 V and $E_{1/2}$(ox) = +1.17 V with reference to a saturated calomel electrode. The compound slowly hydrolyzes in moist air or on prolonged contact with water. On reaction with a ~tenfold excess of norbornadiene, $C_7H_8$, in ether a 1:1 adduct is formed. It's molecular structure is given in **Fig. 7**. The adduct formation causes a distortion from planarity of the thiazyl unit with N(1) undergoing a significant displacement [1].

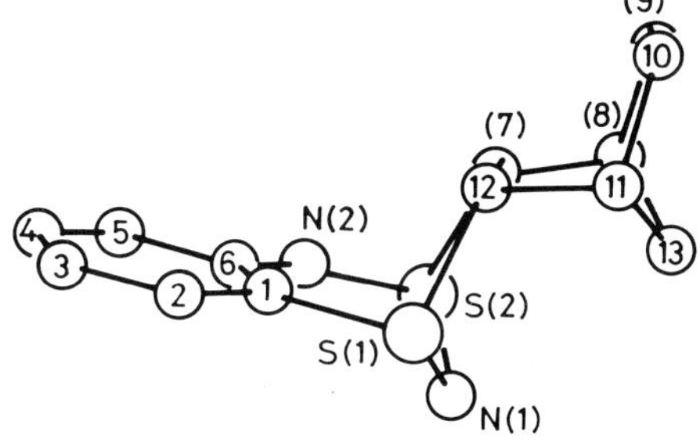


Fig. 7. Molecular structure of 1,3,2,4-benzodithiadiazine-norbornadiene adduct [$\dot{-}$S$\dot{-}$N$\dot{-}$\{1,2-$C_6H_4$\}$\dot{-}$S$\dot{-}$N$\dot{-}$]·$C_7H_8$.

**References:**


[1] Cordes, A. W., Hojo, M., Koenig, H., Noble, M. C., Oakley, R. T., Pennington, W. T. (Inorg. Chem. **25** [1986] 1137/45).

[2] Koenig, H., Oakley, R. T. (J. Chem. Soc. Chem. Commun. **1983** 73/4).


## 9.3 $SN_3C_2$ Ring

### 9.3.1 1,2,3,6-Thiatriazines and 2,1,3,4-Benzothiatriazines

#### 9.3.1.1 Derivatives of 5,6-Dihydro-2H-1,2,3,6-thiatriazine 1-Oxide and 3,5,6,7,8,8a-Hexahydro-1H-2,1,3,4-benzothiatriazine 2-Oxide

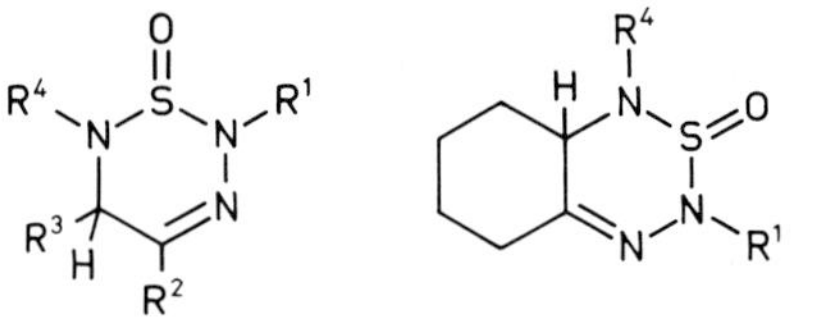

$R^1$, $R^2$, $R^3$, $R^4$ are compiled in Table 4, p. 16.

**[–SO–NR$^1$–N=CR$^2$–CHR$^3$–NR$^4$–] and [–SO–NR$^1$–N={1,2-$C_6H_9$}–NR$^4$–]**

**Formation and Properties.** The compounds, compiled in Table 4, are prepared by the following method [1, 2]: azoalkenes, $R^1$–N=N–CR$^2$=CR$^3$H, are treated in benzene at room temperature with equimolar amounts of N-sulfinyl compounds, $R^4$NSO, and the reaction mixtures are subsequently stirred for 5 h (No. 1 to 3 and 5 to 8), 1 d (No. 9, 10, 11, 13), or 5 d (No. 4 and 12), respectively. Removal of the benzene in vacuum leaves oily residues which are digested with a solvent (see Table 4). Colorless (or in the case of No. 11 ocherous) crystalline solids are obtained by recrystallization of the crude product from the same solvent.

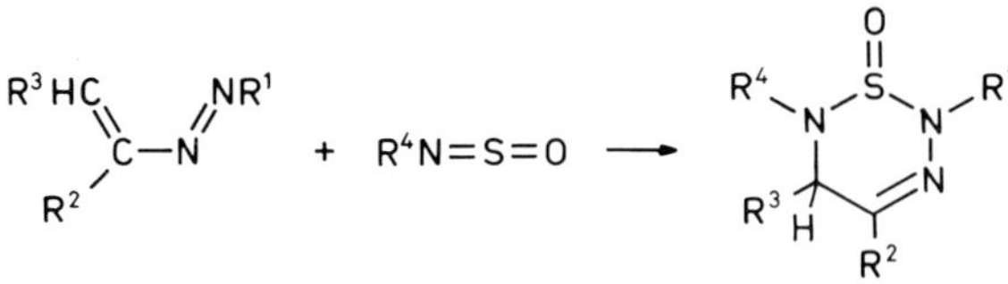

**Spectra.** Infrared and $^1$H NMR data of the compounds 1 to 13 in Table 4 are presented in Table 5 on p. 17 [1, 2]. The $^{13}$C NMR spectra ($CDCl_3$/TMS, $\delta$ in ppm) of [–SO–NR$^1$–N=CR$^2$–CHR$^3$–NR$^4$–] with $R^1 = CH_3$, $R^2R^3 = –(CH_2)_4–$, $R^4 = 4\text{-}CH_3C_6H_4SO_2$ (Table 4, No. 2), and $R^1 = R^2 = CH_3$, $R^3 = H$, $R^4 = 4\text{-}CH_3C_6H_4SO_2$ (Table 4, No. 6) are given below [1]:

No. 2: 21.58 ($CH_3$ ($R^4$)), 24.31, 27.62, 32.76, 34.96 (all –$(CH_2)_4$–), 41.20 ($NCH_3$), 50.8 (C8a), 128.16, 130.05, 134.20, 145.32 (all 6$C_{arom}$), 156.76 (C=N)

No. 6: 21.64 ($CH_3$ ($R^4$)), 21.97 ($H_3\underline{C}$–C=N), 36.72 ($CH_2N$), 40.68 (N–$CH_3$), 128.16, 129.98, 133.55, 145.38 (all 6$C_{arom}$), 148.11 (C=N).

**Chemical Reactions.** The 1,2,3,6-thiatriazine 1-oxides No. 1, 2, 5, 6, 8, and 9 in Table 4, p. 16, rearrange by reaction with trifluoroacetic acid to give ring contraction compounds.

Reaction of compounds No. 1, 2, 8, and 9 with an excess of trifluoroacetic acid in benzene at room temperature, followed by slightly warming for 1 h, gives the products I:

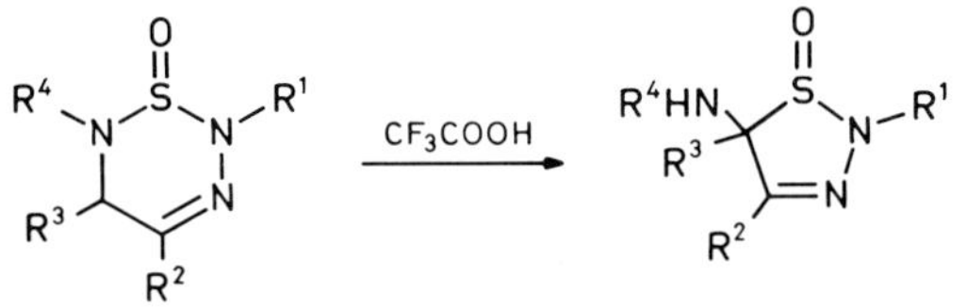

$\xrightarrow{CF_3COOH}$

1, 2, 8, 9      I

| No. | $R^1$ | $R^2$ | $R^3$ | $R^4$ |
|---|---|---|---|---|
| 1 | $CH_3$ | $-(CH_2)_4-$ | | $CH_3SO_2$ |
| 2 | $CH_3$ | $-(CH_2)_4-$ | | $4\text{-}CH_3C_6H_4SO_2$ |
| 8 | $CH_3$ | $CH_3$ | $CH_3$ | $4\text{-}CH_3C_6H_4SO_2$ |
| 9 | $CH_3$ | $CH_3$ | $C_6H_5NHCO$ | $4\text{-}CH_3C_6H_4SO_2$ |

On treatment of compounds No. 5 and 6 with an excess of trifluoroacetic acid, as described above, however, the compounds II form by loss of $H_2O$:

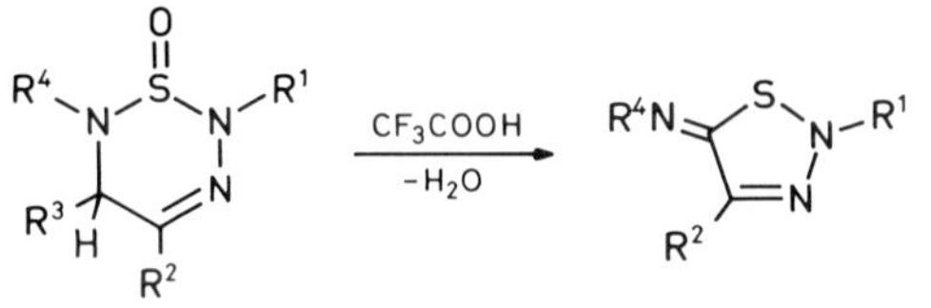

$\xrightarrow[-H_2O]{CF_3COOH}$

5, 6      II

| No. | $R^1$ | $R^2$ | $R^3$ | $R^4$ |
|---|---|---|---|---|
| 5 | $CH_3$ | $CH_3$ | H | $CH_3SO_2$ |
| 6 | $CH_3$ | $CH_3$ | H | $4\text{-}CH_3C_6H_4SO_2$ |

Two reaction mechanisms for the formation of the products I and II are discussed in the paper [2].

If compound 2 is heated with a mixture of ethanol and aqueous HCl, the p-toluenesulfonamide III forms [1, 2].

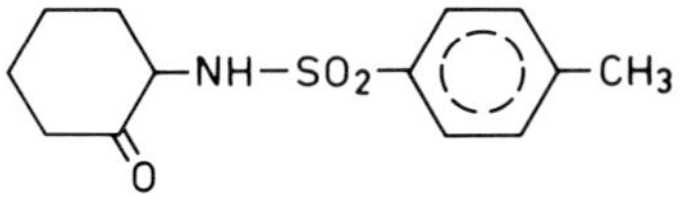

$-NH-SO_2-C_6H_4-CH_3$

III

**References:**

[1] Sommer, S. (Synthesis **1977** 305/7).
[2] Sommer, S., Schubert, U. (Chem. Ber. **111** [1978] 1989/97).

Table 4
Preparation and Melting Point of Derivatives of 5,6-Dihydro-2H-1,2,3,6-thiatriazine 1-Oxide, [–SO–NR[1]–N=CR[2]–CHR[3]–NR[4]–] and 3,5,6,7,8,8a-Hexahydro-1H-2,1,3,4-benzothiatriazine 2-Oxide, [–SO–NR′–N={1,2-$C_6H_9$}–NR[4]–] [1, 2]. References on p. 15.

| No. | $R^1$ | $R^2$ | $R^3$ | $R^4$ | digestion and recryst. solvent | yield in % | m.p. in °C | Ref. |
|---|---|---|---|---|---|---|---|---|
| 1 | $CH_3$ | $-(CH_2)_4-$ | | $CH_3SO_2$ | $CH_3OH$ | 86 | 115 to 117 (dec.) | [1] |
| 2 | $CH_3$ | $-(CH_2)_4-$ | | 4-$CH_3C_6H_4SO_2$ | $CH_3OH$ | 90 | 114 to 116 (dec.) | [1] |
| 3 | $CH_3$ | $-(CH_2)_4-$ | | $C_6H_5CO$ | $CH_3OH$ | 79 | 121 to 123 (dec.) | [1] |
| 4 | $CH_3$ | $-(CH_2)_4-$ | | $C_6H_5$ | $CH_3OH$ | 76 | 115 to 116 (dec.) | [1] |
| 5 | $CH_3$ | $CH_3$ | H | $CH_3SO_2$ | $CH_3OH$ | 73 | 126 to 127 (dec.) | [2] |
| 6 | $CH_3$ | $CH_3$ | H | 4-$CH_3C_6H_4SO_2$ | $CH_3OH$ | 75 | 76 to 77 (dec.) | [1] |
| 7 | $CH_3$ | $CH_3$ | H | $C_6H_5CO$ | $CH_3OH$ | 56 | 67 to 68 | [1] |
| 8 | $CH_3$ | $CH_3$ | $CH_3$ | 4-$CH_3C_6H_4SO_2$ | $CH_3OH$ | 59 | 88 to 89 | [1] |
| 9 | $CH_3$ | $CH_3$ | $C_6H_5NHCO$ | 4-$CH_3C_6H_4SO_2$ | $CHCl_3/CH_3OH$ | 76 | 159 to 161 (dec.) | [1] |
| 10 | $C_6H_5$ | $-(CH_2)_4-$ | | 4-$CH_3C_6H_4SO_2$ | $CH_3COOC_2H_5$ | 84 | 121 to 122 (dec.) | [1] |
| 11 | $C_6H_5$ | $-(CH_2)_4-$ | | $C_2H_5OOC$ | c-$C_6H_{12}$ | 82 | 77 | [1] |
| 12 | $C_6H_5$ | $CH_3$ | $C_6H_5NHCO$ | 4-$CH_3C_6H_4SO_2$ | $CHCl_3/CH_3OH$ | 43 | 139 to 140 (dec.) | [1] |
| 13 | 4-$NO_2C_6H_4$ | $-(CH_2)_4-$ | | 4-$CH_3C_6H_4SO_2$ | $CHCl_3/CH_3OH$ | 96 | 120 to 130 (dec.) | [1] |

Table 5

Infrared and $^{1}H$ NMR Spectra of Derivatives of 5,6-Dihydro-2H-1,2,3,6-thiatriazine 1-Oxide and 3,5,6,7,8,8a-Hexahydro-1H-2,1,3,4-benzothiatriazine 2-Oxide [1, 2].

The composition of compounds No. 1 to 13 is given in Table 4. $^{13}C$ NMR spectra for No. 2 and 6 are compiled on p. 14. References on p. 15.

| No. | IR (in KBr, ν in $cm^{-1}$) ν(C=N) | ν(C=O) | ν(S=O) | $^{1}H$ NMR ($CDCl_3$/TMS, δ in ppm) (for $R^3$ = H the two H atoms in position 5 are designated as $H5^1$ and $H5^2$) |
|---|---|---|---|---|
| 1 | 1641 | | 1110<br>1120 | 1.25 to 2.90 (m, $-(CH_2)_4-$), 3.14 (s, $CH_3$ ($R^4$)), 3.18 (s, $CH_3$ ($R^1$)), 4.12 (m, H8a) |
| 2 | 1637 | | 1130<br>1139 | 1.05 to 2.90 (m, $-(CH_2)_4-$), 2.46 (s, $CH_3$ ($R^4$)), 3.18 (s, $CH_3$ ($R^1$)), 3.75 (m, H8a), 7.39 ($2H_{arom}$), 7.83 ($2H_{arom}$), $A_2B_2$ system |
| 3 | 1649 | 1679 | 1129 | 1.05 to 2.95 (m, $-(CH_2)_4-$), 3.10 (s, $CH_3$ ($R^1$)), 4.48 (m, H8a), 7.2 to 7.8 (m, $5H_{arom}$) |
| 4 | 1640 | | 1095 | 1.15 to 2.80 (m, $-(CH_2)_4-$), 3.14 (s, $CH_3$ ($R^1$)), 4.23 (m, H8a), 7.38 (m, $5H_{arom}$) |
| 5 | 1640 | | | 2.12 (s, $CH_3$ ($R^2$)), 3.07 (s, $CH_3$ ($R^4$)), 3.24 (s, $CH_3$ ($R^1$)), 3.96 (s, ($H5^1$, $H5^2$)) |
| 6 | 1639 | | 1120 | 1.95 (s, $CH_3$ ($R^2$)), 2.44 (s, $CH_3$ ($R^4$)), 3.08 (s, $CH_3$ ($R^1$)), (3.63 (d, $H5^1$), 3.99 (d, $H5^2$), AB system: $^2J(H5^1,H5^2)$ = 17.5 Hz), 7.45 ($2H_{arom}$); 7.79 ($2H_{arom}$), $A_2B_2$ system (acetone-$d_6$) |
| 7 | 1644 | 1660 | 1118 | 2.11 (s, $CH_3$ ($R^2$)), 3.10 (s, $CH_3$ ($R^1$)), (3.94 (d, $H5^1$), 4.42 (d, $H5^2$), AB system: $^2J(H5^1,H5^2)$ = 18.5 Hz), 7.3 to 7.7 (m, $5H_{arom}$) |
| 8 | 1645 | | 1138 | 1.38 (d, $CH_3$ ($R^3$), J = 6.5 Hz), 1.93 (s, $CH_3$ ($R^2$)), 2.43 (s, $CH_3$ ($R^4$)), 3.22 (s, $CH_3$ ($R^1$)), 3.83 (m, H5), (7.35 ($2H_{arom}$), 7.91 ($2H_{arom}$), $A_2B_2$ system) |
| 9 | 1649 | 1672<br>ν(NH) = 3380 | 1148 | 2.11 (s, $CH_3$ ($R^2$)), 2.40 (s, $CH_3$ ($R^4$)), 3.18 (s, $CH_3$ ($R^1$)), 4.83 (s, H5), 6.85 to 7.65 (m, $7H_{arom}$), 7.79 ($2H_{arom}$, part of $A_2B_2$ system), 10.07 (s, NH) |
| 10 | 1649 | | 1146 | 1.15 to 2.95 (m, $-(CH_2)_4-$), 2.43 (s, $CH_3$ ($R^4$)), 3.87 (m, H8a), 7.05 to 7.50 (m, $7H_{arom}$), 7.83 ($2H_{arom}$, part of $A_2B_2$ system) |
| 11 | 1648 | 1725 | 1142 | 1.20 to 3.0 (m, $-(CH_2)_4-$), 1.35 (t, $\underline{H_3}C-CH_2$), 4.26 (m, H8a), 4.37 (q, $H_3C-C\underline{H_2}$), 7.15 to 7.6 (m, $5H_{arom}$) |
| 12 | 1640 | 1672<br>ν(NH) = 3385 | 1099 | 2.23 (s, $CH_3$ ($R^2$)), 2.42 (s, $CH_3$ ($R^4$)), 4.58 (s, H5), 7.0 to 7.6 (m, $12H_{arom}$), 7.94 ($2H_{arom}$, part of $A_2B_2$ system of $R^4$), 8.38 (s, NH) |
| 13 | 1639 | | 1142 | 1.15 to 3.0 (m, $-(CH_2)_4-$), 2.49 (s, $CH_3$ ($R^4$)), 3.88 (m, H8a), (7.42 ($2H_{arom}$), 7.85 ($2H_{arom}$), $A_2B_2$ system of $R^4$), (7.59 ($2H_{arom}$), 8.28 ($2H_{arom}$), $A_2B_2$ system of $R^1$) |

### 9.3.2 1,2,4,6-Thiatriazines

#### 9.3.2.1 1,2,4,6-Thiatriazinyl Radical

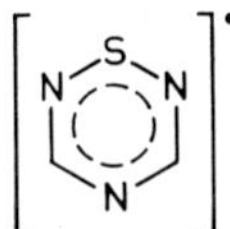


**[┄S┄N┄CH┄N┄CH┄N┄]$^\bullet$** (Reduced formula in the text $SN_3C_2H_2^\bullet$)

The parent radical has not been synthesized or observed until now. Restricted Hartree-Fock (RHF) open-shell calculations (half-electron model) were carried out on a planar $SN_3C_2H_2^\bullet$ fragment with full geometry optimization within $C_{2v}$ symmetry to confirm the parent radical being a 7π-electron radical with a $^2B_1$ ground state. There are three doubly occupied π orbitals, $1b_1$, $1a_2$, and $2b_1$, all of which are bonding contributions. The SOMO $3b_1$ is an antibonding contribution localized primarily over the NSN region but with a substantial contribution being made by the $2p_z$ orbital of the unique nitrogen (see **Fig. 8**). The cofacial approach of two radical units to form the dimer leads to mixing of the π orbitals of the individual components [1].

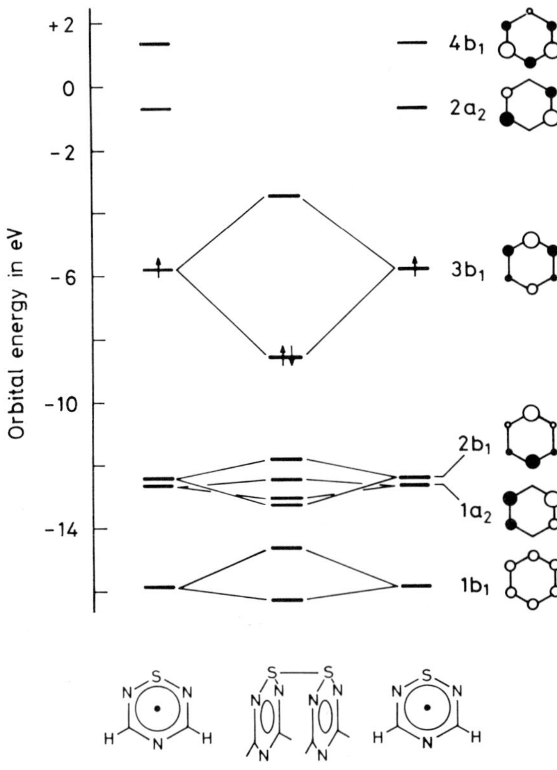


Fig. 8. Molecular orbital energies and contributions of the 1,2,4,6-thiatriazinyl radical. Correlation lines illustrate the mixing of the radical orbitals into those of the dimer $(SN_3C_2H_2)_2$.

### 9.3.2.2 3,5-Disubstituted 1,2,4,6-Thiatriazinyl Radicals

a) $R^1 = Cl$; $R^2 = CCl_3$
b) $R^1 = N(CH_3)_2$; $R^2 = C_6H_5$
c) $R^1 = R^2 = CCl_3$
d) $R^1 = CCl_3$; $R^2 = C_6H_5$
e) $R^1 = R^2 = C_6H_5$

**[-S-N-CR$^1$-N-CR$^2$-N-]$^•$** (Reduced formula in the text $SN_3C_2R^1R^{2•}$)

The radicals form from the corresponding 1-chloro-1H-1,2,4,6-thiatriazine derivatives (see pp. 54/5) by treatment with sodium dust or 2,4,6-triphenyl-3,4-dihydro-2H-1,2,4,5-tetrazinyl (I) (radicals a to d) [2], or with $Sb(C_6H_5)_3$ in $CH_2Cl_2$ (radical e) [1, 3]. The ESR spectra of radicals c and d both exhibit a complicated pattern of about 15 asymmetric lines with the hyperfine coupling constants $a_{N(4)} = 0.36$ mT, $a_{N(2)} = a_{N(6)} = 0.53$ mT. The ESR spectrum of b (given in the paper) consists of 9 clearly defined broadened groups of lines associated with 4 chemically nonequivalent nitrogen atoms; each group is further split into 7 lines with a spacing of 0.09 mT. The g factors of the radicals a to d lie in the region of about 2.006 to 2.007 [2].

I

The 3,5-diphenyl-1,2,4,6-thiatriazinyl radical e in $CH_2Cl_2$ shows an ESR signal of a seven line pattern with equal hyperfine coupling constants to all three nitrogen atoms ($a_N = 0.397$ mT). No additional fine structure due to spin coupling of the phenyl rings to the protons is observed. The deviation of the g-value (2.0059) from the free-electron figure (2.0023) is indicative of a substantial spin density on the sulfur atom. The ESR results thus suggest that the spin orbital in which the unpaired electron resides is equally distributed over N(2), N(6), and N(4), with a large contribution also coming from the sulfur atom [1, 3].

The radicals b, c, and d appear to be rather stable even in the presence of oxygen. By treatment with chlorine, c reforms its starting compound [2]. The remarkably stable radical e forms, in the solid state, a sulfur-sulfur bridged cofacial dimer (see p. 20) [1, 3]. Oxidation of solutions of the radical with $SO_2Cl_2$ regenerated the starting material [3].

### 9.3.2.3 1,1′-Bi-1H-1,2,4,6-thiatriazine

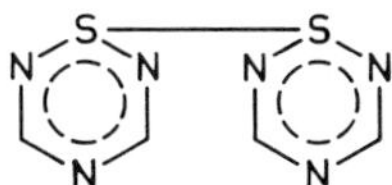

**[-S-N-CH-N-CH-N-]$_2$** (Reduced formula in the text $(SN_3C_2H_2)_2$)

The parent dimer has not been synthesized until now. Restricted Hartree-Fock (open shell) MNDO calculations on a $(SN_3C_2H_2)_2$ model were performed on two cofacial $SN_3C_2H_2$ units, using the geometrical parameters obtained from the free radical calculations. The electronic structure of the dimer, in particular the nature of the weak S–S interaction, can best be analyzed by first considering the electronic makeup of the simple radical $SN_3C_2H_2^•$. The cofacial approach of two radical units to form the dimer leads to mixing of the $\pi$ orbitals of the individual components. The most pronounced interaction is that between the two $3b_1$ orbitals (see Fig. 8) which split apart to form the HOMO and LUMO of the dimer [1].

### 9.3.2.4 3,3′,5,5′-Tetraphenyl-1,1′-bi-1H-1,2,4,6-thiatriazine

R = $C_6H_5$

**([$\dot{-}$S$\dot{-}$N$\dot{-}$C($C_6H_5$)$\dot{-}$N$\dot{-}$C($C_6H_5$)$\dot{-}$N$\dot{-}$])$_2$** (Reduced formula in the text $(SN_3C_2(C_6H_5)_2)_2$)

$(SN_3C_2(C_6H_5)_2)_2$ forms as a dark brown microcrystalline solid in a yield of 85% when solid $Sb(C_6H_5)_3$ is added to a slurry of 1-chloro-3,5-diphenyl-1H-1,2,4,6-thiatriazine (see p. 55) (~1:2 mole ratio) in $CH_3CN$.

The black plates recrystallized from degassed $CH_3CN$ are triclinic, space group $P\bar{1}$-$C_i^1$ (No. 2) with a = 8.894(3), b = 11.058(3), c = 12.814(5) Å, α = 81.76(2)°, β = 86.75(3)°, and γ = 76.00(2)°; Z = 2. V = 1209.9 $Å^3$, $D_x$ = 1.39 g/$cm^3$. R = 0.050 for 1125 reflections. The dimer also crystallizes in the monoclinic space group $P2_1/n$ (standard setting $P2_1/c$)-$C_{2h}^5$ (No. 14) (no crystallizing conditions given), with a = 11.360(3), b = 10.962(3), c = 19.415(7) Å, and β = 99.94(2)°; Z = 4. R = 0.14 for 1051 observed reflections. The molecular structure of the compound is shown in **Fig. 9**.

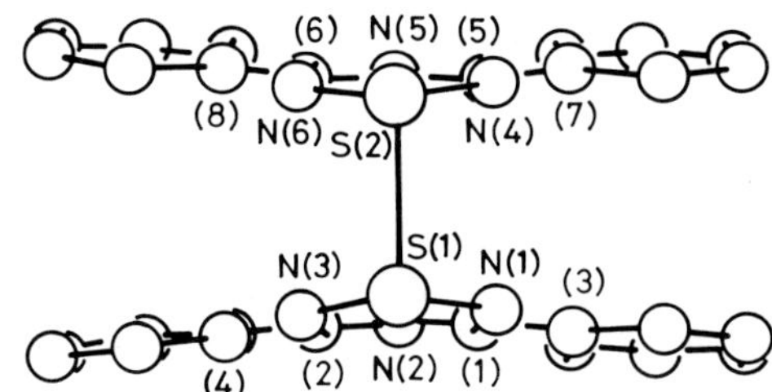


Fig. 9. Molecular structure of $(SN_3C_2(C_6H_5)_2)_2$.

The molecular structure of the compound consists of two $SN_3C_2$ rings with no unusually short S–S contacts. Individually the two $SN_3C_2$ rings exhibit shallow boat conformations. In one ring (containing S(1)), the sulfur atom and the remote nitrogen atom are displaced by 0.135(2) and 0.048(6) Å, respectively, from the mean plane of the four central atoms, which are themselves coplanar to within 0.006(7) Å. In the other ring (containing S(2)), the corresponding displacements are 0.074(2) and 0.053(5) Å, the remaining atoms being coplanar to within 0.010(7) Å. Overall, the two rings are planar to within 0.081(6) and 0.051(5) Å, respectively, and their two mean planes intersect with a dihedral angle of 14°. The shortest contact is 2.666(3) Å for S(1)–S(2), although all pairwise contacts are within the appropriate van der Waals' separation. The two $SN_3C_2$ rings are eclipsed. Selected bond lengths and bond angles are given in Table 6, final atomic coordinates for the nonhydrogen atoms are given in the paper [1].

The black opaque crystals melt at 178 to 181°C. For the IR spectrum (Nujol), 16 absorption bands between 3060 and 340 $cm^{-1}$ are listed in the paper. Mass spectrum (70 eV): m/e = 252 $SN_3C_2(C_6H_5)_2^+$ (45%), 149 $SN_2CC_6H_5^+$ (27%), 103 $C_6H_5CN^+$ (50%), 77 $C_6H_5^+$ (9%), 46 $SN^+$ (100%) [1].

At room temperature the compound is diamagnetic but it dissolves in $CH_2Cl_2$ to give yellow-orange solutions which exhibit the ESR spectrum assigned to its constituent radical $SN_3C_2(C_6H_5)_2^{\bullet}$ (see p. 19) [1, 3].

Table 6

Selected Bond Lengths and Bond Angles in 3,3′,5,5′-Tetraphenyl-1,1′-bi-1H-1,2,4,6-thiatriazine, $(SN_3C_2(C_6H_5)_2)_2$.

| bond length in Å | | | | | |
|---|---|---|---|---|---|
| S(1)–S(2) | 2.666(3) | N(1)–C(1) | 1.316(9) | N(5)–C(6) | 1.33(1) |
| S(1)–N(1) | 1.602(6) | N(2)–C(1) | 1.34(1) | N(6)–C(6) | 1.32(1) |
| S(1)–N(3) | 1.627(6) | N(2)–C(2) | 1.34(1) | C(1)–C(3) | 1.47(1) |
| S(2)–N(4) | 1.614(6) | N(3)–C(2) | 1.32(1) | C(2)–C(4) | 1.47(1) |
| S(2)–N(6) | 1.624(6) | N(4)–C(5) | 1.34(1) | C(5)–C(7) | 1.48(1) |
| | | N(5)–C(5) | 1.33(1) | C(6)–C(8) | 1.47(1) |

| bond angle in ° | | | | | |
|---|---|---|---|---|---|
| S(2)–S(1)–N(1) | 95.7(3) | S(1)–N(3)–C(2) | 116.2(6) | N(2)–C(2)–C(4) | 117.7(8) |
| S(2)–S(1)–N(3) | 96.9(2) | S(2)–N(4)–C(5) | 116.7(6) | N(3)–C(2)–C(4) | 114.2(7) |
| S(1)–S(2)–N(6) | 97.0(3) | C(5)–N(5)–C(6) | 119.6(7) | N(4)–C(5)–N(5) | 127.5(8) |
| S(1)–S(2)–N(4) | 96.6(3) | S(2)–N(6)–C(6) | 116.6(6) | N(4)–C(5)–C(7) | 115.8(8) |
| N(1)–S(1)–N(3) | 110.7(4) | N(1)–C(1)–N(2) | 127.7(8) | N(5)–C(5)–C(7) | 116.8(8) |
| N(4)–S(2)–N(6) | 110.8(3) | N(1)–C(1)–C(3) | 116.3(9) | N(5)–C(6)–N(6) | 128.3(8) |
| S(1)–N(1)–C(1) | 117.2(6) | N(2)–C(1)–C(3) | 115.9(9) | N(5)–C(6)–C(8) | 116.3(8) |
| C(1)–N(2)–C(2) | 119.2(7) | N(2)–C(2)–N(3) | 128.1(8) | N(6)–C(6)–C(8) | 115.5(8) |

**References:**


[1] Hayes, P. J., Oakley, R. T., Cordes, A. W., Pennington, W. T. (J. Am. Chem. Soc. **107** [1985] 1346/51).

[2] Markovskii, L. N., Kornuta, P. P., Katchkovskaya, L. S., Polumbrik, O. M. (Sulfur Letters **1** [1983] 143/5; C.A. **100** [1984] No. 103301).

[3] Cordes, A. W., Hayes, P. J., Josephy, P. D., Koenig, H., Oakley, R. T., Pennington, W. T. (J. Chem. Soc. Chem. Commun. **1984** 1021/2).


### 9.3.2.5 Derivatives of 1,2,4,6-Thiatriazine S (or 1)-Oxide

#### 9.3.2.5.1 3,5-Disubstituted 2H-1,2,4,6-Thiatriazine 1-Oxides

O, S, N, H, N, R², N, R¹

$R^1 = C_6H_5$, $R^2 = CCl_3$; $R^1 = R^2 = CCl_3$;
$R^1 = C_6H_5CH_2NHCO$, $R^2 = C_6H_5$, $(CH_3)_2N$, $C_6H_5CH_2S$

**[-SO-NH-CR¹=N-CR²=N-]**

Compounds 2a ($R^1 = C_6H_5$, $R^2 = CCl_3$) and 2b ($R^1 = R^2 = CCl_3$) were obtained in 87 and 53% yield, respectively, by action of water with the thiatriazines 1a, b. They melt at 149 to 151°C and 121 to 123°C, respectively [1].

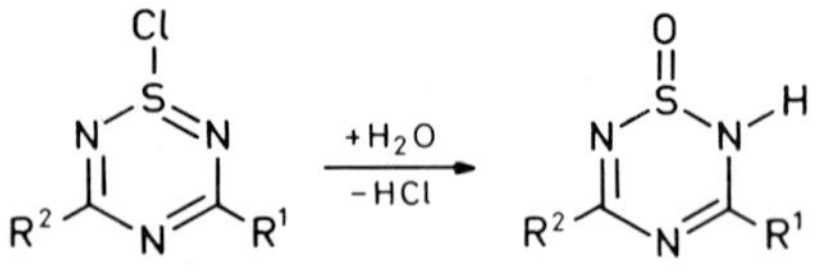

1a, b    2a, b

a) $R^1 = C_6H_5$, $R^2 = CCl_3$
b) $R^1 = R^2 = CCl_3$

Compounds with $R^1 = C_6H_5CH_2NHCO$ and $R^2 = C_6H_5$ (3a), $(CH_3)_2N$ (3b), and $C_6H_5CH_2S$ (3c) form in the reaction of the appropriate 1,2,5-thiadiazole with amidine type binucleophiles $NH{=}C(R^2){-}NH_2$ ($R^2 = C_6H_5$, $(CH_3)_2N$, and $C_6H_5CH_2S$) in $CH_3OH$ at room temperature overnight in yields of 40% (3a), 45% (3b) and 35% (3c). A reaction mechanism was proposed [2]. The following data were measured for compound 3c: melting point 167 to 170°C; $^{13}C$ NMR spectrum ($CDCl_3$/TMS): δ (in ppm) = 33.5 ($CH_2S$), 42.7 ($C_6H_5CH_2$), 146.3 (NHCO), 158.7 ($N{=}\underline{C}{-}C{=}O$), 162.1 (N=C–S); mass spectrum: m/e = 372 $M^+$, 356 $M^+ - O$ [2].

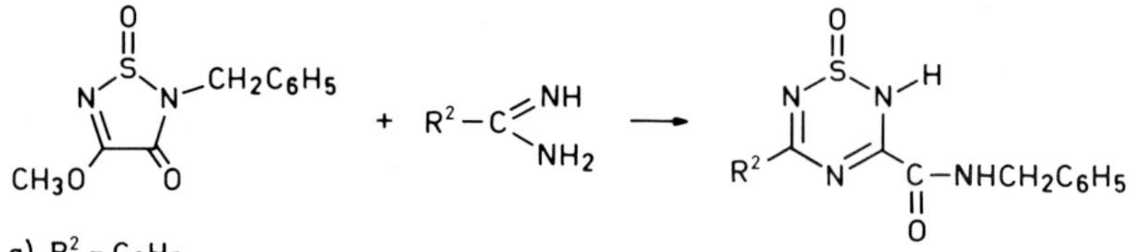

a) $R^2 = C_6H_5$
b) $R^2 = (CH_3)_2N$
c) $R^2 = C_6H_5CH_2S$

3a, b, c

By treating compound 2a with chlorinating agents, e.g., $t\text{-}C_4H_9OCl$ in $CCl_4$ at room temperature or $Cl_2N{-}SO_2C_6H_5$ in boiling benzene, the sulfur atom is oxidized, giving thiatriazine ($S^{VI}$) I [1].

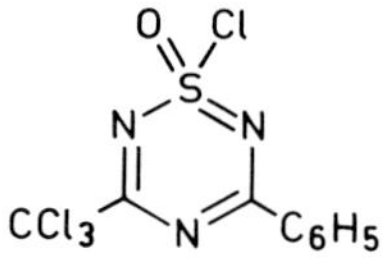

I

**References:**


[1] Kornuta, P. P., Derii, L. I., Markovskii, L. N. (Zh. Org. Khim. **16** [1980] 1308/13; J. Org. Chem. [USSR] **16** [1980] 1130/4).

[2] Karady, S., Amato, J. S., Reamer, R. A., Weinstock, L. M. (Tetrahedron Letters **26** [1985] 6155/8).


### 9.3.2.5.2 2,3,5-Trisubstituted 2H-1,2,4,6-Thiatriazine 1-Oxide

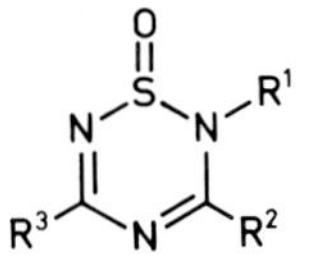

$R^1 = 4\text{-}CH_3C_6H_4SO_2$, $R^2 = H_2N$,
$R^3 = (CH_3)_2N$

**[–SO–NR¹–CR²=N–CR³=N–]**

The compound forms by hydrolysis of the corresponding 1-p-tolylsulfonylimino compound $[{-}S({=}N(SO_2C_6H_4CH_3\text{-}4)){-}N(SO_2C_6H_4CH_3\text{-}4){-}C(NH_2){=}N{-}C(N(CH_3)_2){-}N{-}]$. It melts at 134°C.

IR spectrum (Nujol, ν in $cm^{-1}$): 3410, 3260, 3120 (NH), 1680, 1650, 1600 (C=N, NH), 1285, 1150 ($SO_2$), 1130 (SO); $^1H$ NMR spectrum ($DMSO-d_6/(CH_3)_2CO-d_6$): δ (in ppm) = 2.30 (s, $CH_3$), 2.94 (d, $2CH_3$), 7.15 to 7.73 (m, $C_6H_4$, NH).

**Reference:**

Fischer, E., Teller, M., Kálmán, A., Argay, G. (Tetrahedron **40** [1984] 385/90).

### 9.3.2.5.3 4-Methyl-3,5-bis(tolyloxy)-4H-1,2,4,6-thiatriazine 1-Oxide

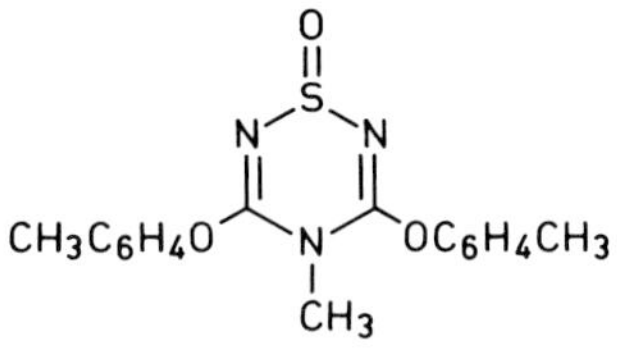

**[-SO-N=C($OC_6H_4CH_3$)-$NCH_3$-C($OC_6H_4CH_3$)=N-]**

The title compound forms in 58% yield by cyclization of the diisobiuret derivative with $SOCl_2$ in $CH_2Cl_2$ at 0°C for 1 h.

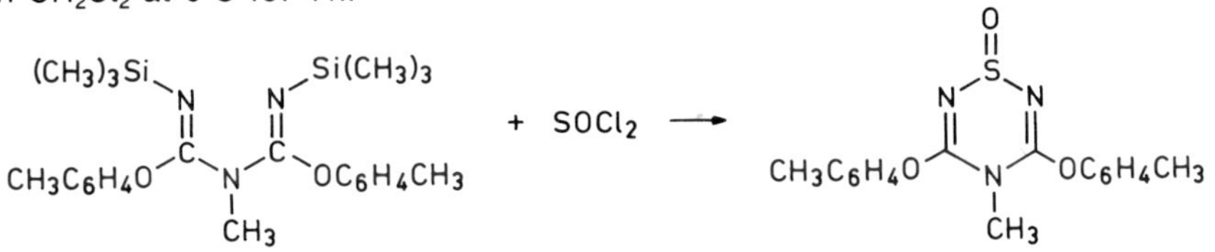

It melts at 232 to 233°C. By treatment of the compound with $3-ClC_6H_4CO_3H$ in $CHCl_3$ at 20°C, the SO group is oxidized to form an $SO_2$ group.

**Reference:**

Cousins, S. J., Ross, B. C., Maw, G. N., Michael, J. D. (Tetrahedron Letters **26** [1985] 1105/8).

### 9.3.2.5.4 2,4,5-Trisubstituted 2H-1,2,4,6-Thiatriazin-3(4H)-one 1-Oxide

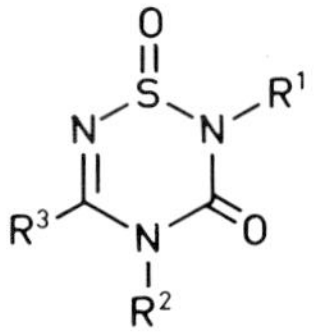

$R^1$, $R^2$, $R^3$ are compiled in Table 7, pp. 24/5.

**[-SO-$NR^1$-CO-$NR^2$-$CR^3$=N-]**

**Preparation. Properties.**

Methods of preparation, yields, properties, and reactions of the compounds are compiled in the scheme p. 26, and Table 7, pp. 24/5. Compounds 1a to 1d are obtained in high yield from the reaction of 1-carbamoyl-2-methylisothioureas with an excess of $SOCl_2$ in the presence of pyridine in $CHCl_3$ at 5°C for 30 min, followed by stirring at room temperature for 5 h. Their

Table 7

Yields, Methods of Preparation, Melting Points, and Spectra of 2,4,5-Trisubstituted 2H-1,2,4,6-Thiatriazin-3(4H)-one 1-Oxides, [–SO–NR[1]–CO–NR[2]–CR[3]=N–] (Reference on p. 27).

| No. | $R^1$ | $R^2$ | $R^3$ | yield in % (method) | m.p. in °C | IR spectra (in Nujol, ν in $cm^{-1}$) ν(CO), ν(SO) | mass spectra (70 eV) m/e | $^1H$ NMR spectra ($CDCl_3$/TMS, δ in ppm) |
|---|---|---|---|---|---|---|---|---|
| 1a | $CH_3$ | H | $SCH_3$ | 82 | 144 | 1690, 1100 | 193 $M^+$,<br>145 $M^+-SO$ | 2.49 (s, $SCH_3$), 3.23 (s, $NCH_3$) |
| 1b | i-$C_3H_7$ | H | $SCH_3$ | 88 | 75 | 1690, 1100 | 221 $M^+$,<br>173 $M^+-SO$ | 1.48 (d, $(CH_3)_2C$), 2.46 (s, $SCH_3$), 4.20 to 4.73 (m, $(CH_3)_2C\underline{H}$) |
| 1c | c-$C_6H_{11}$ | H | $SCH_3$ | 86 | 123 | 1695, 1100 | 261 $M^+$,<br>213 $M^+-SO$ | 1.02 to 2.20 (m, c-$C_6H_{11}$), 2.48 (s, $SCH_3$), 3.88 to 4.40 (m, NH) |
| 1d | $C_6H_5$ | H | $SCH_3$ | 64 | 125 | 1695, 1100 | 255 $M^+$,<br>207 $M^+-SO$ | 2.46 (s, $SCH_3$), 7.73 (s, $C_6H_5$) |
| 2a | $CH_3$ | $CH_3$ | $SCH_3$ | 51 (I)<br>90 (II) | 68 | 1680, 1100 | 207 $M^+$,<br>159 $M^+-SO$ | 2.46 (s, $SCH_3$), 3.18 (s, 2-$CH_3$), 3.39 (s, 4-$CH_3$) |
| 2b | i-$C_3H_7$ | $CH_3$ | $SCH_3$ | 58 (I)<br>95 (II) | 58 | 1680, 1100 | 235 $M^+$,<br>187 $M^+-SO$ | 1.45 (d, $(CH_3)_2C$), 2.49 (s, $SCH_3$), 3.40 (s, 4-$CH_3$), 4.18 to 4.64 (m, $(CH_3)_2C\underline{H}$) |
| 2c | c-$C_6H_{11}$ | $CH_3$ | $SCH_3$ | 59 (I)<br>94 (II) | oil | 1690, 1120 | 275 $M^+$,<br>227 $M^+-SO$ | 1.02 to 2.20 (m, c-$C_6H_{11}$), 2.49 (s, $SCH_3$), 3.37 (s, 4-$CH_3$) |
| 2d | $C_6H_5$ | $CH_3$ | $SCH_3$ | 92 (II) | oil | 1690, 1120 | 269 $M^+$,<br>221 $M^+-SO$ | 2.43 (s, $SCH_3$), 3.67 (s, 4-$CH_3$), 7.26 (s, $C_6H_5$) |

| | | | | | | | | |
|---|---|---|---|---|---|---|---|---|
| 3a | $CH_3$ | $CH_3$ | $N(CH_3)_2$ | 86 | oil | 1680, 1120 | 204 $M^+$, 156 $M^+-SO$ | 2.96 (s, $N(CH_3)_2$), 3.18 (s, 2-$CH_3$), 3.38 (s, 4-$CH_3$) |
| 3b | i-$C_3H_7$ | $CH_3$ | $N(CH_3)_2$ | 85 | oil | 1680, 1100 | 232 $M^+$, 184 $M^+-SO$ | 1.43 (d, $(CH_3)_2C$), 2.97 (s, $N(CH_3)_2$), 3.41 (s, 4-$CH_3$) |
| 3c | c-$C_6H_{11}$ | $CH_3$ | $N(CH_3)_2$ | 81 | 72 | 1680, 1100 | 272 $M^+$, 224 $M^+-SO$ | 0.92 to 2.30 (m, c-$C_6H_{11}$), 2.95 (s, $N(CH_3)_2$), 3.39 (s, 4-$CH_3$) |
| 4b | i-$C_3H_7$ | $NH_2$ | $SCH_3$ | 58 | 155 | 1685, 1120 | 236 $M^+$ | 1.46 (s, $(CH_3)_2C$), 2.32 (s, $SCH_3$), 4.53 (s, br, $NH_2$) |
| 4c | c-$C_6H_{11}$ | $NH_2$ | $SCH_3$ | 63 | 166 | 1685, 1120 | 276 $M^+$ | 1.01 to 2.20 (m, c-$C_6H_{11}$), 2.32 (s, $SCH_3$), 4.43 (s, br, $NH_2$) |
| 5b | i-$C_3H_7$ | $NH_2$ | $N(CH_3)_2$ | 89 | 114 | 1685, 1120 | 233 $M^+$ | 1.46 (s, $(CH_3)_2C$), 3.13 (s, $N(CH_3)_2$), 4.53 (s, br, $NH_2$) |
| 5c | c-$C_6H_{11}$ | $NH_2$ | $N(CH_3)_2$ | 89 | 126 | 1685, 1120 | 273 $M^+$ | 1.00 to 2.20 (m, c-$C_6H_{11}$), 3.09 (s, $N(CH_3)_2$), 4.58 (s, br, $NH_2$) |

4-methyl-substituted derivatives 2a to 2c form by methylation of 1a to 1c with $CH_3I$ in the presence of n-$C_4H_9Li$ in THF at temperatures below −5°C, subsequently being stirred at room temperature for 2 h and refluxing for 4 h (Method I). Compounds 2a to 2d are obtained in higher yields by treatment of 1a to 1d in $CHCl_3$ with an ether solution of diazomethane below 10°C followed by stirring the solution overnight at room temperature (Method II). The 4-amino-substituted heterocycles 4b and 4c form by amination of 1b and 1c with O-(2,4-dinitrophenyl)-hydroxylamine in the presence of n-$C_4H_9Li$ in THF at temperatures below −5 to 0°C, followed by stirring at room temperature for 24 h. Treatment of 2a to 2c, 4b, and 4c with excess dimethylamine in isopropyl alcohol at 5°C and subsequently stirring at room temperature for 1 h effects nucleophilic displacement of the $SCH_3$ group by $N(CH_3)_2$, affording 3a to 3c, 5b, and 5c, respectively.

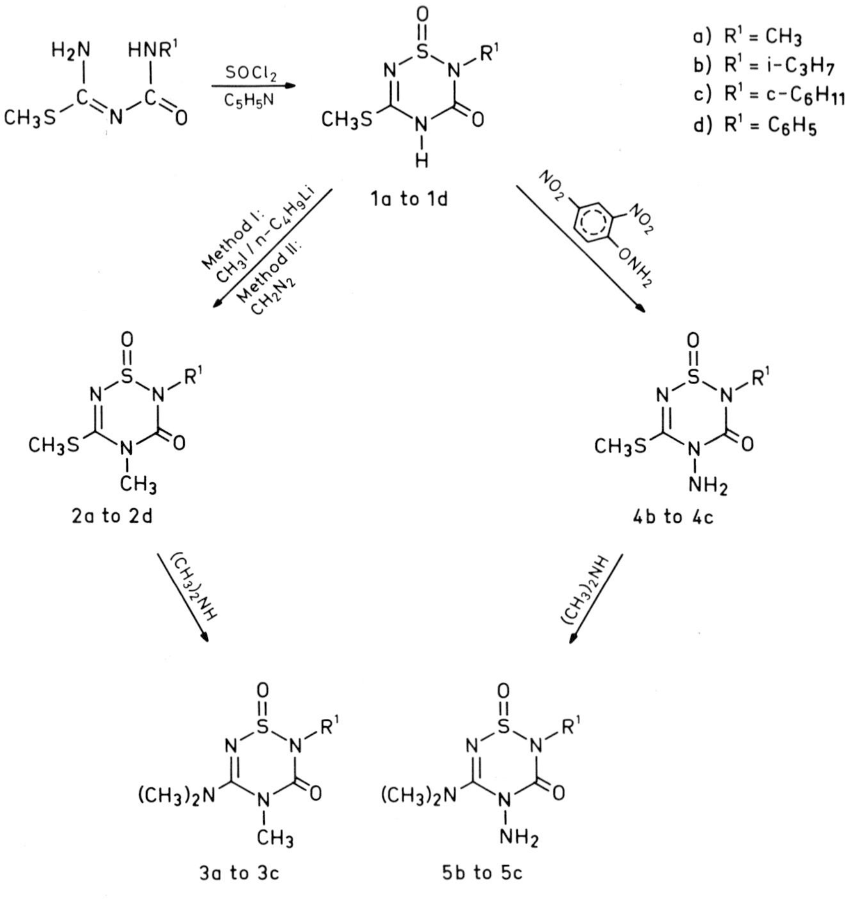

a) $R^1 = CH_3$
b) $R^1 = i\text{-}C_3H_7$
c) $R^1 = c\text{-}C_6H_{11}$
d) $R^1 = C_6H_5$

1a to 1d

2a to 2d

4b to 4c

3a to 3c

5b to 5c

**Reactions.**

If compounds 2a to 2c in Table 7, p. 24, are oxidized with 3-chloroperoxybenzoic acid in $CHCl_3$ at room temperature, the 1,1-dioxide derivatives are obtained. Attempts to oxidize the 4-amino-substituted compounds 5b and 5c to the corresponding 1,1-dioxides using 3-chloroperoxybenzoic acid or trifluoroperoxyacetic acid led to complex mixtures.

Amination of compounds 1b and 1c with O-(2,4-dinitrophenyl)hydroxylamine in the presence of n-$C_4H_9Li$ in THF at room temperature yields the corresponding 4-amino-substituted derivatives 4b and 4c. Treatment of the 4-methyl-substituted heterocycles 2a to 2c and 4-amino-substituted heterocycles 4b and 4c with excess $(CH_3)_2NH$ in isopropyl alcohol at room temperature effected nucleophilic displacement of the $SCH_3$ group, thus affording the corresponding 6-dimethylamino-substituted compounds 3a to 3c, 5b, and 5c. Under similar conditions treatment of compound 2d ($R^1 = C_6H_5$, $R^2 = CH_3$, $R^3 = SCH_3$) with $(CH_3)_2NH$ results in ring cleavage to give 1,1-dimethyl-3-phenylurea, $(CH_3)_2NCONHC_6H_5$.

Methylation of compounds 1a to 1d either with $CH_3I$ in the presence of $C_4H_9Li$ in THF or with $CH_2N_2$ gives 4-methyl-substituted rings 2a to 2d.

2H-1,2,4,6-thiatriazin-3(4H)-one 1-oxides have high potential herbicidal activities.

**Reference:**

Nakayama, Y., Sanemitsu, Y. (J. Heterocycl. Chem. **21** [1984] 1553/6).

### 9.3.2.5.5 Glycosides of 2H-1,2,4,6-Thiatriazine-3(6H)-thione 1-Oxide

$R^1$ and $R^2$ are compiled in Table 8, p. 28.

**[–SO–NR¹–CS–N=CR²–NH–]**

For preparation the appropriate $N^2$-thiocarbamoylcarboxamidines are treated with excess $SOCl_2$ in $CHCl_3$ solution while cooling to give the corresponding 1,2,4,6-thiatriazines 1 to 5, compiled in Table 8 [1, 2].

$$R^2C(NH_2)=N-C(=S)NHR^1 + SOCl_2 \longrightarrow \text{1 to 5} + 2\,HCl$$

1 to 5

Compound 2 forms also on reacting the appropriate $N^2$-thiocarbamoylcarboxamidine with excess $SOCl_2$ in the presence of pyridine in benzene, followed by stirring 1 h at room temperature and 3 h at reflux [1].

Attempted ring contraction of the thiatriazines to triazoles I with elimination of SO was unsuccessful [1, 2]. Thus, refluxing of compound 3 in xylene (~140°C) for 2 h or heating of compound 3 in pyridine at 60 to 80°C yields the starting material and unidentified products [1].

I

**References:**

[1] Ogura, H., Takahashi, H., Sato, O. (Chem. Pharm. Bull. [Tokyo] **29** [1981] 1843/7).
[2] Ogura, H., Takahashi, H., Sato, O. (Nucleic Acids Symp. Ser. No. 8 [1980] s1/s4; C.A. **94** [1981] No. 175410).

Table 8

Properties of Glycosides of 2H-1,2,4,6-Thiatriazine-3(6H)-thione 1-Oxide, [–SO–NR¹–CS–N=CR²–NH–] (AcO = $CH_3COO$, BzO = $C_6H_5COO$) [1, 2].

| No. | $R^1$ | $R^2$ | yield in % | m.p. in °C | $[\alpha]_D^{16}$ (C 1.0, $CH_3OH$) | mass spectrum (75 eV, m/e) | ¹H NMR spectrum (CDCl₃/TMS, δ in ppm) $R^2$ ($CH_3$ or H) | NH | remarks |
|---|---|---|---|---|---|---|---|---|---|
| 1 | 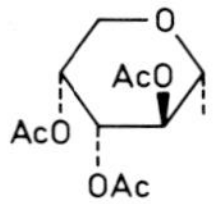 AcO, AcO, OAc | $CH_3$ | 92 | 134 to 135 | −47° | 373 ($M^+$ − SO) | 2.46 (s, 3H) | 7.50 (br, s) | colorless fine needles |
| 2 | 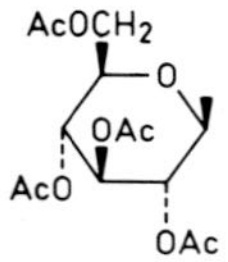 $AcOCH_2$, OAc, AcO, OAc | H | 93 | 147 to 150 | | 431 ($M^+$ − SO) | 8.20 (s, 1H) | 7.10 | colorless needles |
| 3 | 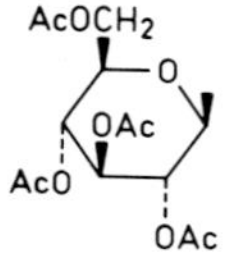 $AcOCH_2$, OAc, AcO, OAc | $CH_3$ | 95 | 157 to 161 | −18° | 445 ($M^+$ − SO) | 2.48 (s, 3H) | 7.08 (br, s) | colorless needles |
| 4 | 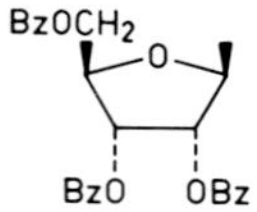 $BzOCH_2$, BzO, OBz | H | 87 | 160 to 163 | | 545 ($M^+$ − SO) | 8.35 (s, 1H) | | colorless solid |
| 5 | 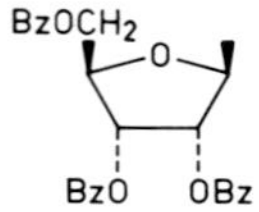 $BzOCH_2$, BzO, OBz | $CH_3$ | 90 | | −83° | 445 (glycoside ion) | 2.02 (s, 3H) | | Rf = 0.60 (TLC on silica gel, benzene-acetone 3:2) |

### 9.3.2.5.6 4-Phenoxy-10H(or 1H)-[1,2,4,6]thiatriazino[4,3-a]benzimidazole 2-Oxide

**[-SO-N=C($OC_6H_5$)-{N-{1,2-$C_6H_4$}-NH-C}=N-]**

The compound is prepared as shown in the scheme below. The reaction of I with II and subsequent hydrolysis of an unisolated and unidentified intermediate were only briefly mentioned [1]. The preparation of III and subsequent hydrolysis by refluxing in $CH_3OH$ were described [2].

R = H, Cl, $CH_3$

X-ray diffraction data show the compound to be monoclinic, space group $P2_1/n$ (standard setting $P2_1/c$)-$C_{2h}^5$ (No. 14) with a = 9.329(2), b = 4.603(5), c = 30.124(7) Å and β = 92.57(2)°; Z = 4, $D_x$ = 1.533 g/cm³ and R = 0.068 for 1089 reflections. **Fig. 10** shows that a rare trigonal-pyramidal O=S(N)(N) moiety with an axially oriented O atom is embedded in the nonplanar thiatriazine ring. The S=O distance of 1.452(7) Å is considerably shorter than that observed in vari-

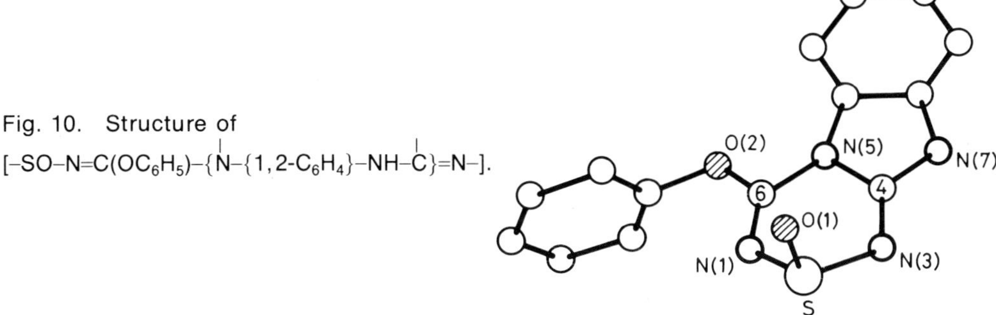


Fig. 10. Structure of [-SO-N=C($OC_6H_5$)-{N-{1,2-$C_6H_4$}-NH-C}=N-].

ous sulfoxides. Selected bond lengths and bond angles are presented in Table 9. For fractional coordinates and mean temperature factors for nonhydrogen atoms, see the paper [1]. Melting point 193°C. IR spectrum (Nujol, ν in $cm^{-1}$): 3200 (NH), 1670 (C=N), 1140 (SO). The chemical shifts of 148.4 and 128.6 ppm for C (5a) and C (9a) atoms in the $^{13}C$ NMR spectrum (solvent and standard not given in the paper) indicate that in solution (as in the solid state) the equilibrium between the 1H and 10H tautomeric forms lies on the side of the 10H compound [2].

Table 9

Selected Bond Lengths and Bond Angles of 4-Phenoxy-10H(or1H)-[1,2,4,6]thiatriazino-[4,3-a]benzimidazole 2-Oxide, [-SO-N=C($OC_6H_5$)-{N-{1,2-$C_6H_4$}-NH-C}=N-] [1].

| bond length in Å | | | | | |
|---|---|---|---|---|---|
| N(1)-S | 1.705(6) | S-O | 1.452(7) | N(5)-C(6) | 1.385(8) |
| N(1)-C(6) | 1.247(9) | N(3)-C(4) | 1.286(10) | C(6)-O(2) | 1.350(7) |
| S-N(3) | 1.669(6) | C(4)-N(5) | 1.415(9) | C(4)-N(7) | 1.359(9) |

| bond angle in ° | | | | | |
|---|---|---|---|---|---|
| S-N(1)-C(6) | 120.1(9) | N(3)-S-O | 107.0(6) | C(4)-N(5)-C(6) | 116.9(10) |
| N(1)-S-N(3) | 102.1(5) | S-N(3)-C(4) | 117.3(9) | N(1)-C(6)-N(5) | 126.1(11) |
| N(1)-S-O | 104.5(6) | N(3)-C(4)-N(5) | 127.0(11) | N(1)-C(6)-O(2) | 124.5(11) |
| | | | | N(5)-C(6)-O(2) | 109.2(10) |

**References:**

[1] Kálmán, A., Argay, G. (Acta Cryst. B **38** [1982] 1370/2).
[2] Fischer, E., Teller, M., Kálmán, A., Argay, G. (Tetrahedron **40** [1984] 385/90).

### 9.3.2.5.7 3-Amino-4H-[1,2,4,6]thiatriazino[2,3-a]benzimidazole 1-Oxide

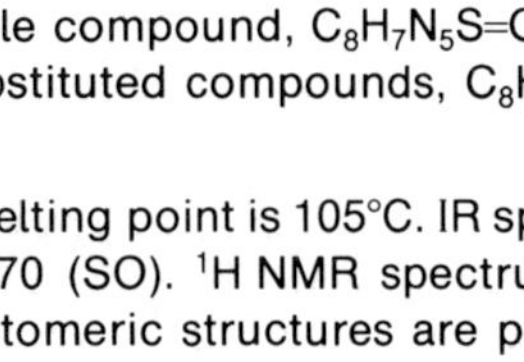

**[-SO-N=C($NH_2$)-NH-{C=N-{1,2-$C_6H_4$}-N}-]**

The title compound, $C_8H_7N_5S{=}O$, is obtained on heating the corresponding arenesulfonyl-imino-substituted compounds, $C_8H_7N_5S{=}NSO_2C_6H_4R$ (R = H, Cl, $CH_3$), in acetic acid at 115 to 117°C.

The melting point is 105°C. IR spectrum (Nujol, ν in $cm^{-1}$): 3420 (NH, $NH_2$), 3330, 3260, 1670 (C=N), 1170 (SO). $^1H$ NMR spectrum (solvent?/TMS): δ = 6.88 to 7.26 ppm (m, aromatic H). Three tautomeric structures are possible.

**Reference:**

Fischer, E., Teller, M., Kálmán, A., Argay, G. (Tetrahedron **40** [1984] 385/90).

### 9.3.2.6 Derivatives of 1-Alkoxy-1,2,4,6-thiatriazine

#### 9.3.2.6.1 3,5-Disubstituted 1-Alkoxy-1H-1,2,4,6-thiatriazines

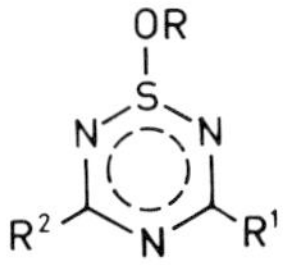

$R = CH_2CH_2Cl$; $R^1 = R^2 = Cl$
$R = CH_3$; $R^1 = CCl_3$; $R^2 = CF_3$, $CCl_3$, $C_6H_5$
$R = C_2H_5$; $R^1 = CCl_3$; $R^2 = C_6H_5$

**[∸S(OR)∸N∸CR$^1$∸N∸CR$^2$∸N∸]**

##### 9.3.2.6.1.1 1-(2-Chloroethoxy)-3,5-dichloro-1H-1,2,4,6-thiatriazine, [∸S($OCH_2CH_2Cl$)∸N∸CCl∸N∸CCl∸N∸]

The compound forms in 98% yield when a solution of 1,3,5-trichloro-1H-1,2,4,6-thiatriazine, [∸SCl∸N∸CCl∸N∸CCl∸N∸] (see p. 54), in $CCl_4$ is treated with an excess of ethylene oxide below 20°C.

The compound (recrystallized from $CCl_4$) melts at 54 to 55.5°C. $^1$H NMR spectrum (in $CCl_4$): δ (in ppm) = 3.55 to 3.77 (m, $CH_2$), 4.02 to 4.23 (m, $CH_2$). Mass spectrum: m/e = 247 $M^+$, 198 $M^+ - CH_2Cl$, 168 $M^+ - OCH_2CH_2Cl$, 151 $M^+ - CNCl_2$, 136 $M^+ - SOCH_2CH_2Cl$, 125 $M^+ - C_2N_2Cl_2$.

The compound reacts in refluxing methanol to give $H_2NCONHC(OCH_3)$=NH.

**Reference:**

Geevers, J., Hackmann, J. T., Trompen, W. P. (J. Chem. Soc. C. **1970** 875/8).

##### 9.3.2.6.1.2 1-Alkoxy-3-alkyl-5-alkyl (or phenyl)-1H-1,2,4,6-thiatriazines

**[∸S(OR)∸N∸CR$^1$∸N∸CR$^2$∸N∸]** R, R$^1$, and R$^2$ are compiled in Table 10, p. 32.

The compounds which are compiled in Table 10 form from the reaction of the appropriately 3,5-substituted 1-chloro-1H-1,2,4,6-thiatriazines, [∸SCl∸N∸CR$^1$∸N∸CR$^2$∸N∸], in benzene with the corresponding sodium alkoxides in ethanol. The sodium chloride was separated, and the mother solution was evaporated at 70 to 80°C/10 to 15 Torr. The compounds were purified by recrystallization or fractional distillation. Yields and properties are given in Table 10. The IR spectra of the 1-alkoxy-1H-1,2,4,6-thiatriazines are characterized by a shift of the absorption band for the stretching vibrations of the ring in the region of 1500 $cm^{-1}$ by 20 to 30 $cm^{-1}$ toward the short wave region compared with the appropriately 3,5-substituted 1-chloro-1H-1,2,4,6-thiatriazines, [∸SCl∸N∸CR$^1$∸N∸CR$^2$∸N∸]. A band at 960 $cm^{-1}$ has been assigned to the stretching vibration of the $OCH_3$ group, an absorption band at 1000 or 890 $cm^{-1}$ to those of the $OC_2H_5$ group.

Table 10

Yields and Properties of 1-Alkoxy-3-alkyl-5-alkyl (or phenyl)-1H-1,2,4,6-thiatriazines, [$\dot{-}$S(OR)$\dot{-}$N$\dot{-}$CR$^1$$\dot{-}$N$\dot{-}$CR$^2$$\dot{-}$N$\dot{-}$].

| R | R$^1$ | R$^2$ | yield in % | m.p. in °C (b.p. in °C/Torr) | properties |
|---|---|---|---|---|---|
| $CH_3$ | $CCl_3$ | $CF_3$ | 27 | (44 to 45/0.03) | $n_D^{20}=1.4952$ |
| $CH_3$ | $CCl_3$ | $CCl_3$ | 54 | 59 to 63 | $^1$H NMR spectrum ($CCl_4$/external hexamethyldisiloxane): δ = 3.88 ppm (s, $CH_3$) |
| $CH_3$ | $CCl_3$ | $C_6H_5$ | 71 | 79 to 81 (from $CH_3OH$) | $^1$H NMR spectrum ($CCl_4$/external hexamethyldisiloxane): δ = 3.68 ppm (s, $CH_3$) |
| $C_2H_5$ | $CCl_3$ | $C_6H_5$ | 55 | 101 to 103 (from $CH_3CN$) | |

**Reference:**

Kornuta, P. P., Derii, L. I., Markovskii, L. N. (Zh. Org. Khim. **16** [1980] 1308/13; J. Org. Chem. [USSR] **16** [1980] 1130/4).

## 9.3.2.7 Derivatives of 1-Arylsulfonylimino-1,2,4,6-thiatriazine

### 9.3.2.7.1 1-Arylsulfonylimino-3,5-dimethoxy-1,1-dihydro-4H- and 2H-1,2,4,6-thiatriazines

NSO$_2$R′ ... $CH_3O$ ... $OCH_3$ ... I

NSO$_2$R′ ... $CH_3O$ ... $OCH_3$ ... II

R = H; R′ = $C_6H_5$, 4-Cl$C_6H_4$, 4-$CH_3C_6H_4$
R = $CH_3$; R′ = 4-Cl$C_6H_4$

**[-S(=NSO$_2$R′)-N=C(OCH$_3$)-NR-C(OCH$_3$)=N-]** (I) and
**[-S(=NSO$_2$R′)-NR-C(OCH$_3$)=N-C(OCH$_3$)=N-]** (II)

The compounds with R = H (Table 11, No. 1 to 3, isomers I and II) form from the reaction of HN=C(OCH$_3$)-N=C(OCH$_3$)-NH$_2$ in $CHCl_3$ or $C_6H_6$ with the appropriate N-sulfinyl-arenesulfonamides, R′SO$_2$NSO (1:2 mole ratio), at room temperature for 4 h. The reaction mixtures are allowed to stand for 48 to 60 h [1, 2]. For a discussion of the course of reaction, see [3]. The compound with R = $CH_3$ (Table 11, No. 4) is obtained from a 73:27% mixture of the isomeric 4-(I) and 2-methyl derivatives (II) when a solution of $CH_2N_2$ in ether is added to a suspension of the appropriate thiatriazine (Table 11, No. 2) in ether for 30 min [1].

The compounds are colorless solids with a surprisingly great stability against hydrolysis. Yields, melting points, and $^1$H NMR data are given in Table 11.

Table 11

Yields, Melting Points, and [1]H NMR Data of 1-Arylsulfonylimino-3,5-dimethoxy-1,1-dihydro-4H- and 2H-1,2,4,6-thiatriazines, [–S(=NSO$_2$R′)–N=C(OCH$_3$)–NR–C(OCH$_3$)=N–] and [–S(=NSO$_2$R′)–NR–C(OCH$_3$)=N–C(OCH$_3$)=N–].
Further information on compounds marked with an asterisk is given at the end of the table.

| No. | R | R′ | yield in % | m.p. in °C (recryst. from) | $^1$H NMR spectrum (DMSO-d$_6$/TMS, δ in ppm) | Ref. |
|---|---|---|---|---|---|---|
| 1 | H | $C_6H_5$ | 27 | 166 (i-$C_3H_7OH$) | 3.70 (s, $OCH_3$), 7.30 to 7.82 (m, $C_6H_5$) [1] | [1, 2] |
| *2 | H | 4-Cl$C_6H_4$ | 10 | 151 ($C_2H_5OH$) | 3.73 (d, $OCH_3$), 3.78 (s, $OCH_3$), 7.35 to 7.75 (m, $C_6H_4$) [1] | [1, 2] |
| *3 | H | 4-$CH_3C_6H_4$ | 76 | 170 ($CH_3OH$) | 2.30 (s, $CH_3$), 3.75 (s, $OCH_3$), 7.13 to 7.65 (m, $C_6H_4$) [1] | [1, 2] |
| *4 | $CH_3$ | 4-Cl$C_6H_4$ | 95 | 181 to 182 ($CH_3OH$) | 3.15 (s, $NCH_3$), 3.28 (s, $NCH_3$), 3.60 (s, $OCH_3$), 3.80 (s, $OCH_3$), 3.95 (s, $OCH_3$) 7.45 to 7.83 (m, $C_6H_4$) | [1] |

* Further information:

**[–S(=NSO$_2$C$_6$H$_4$Cl-4)–N=C(OCH$_3$)–NH–C(OCH$_3$)=N–]** (I) and **[–S(=NSO$_2$C$_6$H$_4$Cl-4)–NH–C(OCH$_3$)=N–C(OCH$_3$)=N–]** (II) (Table 11, No. **2**). The $^1$H NMR spectrum shows that both tautomeric derivatives I and II are in an equilibrium in solution. The doublet at 3.73 ppm (OCH$_3$) is assigned to tautomeric derivative II and the singlet at 3.78 ppm (OCH$_3$) to tautomeric derivative I. The position of the signals in the $^1$H NMR spectrum depends on the temperature; tautomeric derivative III was not detectable spectroscopically.

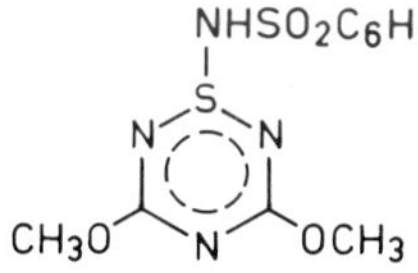

NHSO$_2$C$_6$H$_4$Cl-4
CH$_3$O … OCH$_3$

III

Methylation of the compound by diazomethane results in a mixture of the isomeric 4- and 2-methyl derivatives in the ratio 73:27 [1].

**[–S(=NSO$_2$C$_6$H$_4$CH$_3$-4)–N=C(OCH$_3$)–NH–C(OCH$_3$)=N–]** (I) and **[–S(=NSO$_2$C$_6$H$_4$CH$_3$-4)–NH–C(OCH$_3$)=N–C(OCH$_3$)=N–]** (II) (Table 11, No. **3**). When the reaction is carried out in CCl$_4$, the yield of the product is reduced to 42% [2]. The compound is also obtained from the reaction of 4-CH$_3$C$_6$H$_4$SO$_2$NSO with 4-CH$_3$C$_6$H$_4$SO$_2$N=S=NSO$_2$C$_6$H$_4$CH$_3$-4 at similar conditions in a 10% yield [1, 3]. The substance is very stable; even in boiling water it remains unchanged [1].

An X-ray crystal structure analysis [4] of the compound shows the crystals to be monoclinic, space group $P2_1/c\text{-}C^5_{2h}$ (No. 14) with a = 14.415(1), b = 7.534(1), c = 15.473(2) Å, and β = 120.46(1)°; Z = 4. V = 1448.5 Å$^3$; $D_x$ = 1.515 g/cm$^3$. R = 0.035 for 2698 reflections. The molecular structure is shown in **Fig. 11**, p. 34.

The bonding of the thiatriazine ring exhibits almost perfect mirror symmetry through S(1) and N(4). Nevertheless, the endocyclic torsion angles reveal an asymmetric puckering of the hetero ring. S(IV), situated on top of a distorted trigonal bipyramid, is 0.41 Å out of the best plane of the other five atoms of the hetero ring.

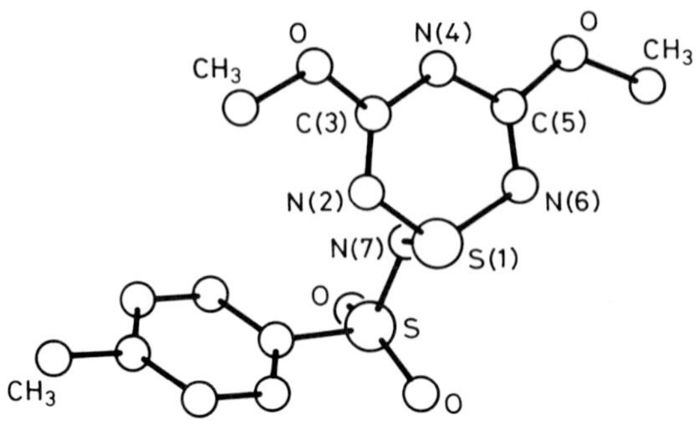


Fig. 11. Molecular structure of
[–S(=$NSO_2C_6H_4CH_3$-4)–N=C($OCH_3$)–NH–C($OCH_3$)=N–].

Tautomer I predominates in the crystal. The proton is fixed between the endocyclic N(4) and the exocyclic N(7) of the imino group by intermolecular interactions in the crystal lattice. Principal bond lengths, bond angles, and torsion angles are given in Table 12. Final coordinates and mean temperature factors for nonhydrogen atoms and parameters for H atoms are given in the paper [4].

Table 12
Selected Bond Lengths, Bond Angles, and Torsion Angles of [–S(=$NSO_2C_6H_4CH_3$-4)–N=C($OCH_3$)–NH–C($OCH_3$)=N–].
Estimated standard deviations in parentheses [4].

| bond length in Å | | | | | |
|---|---|---|---|---|---|
| S(1)–N(2) | 1.658(1) | S(1)–N(7) | 1.617(2) | C(3)–N(4) | 1.365(2) |
| S(1)–N(6) | 1.653(2) | N(2)–C(3) | 1.280(2) | N(4)–C(5) | 1.364(1) |
| | | | | C(5)–N(6) | 1.283(2) |

| bond angle in ° | | | | | |
|---|---|---|---|---|---|
| N(2)–S(1)–N(6) | 106.9(1) | C(3)–N(4)–C(5) | 120.1(2) | N(2)–S(1)–N(7) | 105.3(1) |
| S(1)–N(2)–C(3) | 117.3(2) | N(4)–C(5)–N(6) | 126.4(3) | N(6)–S(1)–N(7) | 101.4(1) |
| N(2)–C(3)–N(4) | 125.7(3) | S(1)–N(6)–C(5) | 117.1(2) | | |

| torsion angle in ° | | | |
|---|---|---|---|
| N(4)–C(3)–N(2)–S(1) | −11.6(2) | N(6)–S(1)–N(2)–C(3) | +25.9(2) |
| N(4)–C(5)–N(6)–S(1) | +5.4(2) | N(6)–C(5)–N(4)–C(3) | +13.6(3) |
| C(5)–N(4)–C(3)–N(2) | −9.9(3) | N(7)–S(1)–N(2)–C(3) | −81.3(2) |
| C(5)–N(6)–S(1)–N(2) | −22.9(2) | N(7)–S(1)–N(6)–C(5) | +87.1(2) |

**[–S(=NSO$_2$C$_6$H$_4$Cl-4)–N=C(OCH$_3$)–NCH$_3$–C(OCH$_3$)=N–]** (I) and **[–S(=NSO$_2$C$_6$H$_4$Cl-4)–NCH$_3$–C(OCH$_3$)=N–C(OCH$_3$)=N–]** (II) (Table 11, No. **4**). Isomer I can be concentrated to 92% by column chromatography through elution with a solvent mixture of 45% $CCl_4$, 45% n-$C_4H_9OH$, and 10% $CH_3OH$; yield 40%; m.p. 182°C. $^1$H NMR spectrum (DMSO-$d_6$/TMS): $\delta$ (in ppm) = 3.28 (s, $NCH_3$), 3.80 (s, $OCH_3$), 7.45 to 7.83 (m, $C_6H_4$). The compound is usable as a histamine $H_2$ antagonist [1].

**References:**


[1] Fischer, E., Rembarz, G., Teller, M. (J. Prakt. Chem. **324** [1982] 920/4).

[2] Fischer, E., Rembarz, G., Teller, M. (Ger. [East] 142338 [1979/80]; C.A. **94** [1981] No. 175179).

[3] Fischer, E., Teller, M., Martin, D. (Wiss. Z. Wilhelm-Pieck-Univ. Rostock Naturwiss. Reihe **31** [1982] 1/4; C.A. **100** [1984] No. 51552).

[4] Kálmán, A., Argay, G., Fischer, E., Teller, M. (Acta Cryst. B **37** [1981] 164/8).


### 9.3.2.7.2 5-Substituted 3-Imino-1,1,3,6-tetrahydro- and 3-Amino-1,1-dihydro-2-(p-tolylsulfonyl)-1-(p-tolylsulfonylimino)-2H-1,2,4,6-thiatriazines

NSO$_2$C$_6$H$_4$CH$_3$-4 / SO$_2$C$_6$H$_4$CH$_3$-4 / H / R / NH

R = C$_6$H$_5$

I

NSO$_2$C$_6$H$_4$CH$_3$-4 / SO$_2$C$_6$H$_4$CH$_3$-4 / R / NH$_2$

R = N(CH$_3$)$_2$, N(C$_2$H$_5$)(CH$_2$C$_6$H$_5$)

II

**[–S(=NSO$_2$C$_6$H$_4$CH$_3$-4)–N(SO$_2$C$_6$H$_4$CH$_3$-4)–C(=NH)–N=CR–NH–]** (I) and
**[–S(=NSO$_2$C$_6$H$_4$CH$_3$-4)–N(SO$_2$C$_6$H$_4$CH$_3$-4)–C(NH$_2$)=N–CR=N–]** (II)

The compounds can exist in several mesomeric structures (the most probable, I and II, are given above) whose statistical weights have not been determined.

**[–S(=NSO$_2$C$_6$H$_4$CH$_3$-4)–N(SO$_2$C$_6$H$_4$CH$_3$-4)–C(NH$_2$)=N–C(N(CH$_3$)$_2$)=N–].** The compound is formed from the reaction of H$_2$N–C(N(CH$_3$)$_2$)=N–C≡N with 4-CH$_3$C$_6$H$_4$SO$_2$N=S=NSO$_2$C$_6$H$_4$CH$_3$-4. The crude product was not purified. No physical data are available. The (impure) compound reacts in moist air within 1 h to give 4-CH$_3$C$_6$H$_4$SO$_2$NH$_2$ and [–SO–N(SO$_2$C$_6$H$_4$CH$_3$-4)–C(NH$_2$)=N–C(N(CH$_3$)$_2$)=N–] (see p. 22) [2].

**[–S(=NSO$_2$C$_6$H$_4$CH$_3$-4)–N(SO$_2$C$_6$H$_4$CH$_3$-4)–C(NH$_2$)=N–C(N(C$_2$H$_5$)(CH$_2$C$_6$H$_5$))=N–].** The compound is formed from equimolar amounts of N≡C–N=C(N(C$_2$H$_5$)(CH$_2$C$_6$H$_5$))–NH$_2$ and 4-CH$_3$C$_6$H$_4$SO$_2$N=S=NSO$_2$C$_6$H$_4$CH$_3$-4 by heating in boiling benzene for 3 d. The red solution is then concentrated by evaporation to half of the original volume. After 14 d at 0 to 5°C the crude product crystallizes and is purified by washing with cold $CH_3OH$; yield 21%; m.p. 125°C. IR spectrum (Nujol, $\nu$ in cm$^{-1}$): 3380, 3200, 3100 (NH, NH$_2$); 1650, 1600 (C=N, NH); 1305, 1290, 1150 (SO$_2$); 980 (N=S), $^1$H NMR spectrum (CDCl$_3$/TMS): $\delta$ (in ppm) = 1.13 (t, CH$_3$), 2.25 (s, 2CH$_3$), 3.44 to 3.78 (q, CH$_2$), 4.73 (s, CH$_2$), 6.93 to 7.75 (m, 2C$_6$H$_4$, C$_6$H$_5$, NH$_2$) [2].

**[–S(=NSO$_2$C$_6$H$_4$CH$_3$-4)–N(SO$_2$C$_6$H$_4$CH$_3$-4)–C(NH$_2$)=N–C(C$_6$H$_5$)=N–].** The compound is formed in a yield of 45% by the reaction of N≡C–N=C(C$_6$H$_5$)–NH$_2$ with equimolar amounts of 4-CH$_3$C$_6$H$_4$SO$_2$N=S=NSO$_2$C$_6$H$_4$CH$_3$-4 in boiling benzene for 40 h, or in a yield of 25% by the reaction of N≡C–N=C(C$_6$H$_5$)–NH$_2$ with 4-CH$_3$C$_6$H$_4$SO$_2$N=S=O (1:2 mole ratio) in boiling benzene for 40 h. M.p. 180 to 182°C with decomposition.

The compound is insoluble in benzene, ethyl acetate, $CHCl_3$, and acetone and is soluble in DMSO. It does not react with cold water, but on boiling with water, $SO_2$ and $4\text{-}CH_3C_6H_4SO_2NH_2$ are split off forming $4\text{-}CH_3C_6H_4SO_2N{=}C(NH_2)N{=}C(C_6H_5)NH_2$ [1]. The compound is hydrolyzed in 96% $C_2H_5OH$ (10 min, reflux) to give $4\text{-}CH_3C_6H_4SO_2N{=}C(NH_2)NHCOC_6H_5$ [2].

**References:**

[1] Seyfried, C. (Diss. München 1966).
[2] Fischer, E., Teller, M., Kálmán, A., Argay, G. (Tetrahedron **40** [1984] 385/90).

## 9.3.2.7.3 2-Arylsulfonylimino-4-alkoxy (and 4-phenoxy)-2,2-dihydro-10H-[1,2,4,6]-thiatriazino[4,3-a]benzimidazoles

$R^1$ and $R^2$ are compiled in Table 13.

**$[\text{-}S(=NSO_2R^1)\text{-}N{=}C(OR^2)\text{-}\{N\text{-}\{1,2\text{-}C_6H_4\}\text{-}NH\text{-}C\}{=}N\text{-}]$**

The title compounds are formed from appropriate 2-amino-1H-benzimidazole-1-carboximidic acid esters (I) and N-sulfinylarenesulfonamides, $4\text{-}R^1C_6H_4SO_2N{=}S{=}O$, in $CHCl_3$. The reaction mixture is refluxed for 1 h and then stirred at room temperature for 1 to 2 d. The crude products are purified by boiling in $CH_3OH$ and recrystallization from benzene. The compounds, yields, and some physical and spectroscopic properties are given in Table 13.

I

II

Table 13
Yields, Melting Points, and Spectral Data of 2-Arylsulfonylimino-4-alkoxy (and 4-phenoxy)-2,2-dihydro-10H-[1,2,4,6]thiatriazino[4,3-a]benzimidazoles,
$[\text{-}S(=NSO_2R^1)\text{-}N{=}C(OR^2)\text{-}\{N\text{-}\{1,2\text{-}C_6H_4\}\text{-}NH\text{-}C\}{=}N\text{-}]$.

| No. | $R^1$ | $R^2$ | yield in % | m.p. in °C | spectra IR (KBr; ν in $cm^{-1}$); $^1H$ NMR (DMSO-$d_6$-$(CD_3)_2CO$/TMS, δ in ppm) |
|---|---|---|---|---|---|
| 1 | $C_6H_5$ | $C_6H_5$ | 59 | 169 to 170 | IR: 3280 (NH); 1695, 1680 (C=N); 1550 (ring); 1285, 1150 ($SO_2$); 1010, 995 (N=S) |
| 2 | $4\text{-}ClC_6H_4$ | $C_6H_5$ | 45 | 184 | IR: 3270 (NH); 1695, 1680 (C=N); 1555 (ring); 1280, 1150 ($SO_2$); 1020, 995 (N=S) |

Table 13 (continued)

| No. | $R^1$ | $R^2$ | yield in % | m.p. in °C | spectra IR (KBr; ν in $cm^{-1}$); $^1H$ NMR (DMSO-$d_6$-$(CD_3)_2CO$/TMS, δ in ppm) |
|---|---|---|---|---|---|
| 3 | 4-$CH_3C_6H_4$ | $CCl_3CH_2$ | 12 | 143 | IR: 3250 (NH); 1695, 1680 (C=N); 1560 (ring); 1280, 1145 ($SO_2$); 1030, 1000 (N=S)<br>$^1H$ NMR: 2.31 (s, $CH_3$), 4.75 to 5.32 (q, $CH_2$), 7.13 to 7.81 (m, 2$C_6H_4$, NH) |
| 4 | 4-$CH_3C_6H_4$ | $C_6H_5$ | 40 | 171 | IR: 3280 (NH); 1680 (C=N); 1550 (ring); 1285, 1130 ($SO_2$); 1000 (N=S)<br>$^1H$ NMR: 2.30 (s, $CH_3$), 7.06 to 7.81 (m, 2$C_6H_4$, $C_6H_5$, NH) |

The compound with $R^1$ = 4-$CH_3C_6H_4$, $R^2$ = $C_6H_5$ (Table 13, No. 4) reacts in boiling aqueous $CH_3OH$ to give the 4-phenoxy-10H-[1,2,4,6]thiatriazino[4,3-a]benzimidazole 2-oxide (II) (see p. 36).

**Reference:**

Fischer, E., Teller, M., Kálmán, A., Argay, G. (Tetrahedron **40** [1984] 385/90).

### 9.3.2.7.4 Derivatives of 2-(p-Tolylsulfonylimino)-1,2-dihydro-2H-pyrazolo[5,1-c][1,2,4,6]-thiatriazin(e)-4(3H)-one and -4(3H)-thione

$C_6H_5$ … $NSO_2C_6H_4CH_3$-4

Y = O; R = H, $C_6H_5$
Y = S; R = H, n-$C_3H_7$

**[–S(=NSO$_2$C$_6$H$_4$CH$_3$-4)–NR–C(=Y)–{N–N=C(C$_6$H$_5$)–CH=C}–NH–]**

The compounds, summarized in Table 14, p. 38, form from the reaction of 3-phenyl-5-amino-1H-pyrazole-1-carboxamides (Y = O) or 3-phenyl-5-amino-1H-pyrazole-1-carbothio-amides (Y = S) dissolved in absolute $CHCl_3$ with 4-$CH_3C_6H_4SO_2N$=S=O (1:2 mole ratio) in $CHCl_3$ at 20 to 45°C. From the resulting red solution a mixture of solids precipitates after 6 to 12 h. The solids are extracted with hot $CHCl_3$ or ethyl acetate to separate p-toluenesulfonamide leaving the title compounds as light yellow crystalline solids. Yields, melting points, and assigned IR spectral data of the synthesized derivatives are given in Table 14; a mechanism of the reaction is suggested in the paper.

$C_6H_5$ … $NH_2$ … NHR … Y + 2 O=S=$NSO_2C_6H_4CH_3$-4 ⟶

$C_6H_5$ … $NSO_2C_6H_4CH_3$-4 … R … Y + 4-$CH_3C_6H_4SO_2NH_2$ + $SO_2$

Table 14
Yields, Melting Points, and IR Spectral Data of Derivatives of 2-(p-Tolylsulfonylimino)-1,2-dihydro-2H-pyrazolo[5,1-c][1,2,4,6]thiatriazin(e)-4(3H)-one and -4(3H)-thione,
[–S(=$NSO_2C_6H_4CH_3$-4)–NR–C(=Y)–{N–N=C($C_6H_5$)–CH=C}–NH–].

| Y | R | yield in % | m.p. in °C | IR spectrum (Nujol, ν in $cm^{-1}$) |
|---|---|---|---|---|
| O | H | 25 | 192 to 196 | 3380, 3250 (NH)<br>1320 ($SO_2$, asym. stretching)<br>1170 ($SO_2$, sym. stretching)<br>1005 (N=S, stretching) |
| O | $C_6H_5$ | 21 | 135 to 136 | 3380 (NH)<br>1310 ($SO_2$, asym. stretching)<br>1160 ($SO_2$, sym. stretching)<br>1005 (N=S, stretching) |
| S | H | 18 | 123 | 3380, 3270 (NH)<br>1310 ($SO_2$, asym. stretching)<br>1160 ($SO_2$, sym. stretching)<br>1005 (N=S, stretching) |
| S | n-$C_3H_7$ | 10 | 139 | 3320 (NH)<br>1005 (N=S, stretching) |

The light yellow crystalline compounds are slightly soluble in common organic solvents; they can be recrystallized from DMSO. In air or in solvents containing moisture, the compounds hydrolyze, splitting off p-toluenesulfonamide to give the corresponding S oxides.

The mass spectra (70 eV, 290°C) of the compounds show a similar fragmentation scheme; the molecular peaks have a relative intensity of only 1%. The $M^+$ ions fragment by splitting off NH from the pyrazole and by splitting off $C_6H_5C_2N$ (m/e = 115) to give the appropriate 1,2,4,6-thiatriazine derivatives.

**Reference:**

Fischer, E., Teller, M., Schröder, H. (Sulfur Letters **1** [1983] 119/25).

### 9.3.2.7.5 3-Amino-1-arylsulfonylimino-1,1-dihydro-4H-[1,2,4,6]thiatriazino[2,3-a]benzimidazoles

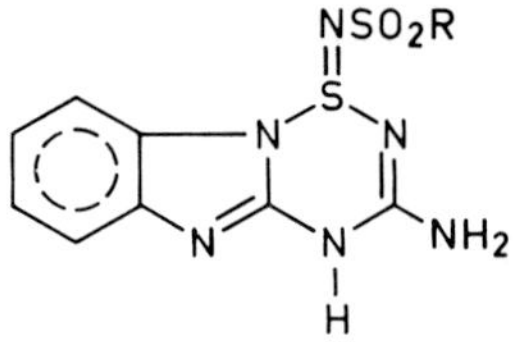

R = $C_6H_5$, 4-$ClC_6H_4$, 4-$CH_3C_6H_4$

**[–S(=$NSO_2R$)–N=C($NH_2$)–NH–{C=N–{1,2-$C_6H_4$}–N}–]**

The title compounds have been obtained from the reaction of 2-guanidino-1H-benzimidazole(I) with appropriate N-sulfinylarenesulfonamides, $RSO_2N{=}S{=}O$ in $CHCl_3$. The reaction mixture is refluxed for 1 h and then stirred at room temperature for 1 to 2 d. The crude products are purified by boiling in $CH_3OH$ and recrystallization from benzene. The compounds prepared with yields and some physical and spectroscopic properties are given in Table 15.

Table 15

Yields, Melting Points, and Spectral Data of 3-Amino-1-arylsulfonylimino-1,1-dihydro-4H-[1,2,4,6]thiatriazino[2,3-a]benzimidazoles, [–S(=$NSO_2R$)–N=C($NH_2$)–NH–{C=N–{1,2-$C_6H_4$}–N}–].

| No. | R | yield in % | m.p. in °C | IR (KBr; ν in $cm^{-1}$); $^1H$ NMR (DMSO-$d_6$–$(CD_3)_2CO$/TMS; δ in ppm) |
|---|---|---|---|---|
| 1 | $C_6H_5$ | 58 | 104 to 107 | IR: 3420, 3350, 3240 (NH, $NH_2$); 1650 (C=N); 1280, 1150 ($SO_2$); 1020, 990 (N=S); 1005, 950 |
| 2 | 4-$ClC_6H_4$ | 56 | 119 to 121 | IR: 3435, 3340, 3235 (NH, $NH_2$); 1650 (C=N); 1280, 1160 ($SO_2$); 1020, 1000 (N=S); 950 |
| 3 | 4-$CH_3C_6H_4$ | 75 | 122 to 123 | IR: 3420, 3340, 3240 (NH, $NH_2$); 1670 (C=N); 1300, 1155 ($SO_2$); 1020, 990 (N=S); 1005, 950<br>$^1H$ NMR: 2.25 (d, $CH_3$), 6.88 to 7.63 (m, 2$C_6H_4$, NH, $NH_2$) |

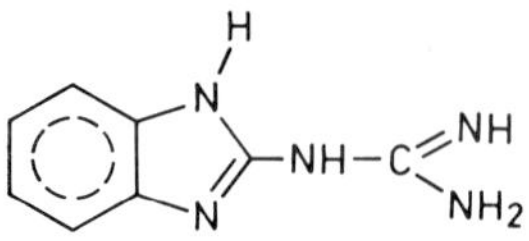

I 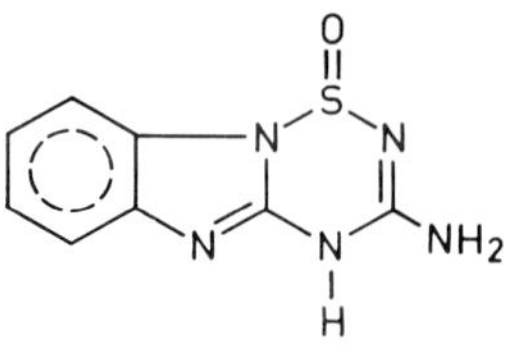    II

Compounds No. 1 and 2 in Table 15 react when heated to 115 to 117°C for 3 min in glacial acetic acid to give 3-amino-4H-[1,2,4,6]thiatriazino[2,3-a]benzimidazole 1-oxide (II) (see p. 30).

**Reference:**

Fischer, E., Teller, M., Kálmán, A., Argay, G. (Tetrahedron **40** [1984] 385/90).

### 9.3.2.8 Derivatives of 1-Dialkylamino-1H-1,2,4,6-thiatriazine

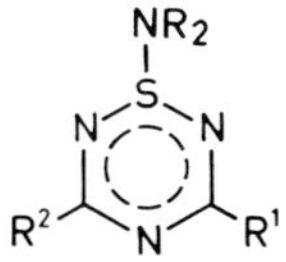

**[∸S($NR_2$)∸N∸$CR^1$∸N∸$CR^2$∸N∸]**

#### 9.3.2.8.1 1-Dialkylamino-3-chloro-5-substituted-1H-1,2,4,6-thiatriazines, [∸S($NR_2$)∸N∸CCl∸N∸$CR^2$∸N∸]

#### 9.3.2.8.1.1 Survey

The 3,5-dichloro-substituted compounds [∸S($NR_2$)∸N∸CCl∸N∸CCl∸N∸] are prepared from [∸SCl∸N∸CCl∸N∸CCl∸N∸] and appropriate secondary amines. The otherwise 5-substituted compounds [∸S($NR_2$)∸N∸CCl∸N∸C$R^2$∸N∸] form from the reaction of [∸S($NR_2$)∸N∸CCl∸N∸CCl∸N∸] with corresponding alcoholates, amines, thiolates, or carbaminates, and with Grignard reagents by nucleophilic substitution.

The cyclic structure of the compounds was concluded from spectral data as well as from their reaction behavior and was established by X-ray crystallography of some derivatives, e.g., [∸S(N($C_3H_7$-i)$_2$)∸N∸CCl∸N∸C(NH$C_6H_{11}$-c)∸N∸].

### 9.3.2.8.1.2 1-Dialkylamino-3,5-dichloro-1H-1,2,4,6-thiatriazines, [∸S($NR_2$)∸N∸CCl∸N∸CCl∸N∸], $NR_2$ is compiled in Table 16.

The compounds listed in Table 16 have been synthesized from 1,3,5-trichloro-1H-1,2,4,6-thiatriazine, [∸SCl∸N∸CCl∸N∸CCl∸N∸] (see p. 53), by treatment with the corresponding secondary amines (1:≧2 mole ratio) in ether at −5°C [1] or −40°C [2, 3, 4]. Yields and melting points are given in Table 16.

Table 16
Yields and Melting Points of 1-Dialkylamino-3,5-dichloro-1H-1,2,4,6-thiatriazines, [∸S($NR_2$)∸N∸CCl∸N∸CCl∸N∸].

| $NR_2$ | yield in % | m.p. in °C (recryst. from) | Ref. |
|---|---|---|---|
| N($C_3H_7$-i)$_2$ | 66 | 116.5 to 118.5 ($CH_3OH$–$H_2O$ 4:1) | [1] |
| | 30 to 37 | 114 to 116/dec. ($CH_3OH$) | [3] |
| | | 116 to 118 | [4] |
| N($C_6H_{11}$-c)$_2$ | 30 | 148 to 149/dec. ($CH_3OH$) | [2, 3] |
| $CH_3$<br>–N (piperidine ring)<br>$CH_3$ | 48 (crude product) | 93 to 94.5 ($CH_3OH$–$H_2O$ 4:1) | [1] |
| | | 91 to 93 | [4] |

Originally, the compounds [∸S($NR_2$)∸N∸CCl∸N∸CCl∸N∸] had been erroneously characterized as having an acyclic structure [1]. The six-membered heterocyclic structure was concluded from IR [2] and NMR spectral data [5, 6] and established by X-ray diffraction data [7] of the diisopropylamino derivative.

Crystals of this compound ($NR_2$ = N($C_3H_7$-i)$_2$) are orthorhombic, space group Pnam–$D_{2h}^{16}$ (No. 62), with a = 13.170(7), b = 7.537(3), and c = 13.137(4) Å; Z = 4. V = 1304 Å$^3$. $D_x$ = 1.371 g/cm$^3$. R = 0.039 for 668 reflections.

The molecular structure is shown in **Fig. 12.**

The thiatriazine ring is not planar, it has a half-chair conformation. The S atom is significantly (by 0.26 Å) out of the best plane for the remaining five atoms. It is situated at the top of a distorted trigonal bipyramid and makes three S–N multiple bonds. The exocyclic N(4) atom is linked pseudoaxially. Selected bond lengths and bond angles are given in Table 17; for endocyclic and exocyclic torsion angles as well as for final coordinates for the nonhydrogen atoms and parameters for H atoms, see the paper [7].

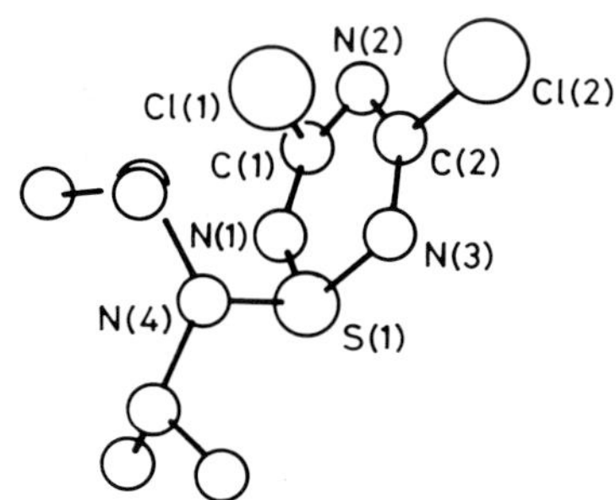


Fig. 12. Molecular structure of [$\overset{..}{-}$S(N($C_3H_7$-i)$_2$)$\overset{..}{-}$N$\overset{..}{-}$CCl$\overset{..}{-}$N$\overset{..}{-}$CCl$\overset{..}{-}$N$\overset{..}{-}$].

Table 17

Selected Bond Lengths and Bond Angles of 1-Diisopropylamino-3,5-dichloro-1H-1,2,4,6-thiatriazine, [$\overset{..}{-}$S(N($C_3H_7$-i)$_2$)$\overset{..}{-}$N$\overset{..}{-}$CCl$\overset{..}{-}$N$\overset{..}{-}$CCl$\overset{..}{-}$N$\overset{..}{-}$].

| bond length | in Å | bond angle | in ° |
|---|---|---|---|
| S(1)–N(1) | 1.655(3) | N(1)–S(1)–N(3) | 105.6(2) |
| N(1)–C(1) | 1.300(4) | S(1)–N(1)–C(1) | 117.0(2) |
| C(1)–N(2) | 1.329(4) | N(1)–C(1)–N(2) | 132.2(3) |
| S(1)–N(4) | 1.590(4) | N(1)–C(1)–Cl(1) | 114.3(2) |
| C(1)–Cl(1) | 1.739(3) | N(2)–C(1)–Cl(1) | 113.6(2) |
| | | C(1)–N(2)–C(2) | 113.8(4) |

ESCA spectroscopic investigations of the compound with $NR_2 = N(C_3H_7\text{-i})_2$ confirm a valency of four for the ring sulfur. The bond energy of the sulfur 2p level was found to be 166.85 eV (with N correction) and 167.05 eV (with C correction) [8]. A rotation barrier of only 38 to 45 kJ/mol has been found for the exocyclic S(1)–N(4) bond in spite of the short distance of 1.59 Å with unambiguous ylidic bond character. Therefore interactions of 3pπ(S)–2pπ(N) or 3dπ(S)–2pπ(N) orbitals should not significantly influence the barrier to rotation of the S(1)–N(4) bond [4].

$^1$H NMR spectrum of the compound with $NR_2 = N(C_3H_7\text{-i})_2$ (15% in $CCl_4$/internal hexamethyldisiloxane): δ (in ppm) = 1.30 (d, $CH_3$), 3.35 (sept, CH) [5]. $^{13}$C NMR spectrum of the compound with $NR_2 = N(C_3H_7\text{-i})_2$ ($CDCl_3$/TMS): δ (in ppm) = 23.26 ($CH_3$), 48.99 (CH), 164.42 (C(ring)) [6].

The IR spectra of the compounds show characteristic absorption bands at 1510 and 1350 $cm^{-1}$ ($NR_2 = N(C_3H_7\text{-i})_2$) or at 1550 and 1370 $cm^{-1}$ ($NR_2 = N(C_6H_{11}\text{-c})_2$) which have been assigned to in-plane vibrations of the ring [2].

A strong absorption at 280 nm ($\varepsilon = 1.6 \times 10^4$) appears in the UV-visible spectrum of the diisopropylamino derivative [2].

The thermal stability of the three derivatives is not very high [4]. The diisopropylamino and dicyclohexylamino derivatives decompose at temperatures above their melting points with evolution of ClCN. The compounds cannot be stored in air without decomposition. Intensely exothermic spontaneous decomposition occurs occasionally; therefore large amounts must be stored and handled with caution [2]. Crystalline diisopropylamino and dicyclohexylamino derivatives are stable for quite a long time when stored in dry argon at low temperatures or when dissolved in anhydrous aprotic solvents [9].

The catalytic hydrogenation of the compounds with noble metal catalysts in absolute organic solvents failed [9], but the hydrogenation of 1-diisopropylamino-3,5-dichloro-1H-1,2,4,6-thiatriazine was carried out successfully in aqueous alcohol in the presence of a Raney nickel catalyst. At a $H_2$ pressure as low as 1 bar, ring cleavage is initiated to give nickel sulfide by coincident HCl elimination with reformation of the nitrile group. This leads to the formation of $(i\text{-}C_3H_7)_2NH_2Cl$ and $NCN{=}CClNH_2$ [9], see also [10]. For the course of reaction of the catalytic hydrogenation, see [9]. If the catalytic hydrogenation of the compound is performed at $H_2$ standard pressure in the presence of Raney nickel and by using water as solvent, the nickel complex I is obtained [9].

```
HC—NH HN—CH
/ ⁄‾‾\ \ / ⁄‾‾\ \
N (     ) Ni (     ) N
\ \__/ / \ \__/ /
HC—NH HN—CH
```

I

Electrolytic reduction of mixtures of $[\dot{-}S(N(C_3H_7\text{-}i)_2)\dot{-}N\dot{-}CCl\dot{-}N\dot{-}CCl\dot{-}N\dot{-}]$ and $CH_3OH$–$H_2O$ (9 : 1) in buffered systems using $NaClO_4$ as conducting salt leads to ring destruction; sulfide and chloride ions, as well as $(i\text{-}C_3H_7)_2NH_2^+$ and $NH_4^+$ salts have been detected [10].

The compounds listed in Table 16, p. 40, hydrolyze in $CH_3OH$–$H_2O$ mixtures to give $SO_2$, $CO_2$, and dialkylamine hydrochloride. Partial hydrolysis products have been found to be cyanamide, biuret, and dialkylcyanoguanidine, depending on the reaction conditions [4, 9, 11]. A proposed hydrolysis mechanism was given [9]. For the diisopropylamino derivative hydrolysis in a $CH_3OH$–$H_2O$ (9 : 1) mixture was followed by observing the $Cl^-$ concentration as a function of time and pH. It was shown that the compound can be handled without decomposition in neutral or acidic media for a short time; in alkaline solution either one or two equivalents of $Cl^-$ per mol of the compound are quickly split off depending on the $OH^-$ concentration [9].

The chlorine atoms of the 1-diisopropyl (or 1-dicyclohexyl)amino-3,5-dichloro-1H-1,2,4,6-thiatriazines can be replaced by nucleophilic substituents with ring conservation; by $NH_3$ and aliphatic amines to form the corresponding monoamino-substituted compounds (see p. 45) [3, 5 to 12]; by the aromatic amines $4\text{-}CH_3OC_6H_4NH_2$ and $4\text{-}CH_3C_6H_4NH_2$ to produce red crystalline compounds containing no Cl and presumed to be N-substituted 1,3,5-triamino-1H-1,2,4,6-thiatriazines (see p. 52); with $C_6H_5NH_2$ or $4\text{-}ClC_6H_4NH_2$ no reaction was observed [12]. With the sodium thiolates RSNa, $R = CH_3$, $C_2H_5$, $i\text{-}C_3H_7$, $n\text{-}C_4H_9$, $C_6H_5CH_2$, $C_6H_5$, in alcohol at room temperature, either the corresponding 5-mono- or the 3,5-disubstituted alkyl (aryl)thio compounds were obtained, depending on the stoichiometric ratio (see pp. 43 and 50) [3, 13]. With sodium N,N-dialkylcarbamodithioates $NaSC(S)NR_2$ with $R = CH_3$, $C_2H_5$ in alcohol, 1-diisopropyl (or dicyclohexyl)amino-3-chloro-5-(N,N-dialkylthiocarbamoylthio)-1H-1,2,4,6-thiatriazines form (see p. 44). Substitution of the second Cl was not successful with an excess of the carbaminates and by heating the reaction mixture [13], see also [3]. The analogous reaction of the 1-diisopropyl- and 1-dicyclohexylamino-3,5-dichloro-1H-1,2,4,6-thiatriazines with N,N-dialkylcarbamothioates, $NaSC(O)NR_2$ with $R = CH_3$, $C_2H_5$, under the same conditions leads not to isolable derivatives but to COS and other decomposition products; in the presence of water, cyanoguanidine, amine hydrochloride, $SO_2$, and COS are formed [13].

The diisopropylamino derivative reacts with the Grignard compound $C_6H_5MgBr$ in ether at room temperature to form $[\dot{-}S(N(C_3H_7\text{-}i)_2)\dot{-}N\dot{-}CCl\dot{-}N\dot{-}C(C_6H_5)\dot{-}N\dot{-}]$, see p. 49 [5].

**References:**


[1] Geevers, J., Hackmann, J. T., Trompen, W. P. (J. Chem. Soc. C **1970** 875/8).

[2] Schramm, W., Voß, G., Rembarz, G., Fischer, E. (Z. Chem. [Leipzig] **14** [1974] 471/2).

[3] Schramm, W., Voß, G., Rembarz, G., Fischer, E. (Ger. [East] 113006 [1974/75]; C.A. **84** [1976] No. 105666).
[4] Fischer, E. (Wiss. Z. Wilhelm-Pieck-Univ. Rostock Math. Naturwiss. Reihe **27** [1978] 609/16).
[5] Michalik, M., Fischer, E., Rembarz, G., Voß, G., Storek, W. (J. Prakt. Chem. **319** [1977] 739/44).
[6] Storek, W., Schramm, W., Voß, G., Rembarz, G., Fischer, E. (Z. Chem. [Leipzig] **15** [1975] 104/5).
[7] Kálmán, A., Argay, G., Fischer, E., Rembarz, G. (Acta Cryst. B **35** [1979] 860/6).
[8] Leonhardt, G., Scheibe, R., Schramm, W., Voß, G., Fischer, E., Rembarz, G. (Z. Chem. [Leipzig] **15** [1975] 193/4).
[9] Fischer, E., Rembarz, G., Klatt, E., Weber, H., Rachimowa, A. A. (Wiss. Z. Wilhelm-Pieck-Univ. Rostock Math. Naturwiss. Reihe **28** [1979] 861/4).
[10] Jeroschewski, P., Voß, G., Fischer, E. (Z. Chem. [Leipzig] **17** [1977] 145).

[11] Fischer, E., Teller, M., Kálmán, A., Argay, G. (Tetrahedron **40** [1984] 385/90).
[12] Schramm, W., Voß, G., Michalik, M., Rembarz, G., Fischer, E. (Z. Chem. [Leipzig] **15** [1975] 19).
[13] Schramm, W., Voß, G., Rembarz, G., Fischer, E. (Z. Chem. [Leipzig] **15** [1975] 57/8).

### 9.3.2.8.1.3 1-Diisopropylamino-3-chloro-5-ethoxy-1H-1,2,4,6-thiatriazine, [$\dot{-}$S(N($C_3H_7$-i)$_2$)$\dot{-}$N$\dot{-}$CCl$\dot{-}$N$\dot{-}$C($OC_2H_5$)$\dot{-}$N$\dot{-}$]

The formation of the compound has not been described but presumably occurs during the reaction of [$\dot{-}$S(N($C_3H_7$-i)$_2$)$\dot{-}$N$\dot{-}$CCl$\dot{-}$N$\dot{-}$CCl$\dot{-}$N$\dot{-}$] with $NaOC_2H_5$. $^1$H NMR spectrum (15% in $CCl_4$/internal hexamethyldisiloxane): δ (in ppm) = 1.20 and 1.24 (d, $NC(CH_3)_2$), 1.24 (t, $OCCH_3$), 3.28 (sept, CH), 4.16 (q, $CH_2$). $^{13}$C NMR spectrum (25% in $CDCl_3$/TMS): δ (in ppm) = 14.11 ($CH_2\underline{C}H_3$), 19.16 ($CH(\underline{C}H_3)_2$), 47.41 (CH), 65.27 ($OCH_2$), 116.23 (C(ring)–O), 165.45 (C(ring)–Cl).

**Reference:**

Michalik, M., Fischer, E., Rembarz, G., Voß, G., Storek, W. (J. Prakt. Chem. **319** [1977] 739/44).

### 9.3.2.8.1.4 1-Diisopropylamino-3-chloro-5-alkyl (or aryl)thio-1H-1,2,4,6-thiatriazines, [$\dot{-}$S(N($C_3H_7$-i)$_2$)$\dot{-}$N$\dot{-}$CCl$\dot{-}$N$\dot{-}$CR$^2\dot{-}$N$\dot{-}$], $R^2 = SCH_3$, $SC_2H_5$, $SC_6H_5$

The title compounds form from the reaction of [$\dot{-}$S(N($C_3H_7$-i)$_2$)$\dot{-}$N$\dot{-}$CCl$\dot{-}$N$\dot{-}$CCl$\dot{-}$N$\dot{-}$] (see above) with equimolar amounts of the corresponding sodium thiolates, RSNa, R = $CH_3$, $C_2H_5$, $C_6H_5$, in alcohol at room temperature [1].

Yields and melting points of the compounds are listed in the following table:

| $R^2$ | yield in % | m.p. in °C |
|---|---|---|
| $SCH_3$ | 80 [1] | 63 to 65 [1] |
| | 85 [2] | 60 to 65 [2] |
| $SC_2H_5$ | 73 [1] | 45 to 47 [1] |
| $SC_6H_5$ | 60 [1] | 71 to 74 [1] |

The compounds form colorless crystals with difficulty. In air they decompose with evolution of $SO_2$. Their IR spectra show characteristic absorption bands in the 1520 to 1500 $cm^{-1}$ and 1350 to 1330 $cm^{-1}$ ranges [1].

For [$\dot{-}$S(N($C_3H_7$-i)$_2$)$\dot{-}$N$\dot{-}$CCl$\dot{-}$N$\dot{-}$C($SCH_3$)$\dot{-}$N$\dot{-}$] spectral data are available: $^1$H NMR (15% in $CDCl_3$/internal hexamethyldisiloxane): δ (in ppm) = 1.25 and 1.26 (d, C($CH_3$)$_2$), 2.34 (s, $SCH_3$), 3.42 (sept, CH) [3]. $^{13}$C NMR (25% in $CDCl_3$/TMS): δ (in ppm) = 20.70 and 23.33 (C($\underline{C}H_3$)$_2$), 23.33 ($SCH_3$), 45.67 and 46.84 (SNCH), 159.23 (C(ring)–S), 163.38 (C(ring)–Cl) [3, 4]. The appearance of two signals for the methine group as well as for the methyl groups was discussed [3]. ESCA spectroscopic investigations confirm a valency of four for the ring sulfur. The bond energy of the sulfur 2p level is 166.65 eV (with N correction) and 166.7 eV (with C correction) [5].

**References:**


[1] Schramm, W., Voß, G., Rembarz, G., Fischer, E. (Z. Chem. [Leipzig] **15** [1975] 57/8).

[2] Schramm, W., Voß, G., Rembarz, G., Fischer, E. (Ger. [East] 113006 [1974/75]; C.A. **84** [1976] No. 105666).

[3] Michalik, M., Fischer, E., Rembarz, G., Voß, G., Storek, W. (J. Prakt. Chem. **319** [1977] 739/44).

[4] Storek, W., Schramm, W., Voß, G., Rembarz, G., Fischer, E. (Z. Chem. [Leipzig] **15** [1975] 104/5).

[5] Leonhardt, G., Scheibe, R., Schramm, W., Voß, G., Fischer, E., Rembarz, G. (Z. Chem. [Leipzig] **15** [1975] 193/4).


### 9.3.2.8.1.5 1-Dialkylamino-3-chloro-5-(N,N-dialkylthiocarbamoylthio)-1H-1,2,4,6-thiatriazines

**[$\dot{-}$S($NR_2$)$\dot{-}$N$\dot{-}$CCl$\dot{-}$N$\dot{-}$C$R^2$$\dot{-}$N$\dot{-}$]**, $NR_2$ and $R^2$ are compiled in the table below.

The compounds form from the reaction of 1-diisopropyl (or dicyclohexyl)amino-3,5-dichloro-1H-1,2,4,6-thiatriazines with sodium N,N-dimethyl (or diethyl)carbamodithioates, NaSC(S)$NR_2$, R = $CH_3$, $C_2H_5$ (1:1 mole ratio) in absolute ethanol. Yields and melting points are listed in the following table:

| $NR_2$ | $R^2$ | yield in % | m.p. in °C |
|---|---|---|---|
| N($C_3H_7$-i)$_2$ | SC(S)N($CH_3$)$_2$ | 85 | 111 to 113 |
| N($C_3H_7$-i)$_2$ | SC(S)N($C_2H_5$)$_2$ | 40 | 145 to 147 |
| N($C_6H_{11}$-c)$_2$ | SC(S)N($C_2H_5$)$_2$ | 62 | 126 to 129 |

The compounds form stable yellow crystals [1].

[$\dot{-}$S(N($C_3H_7$-i)$_2$)$\dot{-}$N$\dot{-}$CCl$\dot{-}$N$\dot{-}$C(SC(S)N($C_2H_5$)$_2$)$\dot{-}$N$\dot{-}$]. UV spectrum (solvent?): $\lambda_{max}$ = 297 nm (ε = 8.2 × $10^4$ L·$mol^{-1}$·$cm^{-1}$) [1]. $^{13}$C NMR spectrum ($CDCl_3$/TMS): δ (in ppm) = 10.60 and 13.72 ($CH_2\underline{C}H_3$), 23.37 and 23.49 (CH($\underline{C}H_3$)$_2$), 48.46 and 50.27 ($\underline{C}H_2CH_3$), 48.64 ($\underline{C}$H($CH_3$)$_2$), 163.31 (C(ring)–Cl), 173.90 (C(ring)–S) [2].

**References:**


[1] Schramm, W., Voß, G., Rembarz, G., Fischer, E. (Z. Chem. [Leipzig] **15** [1975] 57/8).

[2] Storek, W., Schramm, W., Voß, G., Rembarz, G., Fischer, E. (Z. Chem. [Leipzig] **15** [1975] 104/5).

**9.3.2.8.1.6 1-Dialkylamino-3-chloro-5-amino (or mono-or dialkylamino)-1H-1,2,4,6-thiatriazines, [∸S($NR_2$)∸N∸CCl∸N∸$CR^2$∸N∸],** $NR_2$ and $R^2$ are listed in Table 18.

The title compounds compiled in Table 18 are formed from 1-diisopropylamino, 1-diisobutylamino-, 1-dicyclohexylamino-, and 1-(2,6-dimethylpiperidino)-3,5-dichloro-1H-1,2,4,6-thiatriazine by nucleophilic substitution of a Cl atom by reaction with $NH_3$ or primary and secondary amines in ether at room temperature [1 to 4]. Yields and melting points are given in Table 18.

Table 18
Yields and Melting Points of 1-Dialkylamino-3-chloro-5-amino (or mono- or dialkylamino)-1H-1,2,4,6-thiatriazines, [∸S($NR_2$)∸N∸CCl∸N∸$CR^2$∸N∸].

| No. | $NR_2$ | $R^2$ | yield in % | m.p. with dec. in °C | Ref. |
|---|---|---|---|---|---|
| 1 | $N(C_3H_7\text{-}i)_2$ | $NH_2$ | 50 | 138 to 139 | [4] |
| 2 | $N(C_3H_7\text{-}i)_2$ | $NHCH_3$ | 33 | 115 to 120 | [4] |
| 3 | $N(C_3H_7\text{-}i)_2$ | $NHC_2H_5$ | 83 | 98 to 104 | [2] |
| 4 | $N(C_3H_7\text{-}i)_2$ | $NHC_3H_7\text{-}n$ | 55 | 99 to 103 | [4] |
| 5 | $N(C_3H_7\text{-}i)_2$ | $NHCH_2CH{=}CH_2$ | 70 | 80 a) | [4] |
| 6 | $N(C_3H_7\text{-}i)_2$ | $NHC_4H_9\text{-}t$ | 85 | 111 to 115 | [4] |
| 7 | $N(C_3H_7\text{-}i)_2$ | $NHC_6H_{11}\text{-}c$ | 80 | 116 to 119 | [4] |
| | | | 84 | 118 to 120 | [1,5] |
| 8 | $N(C_3H_7\text{-}i)_2$ | $NHCH_2C_6H_5$ | 59 | 115 to 118 | [4] |
| 9 | $N(C_3H_7\text{-}i)_2$ | –N (piperidino ring) | 83 | 98 to 100 | [4] |
| 10 | $N(C_3H_7\text{-}i)_2$ | $N(CH_3)_2$ | — | dec. | [6] |
| 11 | $N(C_3H_7\text{-}i)_2$ | $N(C_2H_5)_2$ | 85 | 145 to 147 b) | [1] |
| 12 | $N(C_3H_7\text{-}i)_2$ | $N(C_2H_5)C_6H_{11}\text{-}c$ | 81 | 109 to 111 | [4] |
| 13 | $N(C_3H_7\text{-}i)_2$ | $N(C_4H_9\text{-}i)_2$ | 80 | 104 to 105 | [2] |
| | | | | 105 to 106 | [6] |
| 14 | $N(C_3H_7\text{-}i)_2$ | $N(CH_2C_6H_5)_2$ | — | 75 to 78 | [6] |
| 15 | $N(C_4H_9\text{-}i)_2$ | $N(C_4H_9\text{-}i)_2$ | — | 65 to 68 | [6] |
| 16 | $N(C_6H_{11}\text{-}c)_2$ | $NHC_6H_{11}\text{-}c$ | 60 | 125 to 130 | [4] |
| 17 | $CH_3$<br>–N (piperidino ring)<br>$CH_3$ | $N(CH_2C_6H_5)_2$ | — | — | [6] |

a) The compound shows an extensive range of decomposition. – b) Without decomposition, yellow crystals.

The nonplanar ring structure of the compounds was concluded from NMR spectral data [2] and established by X-ray diffraction data [5] of the derivative with $NR_2 = N(C_3H_7\text{-}i)_2$, $R^2 = NHC_6H_{11}\text{-}c$ (see below). From the high activation energies (70 to 74 kJ/mol) for the rotation of the dialkylamino group in the 5 position around the C–N bond, it was concluded that polar structures do have a noticeable effect on the ground state of these compounds [6].

The crystal structure of [∸S($N(C_3H_7\text{-}i)_2$)∸N∸CCl∸N∸C($NHC_6H_{11}\text{-}c$)∸N∸] (Table 18, No. 7) has been established by X-ray diffraction analysis. The crystals are monoclinic, space group $P2_1\text{–}C_2^2$ (No. 4), with a = 12.730(3), b = 12.085(3), c = 13.369(4) Å, and $\beta$ = 115.91(2)°; Z = 4 (two molecules in the asymmetric unit); V = 1849.98 Å$^3$, $D_x$ = 1.191 g/cm$^3$, $D_m$ = 1.18 g/cm$^3$. R = 0.050 for 2363 reflections.

The molecular structure is shown in a stereoscopic figure in the original paper [5]. In the crystal two molecular species exist, A and B. The thiatriazine ring in both of the symmetry-independent molecules is clearly nonplanar, each having a twisted half-boat conformation. The hetero-rings in the two molecules have slightly different conformations. Molecule A shows a mean deviation from planarity of 0.035 Å, with the S atom 0.36 Å out of the best plane through the remaining five atoms, whereas for molecule B the corresponding deviation is 0.020 Å, with the S atom 0.27 Å out of the plane. The S atom is situated at the top of a distorted trigonal pyramid. This gives rise to three different S–N bonds [5]. The Cl substituent is axial to the ring system and forms a bond nearly perpendicular to the ring plane [6]. Selected bond lengths and bond angles are given in Table 19. For final coordinates for all atoms, torsion angles, and Newman projections showing the conformations of the characteristic moieties for both molecules A and B, see the paper [5].

Table 19
Bond Lengths and Bond Angles of
1-Diisopropylamino-3-chloro-5-cyclohexylamino-1H-1,2,4,6-thiatriazine,
$[\overset{..}{-}S(N(C_3H_7\text{-}i)_2)\overset{..}{-}N\overset{..}{-}CCl\overset{..}{-}N\overset{..}{-}C(NHC_6H_{11}\text{-}c)\overset{..}{-}N\overset{..}{-}]$
(Table 18, No. 7.)

| | bond length in Å molecule A | bond length in Å molecule B | | bond angle in ° molecule A | bond angle in ° molecule B |
|---|---|---|---|---|---|
| S–N(1) | 1.668(6) | 1.670(5) | N(3)–S–N(1) | 106.8(3) | 107.2(3) |
| N(1)–C(1) | 1.281(10) | 1.289(10) | S–N(1)–C(1) | 113.9(5) | 114.4(5) |
| C(1)–N(2) | 1.322(9) | 1.310(10) | N(1)–C(1)–N(2) | 135.3(6) | 134.2(5) |
| N(2)–C(2) | 1.384(8) | 1.372(7) | C(1)–N(2)–C(2) | 113.9(6) | 115.6(6) |
| C(2)–N(3) | 1.325(9) | 1.323(8) | N(2)–C(2)–N(3) | 126.6(6) | 126.8(7) |
| N(3)–S | 1.621(6) | 1.620(6) | C(2)–N(3)–S | 118.0(4) | 118.6(4) |
| C(1)–Cl | 1.777(7) | 1.769(6) | N(1)–C(1)–Cl | 112.9(5) | 113.0(5) |
| C(2)–N(4) | 1.333(9) | 1.339(10) | N(2)–C(1)–Cl | 111.9(5) | 112.8(5) |
| S–N(5) | 1.645(5) | 1.634(6) | N(2)–C(2)–N(4) | 113.0(6) | 113.9(6) |
| | | | N(3)–C(2)–N(4) | 120.3(6) | 119.2(5) |
| | | | N(1)–S–N(5) | 104.6(3) | 102.4(3) |
| | | | N(3)–S–N(5) | 110.2(3) | 109.0(3) |

ESCA spectroscopic investigations of the derivatives with $NR_2 = N(C_3H_7\text{-}i)_2$, $R^2 = NH_2$ and with $NR_2 = N(C_3H_7\text{-}i)_2$, $R^2 = NHC_6H_{11}\text{-}c$ confirm a valency of four for the ring sulfur. The bond energy of the sulfur 2p level is 166.65 eV (with N correction) and 166.8 eV (with C correction) for the derivative with $R^2 = NH_2$; for the compound with $R^2 = NHC_6H_{11}\text{-}c$ the corresponding values are 166.85 and 166.95 eV, respectively [7].

$^1$H and $^{13}$C NMR data of some derivatives are compiled in Table 20. For the derivative with $NR_2 = N(C_3H_7\text{-}i)_2$, $R^2 = NH_2$ (Table 18, No. 1) though asymmetrically substituted, no splitting of the signals of the methyl and methine groups was observed at 25°C in the $^{13}$C NMR spectrum. This shows that free rotation of the $N(C_3H_7\text{-}i)_2$ group occurs [8].

Table 20

NMR Spectra of 1-Diisopropylamino-3-chloro-5 amino (or mono- or dialkylamino)-1H-1,2,4,6-thiatriazines, [$\because$S(N($C_3H_7$-i)$_2$)$\because$N$\because$CCl$\because$N$\because$CR$^2$$\because$N$\because$] [2].
(Numbers of compounds from Table 18, p. 45.)

| No. | $R^2$ | solvent/standard*) | $^1$H NMR (δ in ppm at ambient temperatures) | Ref. |
|---|---|---|---|---|
| 1 | $NH_2$ | DMSO-$d_6$ | 1.21 (d, C($CH_3$)$_2$), 3.02 to 3.75 (m, CH) | [2] |
| 2 | $NHCH_3$ | $CDCl_3$ | 1.20 and 1.22 (d, C($CH_3$)$_2$), 2.86 (d, $NCH_3$), 3.40 (sept, CH) | [2] |
| 3 | $NHC_2H_5$ | $CDCl_3$ | 1.10 (t, $CCH_3$), 1.19 and 1.22 (d, C($CH_3$)$_2$), 3.05 to 3.60 (m, CH), 3.05 to 3.60 (m, $NCH_2$) | [2] |
| 4 | $NHC_3H_7$-n | $(CD_3)_2CO$ | 0.80 (t, $CCCH_3$), 1.17 (d, C($CH_3$)$_2$), 1.25 to 1.75 (m, $CCH_2C$), 3.00 to 3.38 (m, $NCH_2$), 3.40 (sept, CH) | [2] |
| 13 | N($C_4H_9$-i)$_2$ | $C_6D_5CD_3$ | 0.70 (d, CC($CH_3$)$_2$), 0.90 and 0.97 (d, NC($CH_3$)$_2$), 2.25 to 3.60 (m, NCH), 2.88 to 3.50 (m, $NCH_2$), 2.88 to 3.50 (m, CCH) | [2] |

*) 15% solutions/internal hexamethyldisiloxane each.

| No. | $R^2$ | solvent/TMS | $^{13}$C NMR (δ in ppm) | Ref. |
|---|---|---|---|---|
| 1 | $NH_2$ | DMSO-$d_6$ | 18.98 ($CH_3$), 46.41 (CH), 118.67 (C(ring)–N), 163.04 (C(ring)–Cl) | [8] |
| 4 | $NHC_3H_7$-n | 25% in $CDCl_3$ | 10.91 ($CH_2\underline{C}H_3$), 22.31 ($C\underline{C}H_2C$), 22.99 (C($\underline{C}H_3$)$_2$), 41.73 ($N\underline{C}H_2C$), 46.68 (SNCH), 160.45 (C(ring)–N), 163.22 (C(ring)–Cl) | [2] |
| 13 | N($C_4H_9$-i)$_2$ | 25% in $CDCl_3$ | 19.75 (NC($\underline{C}H_3$)$_2$), 21.20 (CC($\underline{C}H_3$)$_2$), 26.47 and 27.05 ($C\underline{C}H(CH_3)_2$), 46.20 (SNCH), 54.07 and 54.31 ($N\underline{C}H_2C$), 160.50 (C(ring)–N), 163.89 (C(ring)–Cl) | [2] |

For the derivative with $NR_2$ = N($C_4H_9$-i)$_2$, $R^2$ = N($C_4H_9$-i)$_2$ (Table 18, No. 15), extensive $^1$H and $^{13}$C NMR spectroscopic investigations have been performed showing the double chemical nonequivalence of the two protons of the methylene groups of the diisobutylamino group and of the two isobutyls of the diisobutylamino group in position 5. The methylene carbons as well as the methine carbons of the same group are also nonequivalent. The $^1$H and $^{13}$C NMR data are given in Table 21, p. 48. From the coalescence temperature (64 and 67°C) of the methylene C and methine C signals, the free activation energy of rotation was estimated to be 73.6 kJ/mol (17.6 ± 0.2 kcal/mol) [9]. In the $^1$H NMR spectrum of the derivative with $NR_2$ = N($C_3H_7$-i)$_2$, $R^2$ = N($CH_2C_6H_5$)$_2$ (Table 18, No. 14), two AB systems have been found for the protons of the two $CH_2$ groups. This has been explained by a hindrance to the rotation around the C(5)–N bond. The energy barrier for that rotation was estimated to be 73.37 kJ/mol from the coalescence temperature [10].

The derivatives of 1-dialkylamino-1,2,4,6-thiatriazine described here are in general colorless crystalline substances which decompose in air, sometimes explosively. They all melt with decomposition. They are stable when stored in dry argon at low temperatures or when dissolved in aprotic solvents [4]. The compound with $NR_2 = R^2 = N(C_4H_9\text{-}i)_2$ (Table 18, No. 15) decomposes in $C_6H_5Br$ solution at 80°C [9]. From thermogravimetric analyses it was indicated that ClCN is eliminated on heating [6]. The catalytic hydrogenation of the compounds with $NR_2 = N(C_3H_7\text{-}i)_2$ and $R^2 = NH_2$, $NHCH_3$, $NHC_3H_7$-n, $NHC_4H_9$-t, and $NHC_6H_{11}$-c (Table 18, No. 1, 2, 4, 6, 7) with noble metal catalysts failed. The hydrogenation of these compounds was carried out successfully, however, in aqueous alcohol in the presence of a Raney nickel catalyst. At $H_2$ pressure as low as 1 bar the cleavage of the ring is initiated to give NiS by coincident HCl elimination with reformation of the nitrile group. $(i\text{-}C_3H_7)_2NH_2Cl$ and the corresponding N-alkyl-substituted cyanoguanidine are then formed, along with partial hydrolysis products [11], see also [6]. For the course of reaction of the catalytic hydrogenation, see [11]. In the presence of water the thiatriazine system is hydrolyzed. The hydrolysis of the compounds with $NR_2 = N(C_3H_7\text{-}i)_2$ and $R^2 = NH_2$, $NHCH_3$, $NHC_3H_7$-n, $NHC_4H_9$-t, and $NHC_6H_{11}$-c (Table 18, No. 1, 2, 4, 6, 7) leads preferentially to the formation of the corresponding N-alkylcyanoguanidines, N-alkyl-guanyl ureas or biuret along with $SO_2$, and $(i\text{-}C_3H_7)_2NH_2Cl$ [11]. A proposed hydrolysis mechanism was given [11].

Table 21
$^1H$ and $^{13}C$ NMR Spectroscopic Data of $[\dot{-}S(N(C_4H_9\text{-}i)_2)\dot{-}N\dot{-}CCl\dot{-}N\dot{-}C(N(C_4H_9\text{-}i)_2)\dot{-}N\dot{-}]$, (solvent?/TMS) at 25°C [9].

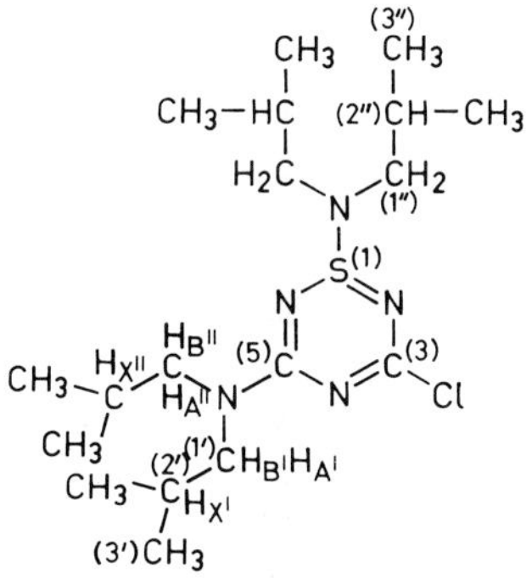

| | δ in ppm | | | | | | | |
|---|---|---|---|---|---|---|---|---|
| position | (1′) | (2′) | (3′) | (1″) | (2″) | (3″) | (3) | (5) |
| $^1H$ | [a] | 2.238[b] | 0.950[c] | 2.775[d] | 1.827[e] | 0.950[c] | — | — |
| $^{13}C$ | 54.25<br>53.95 | 27.11<br>26.47 | 19.92 | 55.07 | 26.24 | 19.92 | 164.24 | 160.33 |

[a] Superposition of two AB parts of the ABX spectra with δ (in ppm) = 3.597 (H(A$^I$)), 3.209 (H(B$^I$)); 3.754 (H(A$^{II}$)), 3.206 (H(B$^{II}$)); $|^2J(H(A^I), H(B^I))| = 14.0$ Hz, $|^2J(H(A^{II}), H(B^{II}))| = 13.6$ Hz; $^3J(H(A^I), H(X^I)) = {}^3J(H(B^I), H(X^I)) = {}^3J(H(A^{II}), H(X^{II})) = {}^3J(H(B^{II}), H(X^{II})) = 7.32$ Hz. – [b] m, X part of the ABX spectrum. – [c] Superposition of all $CH_3$ signals. – [d] d, $^3J(CH, CH_2) = 7.5$ Hz. – [e] t, sept.

For $[\dot{-}S(N(C_3H_7\text{-}i)_2)\dot{-}N\dot{-}CCl\dot{-}N\dot{-}C(NHC_6H_{11}\text{-}c)\dot{-}N\dot{-}]$ (Table 18, No. 7) the hydrolysis was followed by observation of the $Cl^-$ concentration in a $CH_3OH$–$H_2O$ mixture (9:1) as a function of time and pH. It was shown that the compound is relatively stable in alkaline solution. In acid medium there is a fast splitting off of $Cl^-$ because of the relative ease in protonating the compound [11].

The compounds do not react with aliphatic amines to form 1,3,5-triamino-substituted thiatriazines [4].

**References:**

[1] Schramm, W., Voß, G., Rembarz, G., Fischer, E. (Ger. [East] 113006 [1974/75]; C.A. **84** [1976] No. 105666).

[2] Michalik, M., Fischer, E., Rembarz, G., Voß, G., Storek, W. (J. Prakt. Chem. **319** [1977] 739/44).

[3] Kálmán, A., Argay, G., Fischer, E., Rembarz, G. (Acta Cryst. B **35** [1979] 860/6).

[4] Schramm, W., Voß, G., Michalik, M., Rembarz, G., Fischer, E. (Z. Chem. [Leipzig] **15** [1975] 19).

[5] Kálmán, A., Argay, G., Fischer, E., Rembarz, G., Voss, G. (J. Chem. Soc. Perkin Trans. II **1977** 1322/7).

[6] Fischer, E. (Wiss. Z. Wilhelm-Pieck-Univ. Rostock Math. Naturwiss. Reihe **27** [1978] 609/16).

[7] Leonhardt, G., Scheibe, R., Schramm, W., Voß, G., Fischer, E., Rembarz, G. (Z. Chem. [Leipzig] **15** [1975] 193/4).

[8] Storek, W., Schramm, W., Voß, G., Rembarz, G., Fischer, E. (Z. Chem. [Leipzig] **15** [1975] 104/5).

[9] Storek, W., Fischer, E., Michalik, M., Rembarz, G., Voß, G. (Z. Chem. [Leipzig] **16** [1976] 490/1).

[10] Rembarz, G., Fischer, E., Michalik, M. (Wiss. Z. Wilhelm-Pieck-Univ. Rostock Math. Naturwiss. Reihe **28** [1979] 855/9; C.A. **95** [1981] No. 6362).

[11] Fischer, E., Rembarz, G., Klatt, E., Weber, H., Rachimova, A. A. (Wiss. Z. Wilhelm-Pieck-Univ. Rostock Math. Naturwiss. Reihe **28** [1979] 861/4; C.A. **95** [1981] No. 7232).

## 9.3.2.8.1.7 1-Diisopropylamino-3-chloro-5-phenyl-1H-1,2,4,6-thiatriazine, $[\dot{-}S(N(C_3H_7\text{-}i)_2)\dot{-}N\dot{-}CCl\dot{-}N\dot{-}C(C_6H_5)\dot{-}N\dot{-}]$

The title compound is formed in 30% yield from the reaction of equimolar amounts of $[\dot{-}S(N(C_3H_7\text{-}i)_2)\dot{-}N\dot{-}CCl\dot{-}N\dot{-}CCl\dot{-}N\dot{-}]$ (see p. 40) and phenylmagnesium bromide in ether for 20 h at room temperature.

The colorless crystals melt at 58 to 63°C.

$^1$H NMR spectrum (15% in $C_6D_5CD_3$/internal hexamethyldisiloxane): δ (in ppm) = 0.75 and 0.82 (doublets, $CH_3$), 2.95 (sept, NCH).

$^{13}$C NMR spectrum (25% in $CDCl_3$/TMS): δ (in ppm) = 23.28 and 23.53 ($CH_3$), 48.28 (SNCH), 165.89 (C(ring)-Cl), 167.78 (C(ring)-phenyl).

Mass spectrum: m/e = 310 $M^+$, 295 $M^+$-$CH_3$, 275 $M^+$-Cl, 233 $M^+$-$C_6H_5$, 210 $SN_3C_8H_5Cl^+$, 100 $C_6H_{14}N^+$.

**Reference:**

Michalik, M., Fischer, E., Rembarz, G., Voß, G., Storek, W. (J. Prakt. Chem. **319** [1977] 739/44).

### 9.3.2.8.2 1-Dialkylamino-3,5-bis(alkyl (or aryl)thio)-1H-1,2,4,6-thiatriazines, [-S(NR$_2$)-N-CR$^1$-N-CR$^2$-N-], NR$_2$, R$^1$, and R$^2$ are listed in Table 22.

The compounds compiled in Table 22 are formed from the reaction of 1-dialkylamino-3,5-dichloro-1H-1,2,4,6-thiatriazines (see p. 40) with the corresponding sodium thiolates (1:2 mole ratio) in alcohol at room temperature [1]. Yields and melting points of the synthesized derivatives are given in Table 22.

Table 22
Yields and Melting Points of 1-Dialkylamino-3,5-bis(alkyl (or aryl)thio)-1H-1,2,4,6-thiatriazines,
[-S(NR$_2$)-N-CR$^1$-N-CR$^2$-N-].

| $NR_2$ | $R^1 = R^2$ | yield in % | m.p. in °C | Ref. |
|---|---|---|---|---|
| $N(C_3H_7\text{-}i)_2$ | $SCH_3$ | 96 | 104 to 106 | [1] |
| | | 89 | 105 to 106 | [5] |
| $N(C_3H_7\text{-}i)_2$ | $SC_2H_5$ | 87 | 49 to 52 | [1] |
| $N(C_3H_7\text{-}i)_2$ | $SC_3H_7\text{-}i$ | 93 | 75 to 78 | [1] |
| $N(C_3H_7\text{-}i)_2$ | $SC_4H_9\text{-}n$ | 61 | —*) | [1] |
| $N(C_3H_7\text{-}i)_2$ | $SC_6H_5$ | 81 | 152 to 154 | [1] |
| $N(C_3H_7\text{-}i)_2$ | $SCH_2C_6H_5$ | 95 | 72 to 73 | [1] |
| $N(C_6H_{11}\text{-}c)_2$ | $SCH_3$ | 75 | 136 to 139 | [1] |

*) Colorless oil, $n_D^{20} = 1.5438$, purified by column chromatography.

The structure of the compounds was concluded from their NMR spectra (given in Table 24) and confirmed by an X-ray diffraction study of one derivative (see below). In the IR spectra the compounds show characteristic absorption bands in the 1520 to 1500 $cm^{-1}$ and 1350 to 1330 $cm^{-1}$ ranges.

[-S(N(C$_3$H$_7$-i)$_2$)-N-C(SCH$_3$)-N-C(SCH$_3$)-N-] shows a UV spectrum with $\lambda_{max} = 248$ nm, $\varepsilon = 4.6 \times 10^4$ $L \cdot mol^{-1} \cdot cm^{-1}$ [1]. ESCA spectroscopic investigations confirm a valency of four for the ring sulfur [4] and that the compound is not a sulfonium salt [7]. For the bond energy of the sulfur 2p level, 166.8 eV (with N correction and C correction) was obtained [4].

For [-S(N(C$_3$H$_7$-i)$_2$)-N-C(SC$_6$H$_5$)-N-C(SC$_6$H$_5$)-N-] the crystal structure has been established from X-ray diffraction data. Crystals have been obtained by recrystallization from a water-ethanol mixture. They are monoclinic, space group $P2_1/c\text{-}C_{2h}^5$ (No. 14), with a = 8.225(1), b = 16.684(4), c = 16.411(7) Å, and $\beta = 93.84°$; Z = 4. $D_x$ = 1.232 g/cm$^3$, V = 2247 Å$^3$. R = 0.067 for 1714 reflections.

The molecular structure is similar to that of the 3,5-dichloro compound (see p. 40/1). The S atom lies out of the best plane by 0.22 Å. The pseudo threefold axis at N(4) makes an angle less than 90° with the pseudo mirror plane bisecting S(1), N(2), and N(4) [6]. In spite of the shortened S(1)–N(4) distance of 1.61 Å with unambiguous ylidic character of the bonds, rotation barriers of only 38 to 45 kJ/mol have been found. Therefore interactions of 3p$\pi$(S)–2p$\pi$(N) or 3d$\pi$(S)-2p$\pi$(N) orbitals should have no basic influence on the rotation barrier of the considered bond [7]. Selected bond lengths and bond angles are given in Table 23. For endocyclic and exocyclic torsion angles and final coordinates for the non-hydrogen atoms and parameters for H atoms, see the paper [6].

Table 23

Selected Bond Lengths and Bond Angles of 1-Diisopropylamino-3,5-bis(phenylthio)-1H-1,2,4,6-thiatriazine, $[\stackrel{\cdot\cdot}{-}S(N(C_3H_7\text{-}i)_2)\stackrel{\cdot\cdot}{-}N\stackrel{\cdot\cdot}{-}C(SC_6H_5)\stackrel{\cdot\cdot}{-}N\stackrel{\cdot\cdot}{-}C(SC_6H_5)\stackrel{\cdot\cdot}{-}N\stackrel{\cdot\cdot}{-}]$.

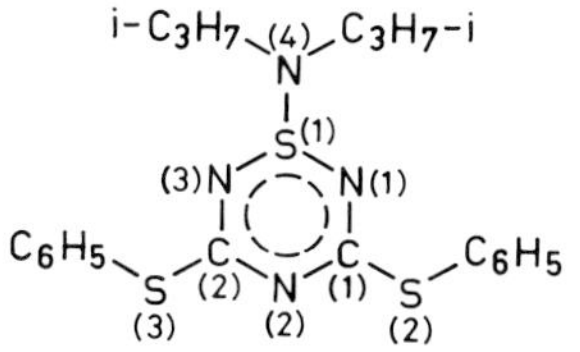

| bond length | in Å | bond angle | in ° |
|---|---|---|---|
| S(1)–N(1) | 1.644(4) | N(1)–S(1)–N(3) | 107.3(2) |
| N(1)–C(1) | 1.318(7) | S(1)–N(1)–C(1) | 116.8(4) |
| C(1)–N(2) | 1.331(7) | N(1)–C(1)–N(2) | 130.7(5) |
| S(1)–N(4) | 1.610(4) | N(1)–C(1)–S(2) | 117.7(4) |
| C(1)–S(2) | 1.762(5) | N(2)–C(1)–S(2) | 111.6(4) |
| S(1)–N(3) | 1.638(4) | C(1)–N(2)–C(2) | 115.6(4) |
| N(3)–C(2) | 1.306(7) | S(1)–N(3)–C(2) | 117.4(4) |
| C(2)–N(2) | 1.343(7) | N(3)–C(2)–N(2) | 130.4(5) |
| C(2)–S(3) | 1.754(5) | N(3)–C(2)–S(3) | 119.7(4) |
| | | N(2)–C(2)–S(3) | 109.9(4) |

The difficult to crystallize compounds give colorless crystals (with exception; see Table 22). They are of higher stability than the monosubstituted alkyl (or aryl)thio derivatives (see p. 24) [1].

Table 24

$^1H$ and $^{13}C$ NMR Spectra of 1-Diisopropylamino-3,5-bis(alkyl (or aryl)thio)-1H-1,2,4,6-thiatriazines, $[\stackrel{\cdot\cdot}{-}S(N(C_3H_7\text{-}i)_2)\stackrel{\cdot\cdot}{-}N\stackrel{\cdot\cdot}{-}CR^1\stackrel{\cdot\cdot}{-}N\stackrel{\cdot\cdot}{-}CR^2\stackrel{\cdot\cdot}{-}N\stackrel{\cdot\cdot}{-}]$.

| $R^1 = R^2$ | $^1H$ (15% in $CDCl_3$/internal hexamethyldisiloxane; δ in ppm) | Ref. | $^{13}C$ (25% in $CDCl_3$/TMS; δ in ppm) | Ref. |
|---|---|---|---|---|
| $SCH_3$ | 1.20 (d, $CCH_3$)<br>2.29 (s, $SCH_3$)<br>3.38 (sept, CH) | [3] | 12.51 ($SCH_3$)<br>23.09 ($CH(\underline{C}H_3)_2$)<br>47.30 ($\underline{C}H(CH_3)_2$)<br>174.08 (ring C(3) and C(5)) | [2] |
| $SC_2H_5$ | 1.20 (d, $C(CH_3)_2$)<br>1.24 (t, $SCCH_3$)<br>2.85 and 2.90 (m, $CH_2$)<br>3.29 (sept, CH)[a)] | [3] | 14.84 ($CH_2\underline{C}H_3$)<br>23.48 ($CH(\underline{C}H_3)_2$)<br>24.01 ($SCH_2$)<br>47.65 (SNCH)<br>174.38 (ring C(3) and C(5)) | [3] |
| $SC_3H_7$-i | 1.20 (d, $CSC(CH_3)_2$)<br>1.28 (d, $SNC(CH_3)_2$)<br>3.40 (sept, CSCH)<br>3.69 (sept, SNCH) | [3] | 22.85 and 23.43 ($SCH(\underline{C}H_3)_2$)<br>23.50 ($NCH(\underline{C}H_3)_2$)<br>34.50 (SCH)<br>47.70 (SNCH)<br>174.54 (ring C(3) and C(5)) | [3] |

Table 24 (continued)

| $R^1 = R^2$ | $^1H$ (15% in $CDCl_3$/internal hexamethyldisiloxane; δ in ppm) | Ref. | $^{13}C$ (25% in $CDCl_3$/TMS; δ in ppm) | Ref. |
|---|---|---|---|---|
| $SC_4H_9$ | 0.92 (d, $SCC(CH_3)_2$)<br>1.19 (d, $SNC(CH_3)_2$)<br>1.82 (m, SCCH)<br>2.74 and 2.80 (dq, $CSCH_2$)<br>3.38 (sept, SNCH) | [3] | — | |
| $SCH_2C_6H_5$ | 0.81 (d, $NC(CH_3)_2$)[b]<br>3.05 (sept, SNCH)<br>4.05 and 4.19 (m, $CSCH_2$) | [3] | 23.20 ($NCH(\underline{C}H_3)_2$)<br>33.90 ($SCH_2$), 47.70 (SNCH)<br>173.84 (ring C(3) and C(5)) | [3] |

[a] The $SC_2H_5$ group appears as an $ABX_3$ spectrum with $^2J_{AB} = 7.5$ Hz. [b] 15% solution in $C_6D_5Br$. The $SCH_2$ group appears as an AB spectrum with $^2J_{AB} = 13.9$ Hz.

**References:**


[1] Schramm, W., Voß, G., Rembarz, G., Fischer, E. (Z. Chem. [Leipzig] **15** [1975] 57/8).

[2] Storek, W., Schramm, W., Voß, G., Rembarz, G., Fischer, E. (Z. Chem. [Leipzig] **15** [1975] 104/5).

[3] Michalik, M., Fischer, E., Rembarz, G., Voß, G., Storek, W. (J. Prakt. Chem. **319** [1977] 739/44).

[4] Leonhardt, G., Scheibe, R., Schramm, W., Voß, G., Fischer, E., Rembarz, G. (Z. Chem. [Leipzig] **15** [1975] 193/4).

[5] Schramm, W., Voß, G., Rembarz, G., Fischer, E. (Ger. [East] 113006 [1974/75]; C.A. **84** [1976] No. 105666).

[6] Kálmán, A., Argay, G., Fischer, E., Rembarz, G. (Acta Cryst. B **35** [1979] 860/6).

[7] Fischer, E. (Wiss. Z. Wilhelm-Pieck-Univ. Rostock Math. Naturwiss. Reihe **27** [1978] 609/16; C.A. **93** [1980] No. 94303).


### 9.3.2.8.3 1-Dialkylamino-3,5-bis(arylamino)-1H-1,2,4,6-thiatriazines, [$\overline{-S(NR_2)}$$\dot{-}$N$\dot{-}$CR$^1$$\dot{-}$N$\dot{-}$CR$^2$$\dot{-}$N$\dot{-}$]

$NR_2 = N(C_3H_7\text{-}i)_2$, $N(C_6H_{11}\text{-}c)_2$; $R^1 = R^2 = NHC_6H_4CH_3\text{-}4$, $NHC_6H_4OCH_3\text{-}4$

The compounds are supposed to form from the reaction of 1-diisopropyl (or dicyclohexyl)-amino-3,5-dichloro-1H-1,2,4,6-thiatriazine (see p. 40) with $4\text{-}CH_3C_6H_4NH_2$ or $4\text{-}CH_3OC_6H_4NH_2$ in excess in ether at room temperature. The red crystalline substances do not contain Cl.

**Reference:**


Schramm, W., Voß, G., Michalik, M., Rembarz, G., Fischer, E. (Z. Chem. [Leipzig] **15** [1975] 19).


### 9.3.2.8.4 1-Dialkylamino-3-trichloromethyl-5-phenyl-1H-1,2,4,6-thiatriazines, [$\dot{-}$S($NR_2$)$\dot{-}$N$\dot{-}$C($CCl_3$)$\dot{-}$N$\dot{-}$C($C_6H_5$)$\dot{-}$N$\dot{-}$], $NR_2 = N(CH_3)_2$, $N(C_2H_5)_2$

The title compounds form from the reaction of [$\dot{-}$SCl$\dot{-}$N$\dot{-}$C($CCl_3$)$\dot{-}$N$\dot{-}$C($C_6H_5$)$\dot{-}$N$\dot{-}$] (see p. 55) in benzene with $(CH_3)_2NH$ or $(C_2H_5)_2NH$ (1:2 mole ratio) at 18 to 20°C for 1 h. The amine hydrochlorides were separated, and the mother solution was evaporated at 70 to 80°C/10 to 15 Torr. The compounds were purified by recrystallization. Yields and properties are given in the following table:

| $NR_2$ | yield in % | m.p. in °C | $^1H$ NMR spectrum ($CCl_4$/external hexamethyldisiloxane) |
|---|---|---|---|
| $N(CH_3)_2$ | 63 | 84 to 85 (from hexane) | $\delta = 3.03$ ppm (s, $CH_3$) |
| $N(C_2H_5)_2$ | 81 | 72 to 73 (from petroleum ether) | |

The IR spectra of the compounds are characterized by a shift of the absorption band for the stretching vibrations of the ring in the region of 1500 $cm^{-1}$ by 20 to 30 $cm^{-1}$ toward the short wave region compared with [$\dot{-}$SCl$\dot{-}$N$\dot{-}$C($CCl_3$)$\dot{-}$N$\dot{-}$C($C_6H_5$)$\dot{-}$N$\dot{-}$].

**Reference:**

Kornuta, P. P., Derii, L. I., Markovskii, L. N. (Zh. Org. Khim. **16** [1980] 1308/13; J. Org. Chem. [USSR] **16** [1980] 1130/4).

### 9.3.2.9 Derivatives of 1-Chloro-1H-1,2,4,6-thiatriazine

#### 9.3.2.9.1 3,5-Disubstituted 1-Chloro-1H-1,2,4,6-thiatriazines

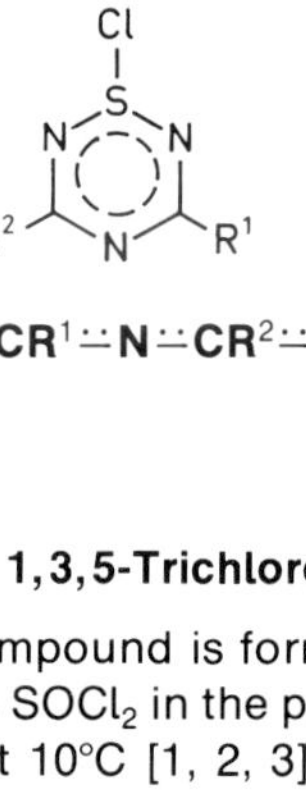

$R^1 = R^2 = Cl$, $CCl_3$, $C_6H_5$
$R^1 = Cl$; $R^2 = CCl_3$, $C_6H_5$
$R^1 = CCl_3$; $R^2 = CF_3$, $C_6H_5$

**[$\dot{-}$SCl$\dot{-}$N$\dot{-}$CR$^1$$\dot{-}$N$\dot{-}$CR$^2$$\dot{-}$N$\dot{-}$]**

##### 9.3.2.9.1.1 1,3,5-Trichloro-1H-1,2,4,6-thiatriazine, [$\dot{-}$SCl$\dot{-}$N$\dot{-}$CCl$\dot{-}$N$\dot{-}$CCl$\dot{-}$N$\dot{-}$]

The title compound is formed from the complex reaction of sodium dicyanamide, $NaN(CN)_2$, with excess $SOCl_2$ in the presence of dimethylformamide and a tetraalkylammonium halogenide at about 10°C [1, 2, 3]. The yield is 60% [3] to 65% [1].

Originally, the structure of the compound was assumed to be that of an acyclic cyanochloroformamidine [1], but during reexamination the cyclic structure was concluded from spectral data [2, 4] and from its reaction behavior (see below). ESCA investigations without a doubt show the sulfur to have a valency of four. A maximum at 167.1 eV for the sulfur 2p level was obtained including a charging correction. For the $C_{1s}$ value, 288.65 eV was obtained without a correction [4].

The compound shows characteristic bands at 1480 and 1330 $cm^{-1}$ in the IR spectrum [2]. $^{13}C$ NMR spectrum (solvent?/TMS): $\delta = 169.53$ ppm. Mass spectrum: m/e = 203 $M^+$, 168 $SN_3C_2Cl_2^+$, 133 $SN_3C_2Cl^+$, 122 $N_2C_2Cl_2^+$, 107 $SN_2CCl^+$, 93 $SNCCl^+$, 87 $N_2C_2Cl^+$, 67 $SCl^+$ [4]. The dipole moment was found to be 0.9 D [4].

1,3,5-Trichloro-1H-1,2,4,6-thiatriazine is an extremely moisture-sensitive, pale yellow, oily liquid [1, 4], b.p. 70 to 73°C/2 Torr [1, 3], which solidifies into a heavy white crystalline product, m.p. 32 to 34°C [1]. The reaction with water leads to a variety of products depending upon conditions [1]. In every case, however, the heterocyclic system is destroyed producing $CO_2$, $NH_4Cl$, and $SO_2$ [1, 6]. Hydrolysis with water at 0°C resulted in a mixture of $SO_2$, $NH_2CCl{=}NCONH_2$, and $HN(CONH_2)_2$. Reaction with 36% hydrochloric acid in dry ether at −10°C gives N-cyano-chloroformamidine dihydrochloride, $N{\equiv}C{-}N{=}CClNH_2 \cdot 2HCl$ in 62.6% yield. The same reaction at 0°C produces in addition $H_2NCON{=}CClNH_2$ and $HN(CONH_2)_2$. When the heterocycle is dissolved in cold glacial acetic acid and the reaction mixture kept below 25°C, 1-acetylbiuret, $H_2NCONHCONHCOCH_3$ is obtained along with $CH_3COCl$ and $SO_2$. With the alcohols ROH ($R=CH_3$, and $C_2H_5$) in dry $CCl_4$ at less than 25°C, the corresponding O-alkylisobiuret hydrochlorides, $H_2NCONHC(OR){=}NH \cdot HCl$ are formed; reactions with higher alcohols lead to biuret, $HN(CONH_2)_2$, and alkyl chlorides [1].

1,3,5-Trichloro-1H-1,2,4,6-thiatriazine is also the starting compound for a large number of thiatriazine derivatives, since the chloro atoms are replaceable stepwise (starting with the Cl at sulfur) by nucleophiles with conservation of the heterocyclic ring. The compound reacts with an excess of ethylene oxide in $CCl_4$ below 20°C to give 1-chloroethoxy-3,5-dichloro-1H-1,2,4,6-thiatriazine, [∸S($OCH_2CH_2Cl$)∸N∸CCl∸N∸CCl∸N∸], in 98% yield [1] (see p. 31). The secondary amines, exclusively with bulky groups, $(i\text{-}C_3H_7)_2NH$ [1, 2, 3], 2,6-dimethylpiperidine [1], and $(c\text{-}C_6H_{11})_2NH$ [2, 3] in $(C_2H_5)_2O$ at −5°C [1] or −40°C [2, 3] react to yield the corresponding 1-dialkylamino-3,5-dichloro-1H-1,2,4,6-thiatriazines, see p. 40. No reaction takes place with primary amines [5].

**References:**


[1] Geevers, J., Hackmann, J. T., Trompen, W. P. (J. Chem. Soc. C **1970** 875/8).

[2] Schramm, W., Voß, G., Rembarz, G., Fischer, E. (Z. Chem. [Leipzig] **14** [1974] 471/2).

[3] Schramm, W., Voß, G., Rembarz, G., Fischer, E. (Ger. [East] 113006 [1974/75]; C.A. **84** [1976] No. 105666).

[4] Voß, G., Fischer, E., Rembarz, G., Schramm, W. (Z. Chem. [Leipzig] **16** [1976] 358/9).

[5] Fischer, E. (Wiss. Z. Wilhelm-Pieck-Univ. Rostock Math. Naturwiss. Reihe **27** [1978] 609/16; C.A. **93** [1980] No. 94303).

[6] Fischer, E., Rembarz, G., Klatt, E., Weber, H., Rachimowa, A. A. (Wiss. Z. Wilhelm-Pieck-Univ. Rostock Math. Naturwiss. Reihe **28** [1979] 861/4; C.A. **95** [1981] No. 7232).


### 9.3.2.9.1.2 1,3-Dichloro-5-trichloromethyl-1H-1,2,4,6-thiatriazine, [∸SCl∸N∸CCl∸N∸C($CCl_3$)∸N∸], and 1,3-Dichloro-5-phenyl-1H-1,2,4,6-thiatriazine, [∸SCl∸N∸CCl∸N∸C($C_6H_5$)∸N∸]

The title compounds are formed in 40 and 45% yield, respectively, in a cyclization reaction, when a mixture of the N-cyanoamidines, $H_2NCR^2{=}N{-}C{\equiv}N$ with $R^2=CCl_3$, $C_6H_5$, and a fivefold excess of $SCl_2$ is heated for several hours [1].

[∸SCl∸N∸CCl∸N∸C($CCl_3$)∸N∸] is a liquid, b.p. 81 to 86°C/0.04 Torr, $n_D^{20}=1.6193$. The IR spectrum (in $CCl_4$) shows absorption bands at 1518 and 1381 $cm^{-1}$, characteristic for the stretching vibrations of the ring. UV-visible spectrum (in n-hexane): $\lambda_{max}=278$ nm (log $\varepsilon=3.79$) [1]. When treated with sodium dust or with 2,4,6,-triphenyl-3,4-dihydro-2H-1,2,4,5-tetrazinyl (compound I, p. 19) the compound is transformed into the radical [∸S∸N∸CCl∸N∸C($CCl_3$)∸N∸]$^\bullet$ (see p. 19) [2].

[$\dot{-}$SCl$\dot{-}$N$\dot{-}$CCl$\dot{-}$N$\dot{-}$C($C_6H_5$)$\dot{-}$N$\dot{-}$] is a solid, m.p. 93 to 95°C (recrystallized from petroleum ether). The IR spectrum (in $CCl_4$) shows absorption bands at 1494 and 1406 $cm^{-1}$ characteristic for the stretching vibrations of the ring. UV-visible spectrum (n-hexane): $\lambda_{max}$= 259 nm (log $\varepsilon$ = 4.26). Cl NQR spectrum: 35.341 (C-$^{35}$Cl), 27.853 (C-$^{37}$Cl), and 27.163 MHz (S-$^{35}$Cl) [1].

**References:**

[1] Kornuta, P. P., Derii, L. I., Romanenko, E. A. (Khim. Geterotsikl. Soedin. **1978** 273; Chem. Heterocycl. Compounds [USSR] **1978** 226).

[2] Markovskii, L. N., Kornuta, P. P., Katchkovskaya, L. S., Polumbrik, O. M. (Sulfur Letters **1** [1983] 143/5; C.A. **100** [1984] No. 103301).

### 9.3.2.9.1.3 Other 1-Chloro-3,5-disubstituted-1H-1,2,4,6-thiatriazines, [$\dot{-}$SCl$\dot{-}$N$\dot{-}$CR$^1\dot{-}$N$\dot{-}$CR$^2\dot{-}$N$\dot{-}$], R$^1$ and R$^2$ are compiled in Table 25

The title compounds listed in Table 25 form by a cyclization reaction of the corresponding carboxamidines, HN=CR$^1$–N=CR$^2$NH$_2$, in $C_6H_6$ with a fivefold excess of $SCl_2$ in $C_6H_6$ at 80°C for 1.5 to 2 h (Table 25, No. 1, 2, 3) [1], or by heating a mixture of $C_6H_5C(=NH)NH_2$ and $S_3N_3Cl_3$ (5:2 mole ratio) in $CH_3CN$ at reflux for 16 h (Table 25, No. 4) [2, 3]. Yields and some property data of the compounds are given in Table 25. (Another compound, with R$^1$ = N($CH_3$)$_2$, R$^2$ = $C_6H_5$, is mentioned as the starting compound for the preparation of the radical $SN_3C_2(N(CH_3)_2)(C_6H_5)$, see p. 19.)

Table 25

Yields, Melting (Boiling) Points, and some Property Data of 3,5-Disubstituted 1-Chloro-1H-1,2,4,6-thiatriazines, [$\dot{-}$SCl$\dot{-}$N$\dot{-}$CR$^1\dot{-}$N$\dot{-}$CR$^2\dot{-}$N$\dot{-}$].

| No. | R$^1$ | R$^2$ | yield in % | m.p. in °C (b.p. in °C/Torr) | properties | Ref. |
|---|---|---|---|---|---|---|
| 1 | $CCl_3$ | $CF_3$ | 55 | (41 to 42/0.03) | $D_4^{20}$ = 1.7430 g/cm$^3$, $n_D^{20}$ = 1.5211 | [1] |
| 2 | $CCl_3$ | $CCl_3$ | 38 | 57 to 59 (94 to 95/0.03) | $^{35}$Cl NQR spectrum (at 77 K): 31.40 (s, S–Cl), 40.06, 40.09, and 40.57 MHz (t, $CCl_3$) | [1] |
| 3 | $CCl_3$ | $C_6H_5$ | 80 | 91 to 93 | $^{35}$Cl NQR spectrum (at 77 K): 26.57 (s, S–Cl), 39.14, 40.00, and 40.38 MHz (t, $CCl_3$) | [1] |
| 4 | $C_6H_5$ | $C_6H_5$ | 34 | 162 to 167 | IR spectrum (Nujol): 26 absorption bands between 3060 and 230 $cm^{-1}$. MS (70 eV): m/e (relative abundance in %) = 252 M$^+$–Cl (34), 149 $SN_2CC_6H_5^+$ (21), 103 $C_6H_5CN^+$ (50), 77 $C_6H_5^+$ (19), 46 SN$^+$ (100) [3] | [2, 3] |

An X-ray crystal structure analysis on the yellow transparent needles (recrystallized from $CH_3CN$) of [$\dot{-}$SCl$\dot{-}$N$\dot{-}$C($C_6H_5$)$\dot{-}$N$\dot{-}$C($C_6H_5$)$\dot{-}$N$\dot{-}$] (Table 25, No. 4) shows the compound to be orthorhombic, space group $P2_12_12_1$–$D_2^4$ (No. 19) with a = 6.773(2), b = 11.114(2), and c = 17.571(2) Å; Z = 4. V = 1323 Å$^3$, $D_x$ = 1.45 g/cm$^3$. R = 0.044 for 1142 independent reflections. The molecular structure of the compound is shown in **Fig. 13**, p. 56. Selected bond lengths and bond angles are given in Table 26, p. 56.

Table 26

Selected Bond Lengths and Bond Angles of 1-Chloro-3,5-diphenyl-1H-1,2,4,6-thiatriazine, [$\overline{\cdot\cdot}$SCl$\overline{\cdot\cdot}$N$\overline{\cdot\cdot}$C($C_6H_5$)$\overline{\cdot\cdot}$N$\overline{\cdot\cdot}$C($C_6H_5$)$\overline{\cdot\cdot}$N$\overline{\cdot\cdot}$] [2].

| bond length | in Å | bond angle | in ° |
|---|---|---|---|
| S–Cl | 2.283(1) | Cl–S–N(1) | 100.5(1) |
| S–N(1) | 1.590(3) | Cl–S–N(3) | 101.9(1) |
| S–N(3) | 1.583(3) | N(1)–S–N(3) | 117.6(2) |
| N(1)–C(1) | 1.332(4) | C(1)–N(2)–C(2) | 119.8(3) |
| N(2)–C(1) | 1.348(4) | S–N(3)–C(2) | 117.7(2) |
| N(2)–C(2) | 1.332(4) | N(1)–C(1)–N(2) | 126.2(3) |
| N(3)–C(2) | 1.348(4) | N(2)–C(2)–N(3) | 125.9(3) |
| C(1)–C(3) | 1.468(4) | S–N(1)–C(1) | 117.6(2) |
| C(2)–C(4) | 1.464(5) | | |

The compound consists of a six-membered $SN_3C_2$ ring which adopts a shallow boat-like conformation, with the sulfur and N(3) nitrogen atoms lying 0.24 and 0.05 Å, respectively, away from the mean plane of the other four atoms [2].

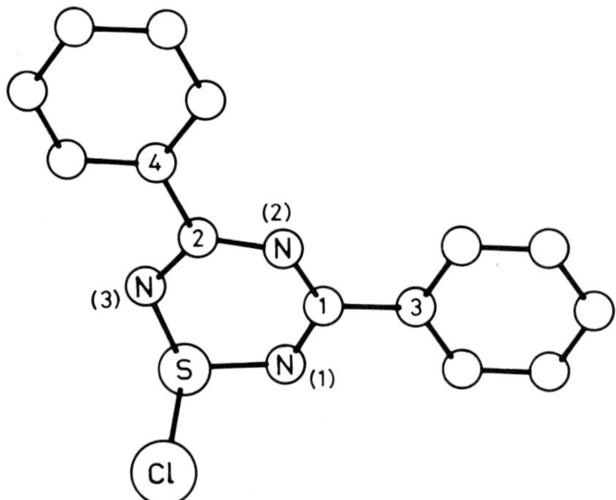


Fig. 13. Molecular structure of [$\overline{\cdot\cdot}$SCl$\overline{\cdot\cdot}$N$\overline{\cdot\cdot}$C($C_6H_5$)$\overline{\cdot\cdot}$N$\overline{\cdot\cdot}$C($C_6H_5$)$\overline{\cdot\cdot}$N$\overline{\cdot\cdot}$] [2].

In the IR spectra of the compounds there are absorption bands in the ranges 1550 to 1510 $cm^{-1}$ and 1440 to 1400 $cm^{-1}$ and at 1020 $cm^{-1}$ which have been assigned to the stretching vibrations of the thiatriazine ring [1]. For some $^{35}$Cl NQR and mass spectral data, see Table 25, p. 55.

The compounds are readily soluble in most inert solvents. They are extremely reactive; the chlorine atom at sulfur is readily substituted by the action of nucleophiles. The reaction of the compound with $R^1 = CCl_3$, $R^2 = C_6H_5$ (Table 25, No. 3) with $(CH_3)_2NH$ and $(C_2H_5)_2NH$ leads to the corresponding 1-dialkylamino-1H-1,2,4,6-thiatriazines (see p. 52). Reaction of the compounds with $R^1 = CCl_3$, $R^2 = CF_3$, $CCl_3$, or $C_6H_5$ (Table 25, No. 1, 2, 3) with $NaOCH_3$ or $NaOC_2H_5$ yields the corresponding 1-alkoxy-1H-1,2,4,6-thiatriazines (see p. 31). By the action of water the compounds No. 2 and 3 in Table 25 are readily hydrolyzed to give 2H-1,2,4,6-thiatriazine 1-oxides (see p. 21) [1]. When treated with sodium dust or the 2,4,6-triphenyl-3,4-dihydro-2H-1,2,4,5-tetrazinyl radical (see compound I, p. 19), the 3,5-bis(trichloromethyl) (Table 25, No. 2) and 3-trichloromethyl-5-phenyl (Table 25, No. 3) derivatives are completely reduced to give the corresponding 1,2,4,6-thiatriazinyl radicals (see p. 19) [4]. The same transformation can be performed on the 3,5-diphenyl derivative (Table 25, No. 4) by treating it with $Sb(C_6H_5)_3$ in deoxygenated $CH_2Cl_2$ [2, 3]. In $CH_3CN$ this reduction leads to the dimer 3,3′,5,5′-tetraphenyl-1,1′-bi-1H-1,2,4,6-thiatriazine, $(SN_3C_2(C_6H_5)_2)_2$ (see p. 20) [3].

**References:**


[1] Kornuta, P. P., Derii, L. I., Markovskii, L. N. (Zh. Org. Khim. **16** [1980] 1308/13; J. Org. Chem. [USSR] **16** [1980] 1130/4).

[2] Cordes, A. W., Hayes, P. J., Josephy, D., Koenig, H., Oakley, R. T., Pennington, W. T. (J. Chem. Soc. Chem. Commun. **1984** 1021/2).

[3] Hayes, P. J., Oakley, R. T., Cordes, A. W., Pennington, W. T. (J. Am. Chem. Soc. **107** [1985] 1346/51).

[4] Markovskii, L. N., Kornuta, P. P., Katchkovskaya, L. S., Polumbrik, O. M. (Sulfur Letters **1** [1983] 143/5; C.A. **100** [1984] No. 103301).


## 9.3.2.10 Derivatives of 1-Iodo-1H-1,2,4,6-thiatriazine

### 9.3.2.10.1 1-Iodo-3,5-diphenyl-1H-1,2,4,6-thiatriazine

**[∸SI∸N∸C($C_6H_5$)∸N∸C($C_6H_5$)∸N∸]**

The compound is prepared by reaction of 3,3′,5,5′-tetraphenyl-1,1′-bi-1H-1,2,4,6-thiatriazine, $(SN_3C_2(C_6H_5)_2)_2$, see p. 20, with $I_2$. Black platelets are obtained from $CH_3CN$ solution. The crystal structure has been determined by X-ray diffraction. The crystals are monoclinic, space group C2/c-$C^6_{2h}$ (No. 15) with a = 24.527(3), b = 5.117(2), c = 22.495(3) Å, β = 93.85(1)°, V = 2817(2) Å$^3$, Z = 8, $D_x$ = 1.79 g/cm$^3$. R = 0.031 for 1524 unique observed reflections. The molecular structure of the compound is shown in **Fig. 14**. Selected bond lengths and bond angles are given in Table 27, p. 58. The $N_3C_2$ ring segment is planar within 0.048(5) Å and the S atom is displaced 0.256(1) Å from this plane.

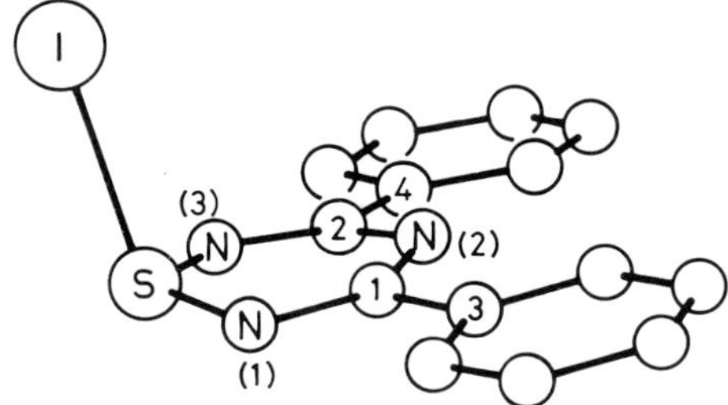


Fig. 14. Molecular structure of [∸SI∸N∸C($C_6H_5$)∸N∸C($C_6H_5$)∸N∸].

Table 27
Bond Lengths and Bond Angles of 1-Iodo-3,5-diphenyl-1H-1,2,4,6-thiatriazine, $[\dot{-}SI\dot{-}N\dot{-}C(C_6H_5)\dot{-}N\dot{-}C(C_6H_5)\dot{-}N\dot{-}]$.

| bond length | in Å | bond angle | in ° |
|---|---|---|---|
| S–I | 2.665(2) | I–S–N(1) | 103.5(2) |
| S–N(1) | 1.604(4) | I–S–N(3) | 101.4(2) |
| S–N(3) | 1.603(4) | N(1)–S–N(3) | 110.5(2) |
| N(1)–C(1) | 1.341(6) | S–N(1)–C(1) | 116.8(4) |
| N(3)–C(2) | 1.336(6) | S–N(3)–C(2) | 116.5(4) |
| N(2)–C(1) | 1.331(6) | C(1)–N(2)–C(2) | 119.7(4) |
| N(2)–C(2) | 1.344(6) | N(1)–C(1)–N(2) | 126.5(5) |
| C(1)–C(3) | 1.475(7) | N(3)–C(2)–N(2) | 127.1(5) |
| C(2)–C(4) | 1.481(7) | N(1)–C(1)–C(3) | 115.7(4) |
| | | N(2)–C(1)–C(3) | 117.8(4) |
| | | N(3)–C(2)–C(4) | 116.7(4) |
| | | N(2)–C(2)–C(4) | 116.1(5) |

**Reference:**

Cordes, A. W., Craig, S. L., Condren, M. S., Oakley, R. T., Reed, R. W. (Acta Cryst. C **42** [1986] 922/3).

### 9.3.2.11 Derivatives of 1-Alkyl (or Aryl)-1H-1,2,4,6-thiatriazine

R = Alkyl, Aryl; $R^1$ and $R^2$ = Aryl
$R = R^1 = C_6H_5$, $R^2 = N(C_6H_5)_2$

**$[\dot{-}SR\dot{-}N\dot{-}CR^1\dot{-}N\dot{-}CR^2\dot{-}N\dot{-}]$**

#### 9.3.2.11.1 1-Alkyl (or Aryl)-3,5-diaryl-1H-1,2,4,6-thiatriazines, $[\dot{-}SR\dot{-}N\dot{-}CR^1\dot{-}N\dot{-}CR^2\dot{-}N\dot{-}]$, R, $R^1$, $R^2$ are compiled in Table 28

The 1H-1,2,4,6-thiatriazine derivatives compiled in Table 28 have been obtained in the past only by cyclization reactions. For the first time some compounds (No. 4 to 6 in Table 28) have been isolated from treatment of N-bromobenzamidine, $C_6H_5C(NH_2)$=NBr, with $NaSC_2H_5$ (2:3 mole ratio) in $H_2O$ at reflux temperature for about 30 min or with $Pb(SC_2H_5)_2$, $Pb(SC_3H_7\text{-}i)_2$, or $Pb(SC_4H_9\text{-}n)_2$ in a 2:1 mole ratio in $CH_3COOC_2H_5$ at reflux temperature for 1.25 to 2.5 h [1]. The compounds No. 1 to 3 and 7 to 11 in Table 28 form from $H_2NCR^1$=NSR=$NCR^2$=$NH_2^+Br^-$ reacting at room temperature with diluted aqueous KOH in excess (except compound No. 11 in Table 28) or 1 N methanolic KOH in stoichiometric amounts. Even when left standing, the cyclization reaction takes place in solutions of the hydrobromides. When heating the hydrobromide or its free base to 110 to 120°C for 2 h, the corresponding thiatriazine derivative is formed by splitting off $NH_4Br$ or $NH_3$, respectively [2]. The compounds No. 1 and 4 in Table 28 have been isolated in very small yields (1 to 2%) from the reaction of sulfur diimides, $CH_3(C_6H_5CH_2)S(=NH)_2$ or $(C_2H_5)_2S(=NH)_2$, and $C_2H_5(C_6H_5CH_2)S(=NH)_2$, respectively, with $ClCH_2N{=}CClC_6H_5$, in $CH_2Cl_2$ in the presence of $N(C_2H_5)_3$ [3]. Yields, melting points, and descriptions of the compounds are given in Table 28.

Table 28

Yields and Melting Points of 1-Alkyl (or aryl)-3,5-diaryl-1H-1,2,4,6-thiatriazines, [∸SR∸N∸CR¹∸N∸CR²∸N∸].

| No. | R | R¹ | R² | yield in % | m.p. in °C (recrystallized from) | remarks | Ref. |
|---|---|---|---|---|---|---|---|
| 1 | $CH_3$ | $C_6H_5$ | $C_6H_5$ | 56[a)] | 166 to 166.5 ($CH_3OH$) | yellow tetrahedra | [2] |
| 2 | $CH_3$ | $C_6H_5$ | 4-$ClC_6H_4$ | 62 | 122.5 to 123.5 ($CH_3OH$) | yellow needles | [2] |
| 3 | $CH_3$ | 4-$ClC_6H_4$ | 4-$ClC_6H_4$ | 66 | 175 to 175.5 ($CH_3OH$) | yellow needles | [2] |
| 4 | $C_2H_5$ | $C_6H_5$ | $C_6H_5$ | 45<br>39 | 157 to 160<br>162 ($C_2H_5OH$) | yellow crystals<br>long yellow needles | [1] |
| 5 | i-$C_3H_7$ | $C_6H_5$ | $C_6H_5$ | 51 | 106 ($CH_3OH$) | yellow needles | [1] |
| 6 | n-$C_4H_9$ | $C_6H_5$ | $C_6H_5$ | 44 | 60 (60% $C_2H_5OH$) | yellow needles | [1] |
| 7[b)] | $C_6H_5$ | $C_6H_5$ | $C_6H_5$ | 69 | 88.5 to 89 ($CH_3OH$) | yellow needles | [2] |
| 8 | $C_6H_5$ | $C_6H_5$ | 4-$ClC_6H_4$ | 42 | 95 to 97 ($CH_3OH$) | yellow needles | [2] |
| 9 | $C_6H_5$ | 4-$ClC_6H_4$ | 4-$ClC_6H_4$ | 72 | 139 to 139.5 ($CH_3OH$) | small yellow needles | [2] |
| 10 | 3-$NO_2C_6H_4$ | $C_6H_5$ | $C_6H_5$ | 68 | 113 to 117 ($CH_3OH$ or ligroine) | dark yellow, lustrous plates, sensitive to light | [2] |
| 11 | 4-$NO_2C_6H_4$ | $C_6H_5$ | $C_6H_5$ | 47 | 195 to 195.5 ($CCl_4$) | small yellow, elongated rectangular leaves | [2] |

a) The yield can be increased to 90 or 75% by thermolyzing $H_2NC(C_6H_5){=}NS(CH_3){=}NC(C_6H_5){=}NH$ or its hydrobromide at 110 to 120°C for 2 h (see p. 58) [2]. – b) The compound is also formed by treating $H_2NC(C_6H_5){=}NS(C_6H_5){=}NC(C_6H_5){=}NH_2^+Br^-$ with excess diluted aqueous KOH [2].

1H-1,2,4,6-Thiatriazine derivatives are yellow crystalline solids of unlimited stability. They are easily chromatographed on $SiO_2$ thin-layer plates with $C_6H_6$-$CHCl_3$ (1:1) ($R_F$ values of ~0.6). In the IR spectrum of the compounds there is a very strong absorption band at ca. 1370 $cm^{-1}$ which has been assigned to a combined ring vibration [2]. UV spectrum of the compound with $R^1=R^2=C_6H_5$, $R=C_2H_5$ (Table 28, No. 4) ($5\times10^{-5}$M c-$C_6H_{12}$ solution): $\lambda_{max}=268$ nm ($\varepsilon=3100$) and 383 nm ($\varepsilon=1500$) [1]. Mass spectra of the derivatives with $R^1=C_6H_5$, $R^2=C_6H_5$, 4-$ClC_6H_4$, $R=CH_3$, $C_2H_5$ (Table 28, No. 2 and 4), show peaks of the molecular ions and the fragments $M^+$–R, $SN_2CC_6H_5^+$, and $SN^+$ [3].

The compounds are weak bases, reacting with HCl in ether to give colorless monochlorides. In 75% ethanol, pK = 3.2 was measured [2]. By hydrolysis the ring system is split, leading to N-containing products and corresponding sulfinic acids, $RSO_2H$ [4]. The compounds with $R^1=R^2=C_6H_5$, $R=C_2H_5$, i-$C_3H_7$, n-$C_4H_9$ (Table 28, No. 4 to 6) are oxidized nearly quantitatively by $MnO_4^-$ in acetone-$H_2O$ to give the corresponding 1H-1,2,4,6-thiatriazine 1-oxides. Addition compounds, 1 base : 2 $AgNO_3$, of low solubility are formed by reaction with $AgNO_3$ in $CH_3OH$. The same compounds (Table 28, No. 4 to 6) are hydrolyzed by boiling even for a short time or by standing with mineral acids for several hours to give, depending on the conditions, $(C_6H_5CO)_2NH$, $C_6H_5C(NH_2)=NC(O)C_6H_5$, and further decomposition products. Treatment with Zn in acids leads to destruction of the ring with formation of the corresponding thiols and benzamidine, $C_6H_5C(=NH)NH_2$ [1].

1-Ethyl-3,5-diphenyl-1H-1,2,4,6-thiatriazine (Table 28, No. 4) is slightly soluble in $CHCl_3$, $CCl_4$, benzene, acetone, ethyl acetate, pyridine, and in diluted mineral acids, is moderately soluble in ether and ligroine, and is nearly insoluble in water [1]. The corresponding nitrophenyl derivative (Table 28, No. 11) is readily soluble in benzene, moderately so in ether, difficult to dissolve in $C_2H_5OH$, and insoluble in ligroine [2].

**References:**


[1] Goerdeler, J., Loevenich, D. (Chem. Ber. **87** [1954] 1079/82).
[2] Goerdeler, J., Wedekind, B. (Chem. Ber. **95** [1962] 147/53).
[3] Haake, M., Fode, H., Ahrens, K. (Z. Naturforsch. **28b** [1973] 539/40).
[4] Goerdeler, J. (Angew. Chem. **66** [1954] 306/7).


### 9.3.2.11.2 1,3-Diphenyl-5-diphenylamino-1H-1,2,4,6-thiatriazine, [$\overset{..}{-}S(C_6H_5)\overset{..}{-}N\overset{..}{-}C(C_6H_5)\overset{..}{-}N\overset{..}{-}C(N(C_6H_5)_2)\overset{..}{-}N\overset{..}{-}$]

The compound is obtained in 25% yield by cyclization of $(C_6H_5)_2NC(NH_2)=NS(C_6H_5)=NC(C_6H_5)=NH_2^+Br^-$ with $CH_3ONa$ in $CH_3OH$ at reflux for 1 h. The light yellow rhombuses obtained by recrystallization from $CH_3OH$ melt at 174.5°C. They discolor upon action of light. The substance is moderately soluble in $CH_2Cl_2$, benzene, acetone, and ethyl acetate, and is difficult to dissolve in ether and petroleum ether.

**Reference:**


Goerdeler, J., Doerk, K. (Chem. Ber. **95** [1962] 154/7).


## 9.3.2.12 Derivatives of 2,5-Dihydro-1H-1,2,4,6-thiatriazine

### 9.3.2.12.1 1,3-Diphenyl-5,5-bis(trifluoromethyl)-2,5-dihydro-1H-1,2,4,6-thiatriazine and 1,2,3-Triphenyl-5,5-bis(trifluoromethyl)-2,5-dihydro-1H-1,2,4,6-thiatriazine

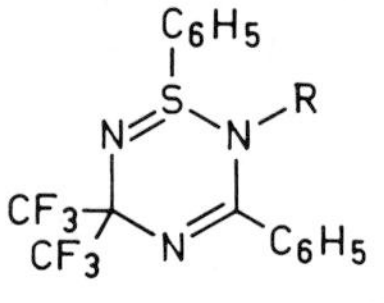

$R = H, C_6H_5$

**[$=S(C_6H_5)-NR-C(C_6H_5)=N-C(CF_3)_2-N=$]**

The title compounds have been obtained in 39.5 and 27% yield, respectively, when solutions of equimolar amounts of $(CF_3)_2ClC-N=SCl(C_6H_5)$ and the benzamidines $C_6H_5C(NH_2)=NR$, $R = H, C_6H_5$, in ether and a solution of double molar amounts of $N(C_2H_5)_3$ in ether-benzene (1:1) were added simultaneously at −30°C with stirring over 15 min. The mixture was then stirred at −10°C for 1.5 h and heated to 20°C. The compounds are recrystallized from hexane-ethanol 3:1.

[=S($C_6H_5$)–NH–C($C_6H_5$)=N–C($CF_3$)$_2$–N=] melts at 144 to 145°C. $^{19}$F NMR spectrum (solvent?/ internal $C_6H_5CF_3$), measured at −20°C: δ (in ppm) = −13.20 (q, $CF_3$), −18.10 (q, $CF_3$), $^4J(F,F)$ = 10.5 Hz; −14.95 (q, $CF_3$), −17.05 (q, $CF_3$), $^4J(F,F)$ = 10.0 Hz. The observed splitting of the signals is due to freezing of two of the possible tautomeric prototropic forms.

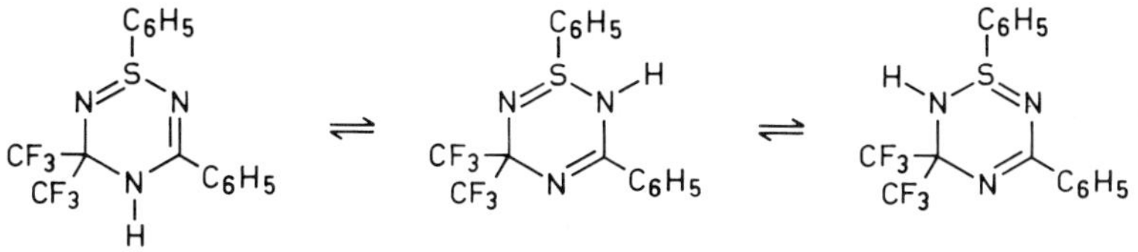

[=S($C_6H_5$)–N$C_6H_5$–C($C_6H_5$)=N–C($CF_3$)$_2$–N=] melts at 105 to 108°C. $^{19}$F NMR spectrum (solvent?/ internal $C_6H_5CF_3$), measured at 25°C: δ (in ppm) = −5.55 (q, $CF_3$), −7.95 (q, $CF_3$), $^4J(F,F)$ = 9.5 Hz. From the presence of one set of signals it was concluded that only one of the possible

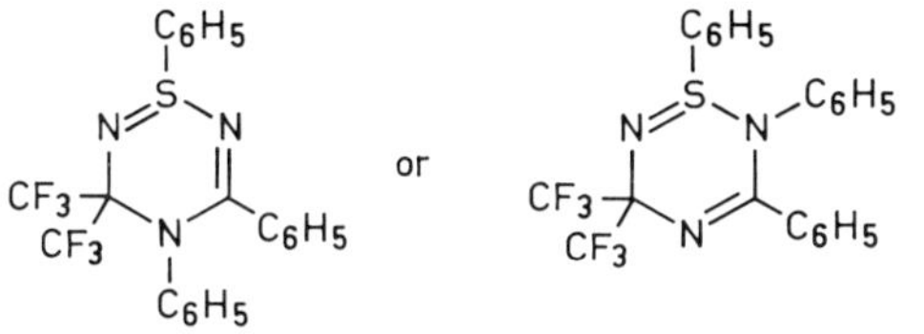

or

isomers is formed and not a mixture.

**Reference:**

Shermolovich, Yu. G., Talanov, V. S., Pirozhenko, V. V., Markovskii, L. N. (Zh. Org. Khim. **18** [1982] 2539/47; J. Org. Chem. [USSR] **18** [1982] 2240/7).

## 9.4 $SN_2C_3$ Ring

### 9.4.1 1,2,6-Thiadiazines

#### 9.4.1.1 4,5-Dihydro-3H-1,2,6-thia(S$^{IV}$)diazine and Derivatives

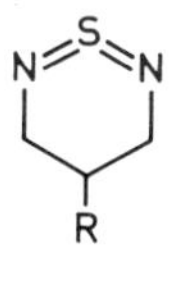

R = H, OH, $OOCCH_3$, $C_6H_4Cl$-4

**[=S=N–$CH_2$–CHR–$CH_2$–N=]** (Reduced formula in the text $SN_2C_3H_5R$)

##### 9.4.1.1.1 Survey

The six-membered $SN_2C_3$ ring forms when the sulfur diimide 4-$CH_3C_6H_4SO_2$–N=S=N–$SO_2C_6H_4CH_3$-4 reacts with diamines $NH_2CH_2CHRCH_2NH_2$ (R = H, OH), or in the reaction of $SF_4$ with $[(C_6H_5)_3P–NH(CH_2)_3NH–P(C_6H_5)_3]^{2+}$, see below.

The structure of the ring in its 4-(4-chlorophenyl) derivative was elucidated by X-ray studies. The ring exists in a half-boat conformation. The N=S=N group and the neighboring C atoms are coplanar, the substituted C(2) atom is situated above this plane. Bond lengths and bond angles are given in **Fig. 15**, p. 62. Bond angles at the nitrogen atoms, as well as the N–C distances, indicate $sp^2$-hybridization at the N atoms of the $SN_2$ group. The bond angles of the carbon atoms bound to nitrogen are 4.5° wider than expected for an ideal tetrahedron [1].

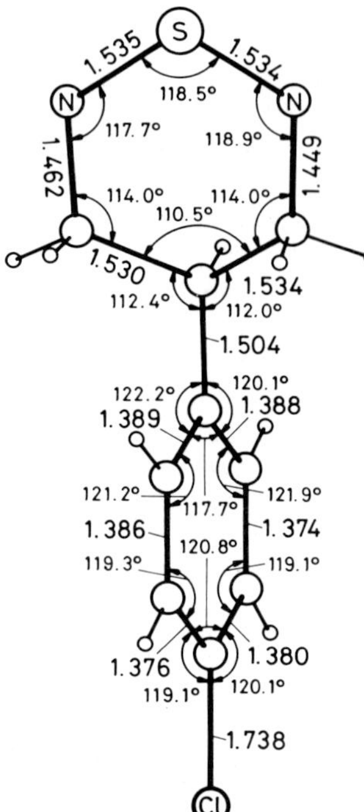


Fig. 15. Molecular structure of $SN_2C_3H_5(C_6H_4Cl\text{-}4)$. Bond lengths in Å; bond angles in °. Standard deviations are 0.003 to 0.005 Å (except C–H with 0.03 to 0.04 Å) and 0.2° to 0.3°.

$^1H$ NMR, UV, and assigned IR spectra of some $SN_2C_3$ ring derivatives are listed in Table 29.

Table 29

Spectra of 4,5-Dihydro-3H-1,2,6-thia($S^{IV}$)diazine and Derivatives [=S=N–$CH_2$–CHR–$CH_2$–N=]

| compound | $SN_2C_3H_6$ | $SN_2C_3H_5(OH)$ | $SN_2C_3H_5(OOCCH_3)$ |
|---|---|---|---|
| $^1H$ NMR spectra (δ in ppm, J in Hz) | δ = 1.89 (quint, β-$CH_2$), δ = 3.85 (t, α-$CH_2$) $^3J$ = 5.0 (in $CDCl_3$/TMS) | δ = 3.5 to 4.3, (m, α-$CH_2$) δ = 4.53 (s, OH) (solvent?/TMS) | δ = 4.0 (m, α-$CH_2$) δ = 5.2 (β-$CH_2$) (solvent?/TMS) |
| UV spectra ($\lambda_{max}$ in nm, ε in $L\cdot mol^{-1}\cdot cm^{-1}$) | 277.7 (log ε = 3.61) (in ethanol) | 280 (log ε = 3.57) (in dioxane) | — |
| IR spectra (ν in $cm^{-1}$)*) | 1135 ($\nu_{asym}$ N=S=N) 1020 ($\nu_{sym}$ N=S=N) (liquid film) | 1135 ($\nu_{asym}$ N=S=N) 1030 ($\nu_{sym}$ N=S=N) (liquid film) | 1165 ($\nu_{asym}$ N=S=N) 1045 ($\nu_{sym}$ N=S=N) (liquid film) |
| Ref. | [3, 4] | [4] | [4] |

*) Further IR absorption bands (in $cm^{-1}$) were tentatively assigned. $SN_2C_3H_6$: 2950 and 2850 ($\nu CH_2$); 1440 ($\delta CH_2$). $SN_2C_3H_5(OH)$: 3300 (νOH), 2900 and 2800 ($\nu CH_2$), 1425 ($\delta CH_2$). $SN_2C_3H_5(OOCCH_3)$: 1740 $cm^{-1}$ (νC=O) [4].

Chemical reactions are described for the particular compounds, see pp. 63/4.

### 9.4.1.1.2 $SN_2C_3H_6$

The compound forms in 38% yield from the reaction of a solution of $NH_2CH_2CH_2CH_2NH_2$ in $CCl_4$ with a suspension of 4-$CH_3C_6H_4SO_2$–N=S=N–$SO_2C_6H_4CH_3$-4 in $CCl_4$ (1:1 mole ratio) [3, 4]. $SN_2C_3H_6$ is also obtained by slowly passing $SF_4$ into a suspension of $[(C_6H_5)_3PNH(CH_2)_3NHP(C_6H_5)_3]^{2+}(Cl^-)_2$ (1:1 mole ratio) in $CH_2Cl_2/(C_2H_5)_3N$ at −10 to 0°C, and stirring at room temperature, yield 15.2% [8]. The yellowish oil [3, 4] is purified by distillation (b.p. 28°C/6 to 8 Torr) [3, 4, 8]. Vertical ionization energies of $SN_2C_3H_6$ were determined by photoelectron spectroscopy: $E_i$ = 9.25, 9.64, 10.88, 11.71, and 12.3 eV. They were tentatively assigned by comparison with chemically related compounds and CNDO calculations for the planar and the chair-formed molecule [5, 6]. NMR, UV, and IR spectra are described on p. 62. The molecular ion in the mass spectrum appears at m/e = 102 [4, 7, 8]. Several schemes for the ring fragmentation in the mass spectrometer were proposed [7].

$SN_2C_3H_6$ is thermally unstable and decomposes rapidly even at room temperature. ESR spectra taken during the decomposition at 100°C indicate the presence of $SN_2^{\bullet -}$ (quintet, g = 2.0107, $a_x$ = 5.1 ± 0.1 G), and $SN^\bullet$ (triplet, g = 2.0059, $a_x$ = 9.1 ± 0.1 G) radicals. At slightly higher temperatures, sudden decomposition occurs, and no more radicals are observed [8].

Discrepancies exist for the reaction with water. Some authors [3, 4] claim that $SN_2C_3H_6$ does not readily react with cold water and only on addition of acids is it rapidly hydrolyzed to the diamine [3, 4]. Other authors [8] report a very violent reaction with water. Slow hydrolysis produces the $HSO_3^-$ salts of the corresponding diamines [8]. The reaction with an equimolar amount or an excess of $CCl_3CHO$ (chloral) in absolute $CCl_4$ at room temperature yields O=S=N$(CH_2)_3$N=$CHCCl_3$ [4].

### 9.4.1.1.3 $SN_2C_3H_5(OH)$

The compound is prepared by adding pure 4-$CH_3C_6H_4SO_2$–N=S=N–$SO_2C_6H_4CH_3$-4 to a solution of $NH_2CH_2CHOHCH_2NH_2$ in $CHCl_3$ (1:1 mole ratio). When the precipitated 4-$CH_3C_6H_4SO_2NH_2$ is removed and the solvent evaporated, crude $SN_2C_3H_5OH$ is obtained almost quantitatively. The yellow oil is purified by distillation in high vacuum. Boiling point 108°C/$10^{-4}$ Torr [4].

NMR, UV, and IR spectra are given on p. 62.

The reaction of $SN_2C_3H_5(OH)$ in pyridine with pure $(CH_3CO)_2O$ yields the acetoxy derivative $SN_2C_3H_5(OOCCH_3)$ (see below) [4].

### 9.4.1.1.4 $SN_2C_3H_5(OOCCH_3)$

This acetoxy derivative forms when $(CH_3CO)_2O$ is slowly added to $SN_2C_3H_5(OH)$ in pyridine at room temperature (1:1 mole ratio). The red oil could not be obtained in a pure state, even after distillation at $10^{-4}$ Torr. Yield 59%. For NMR, UV, and IR spectra, see p. 62 [4].

### 9.4.1.1.5 $SN_2C_3H_5(C_6H_4Cl\text{-}4)$

The preparation of this derivative was not described, only the structure of it. $SN_2C_3H_5(C_6H_4Cl$-4) crystals, obtained from ethanol, are monoclinic, space group $P2_1/n$ (standard setting $P2_1/c$)–$C_{2h}^5$ (No. 14), lattice parameters a = 12.58(1), b = 10.95(6), c = 7.27(7) Å,

$\beta = 103.0(9)°$; $Z = 4$. Measured density $D_m = 1.44$ g/cm$^3$, calculated density $D_x = 1.45$ g/cm$^3$. $R = 0.082$; with consideration of H atoms $R = 0.041$ [1]. Dipole moment (measured) $\mu = 2.21$ D [2]. The structure of the $SN_2C_3$ ring is described on p. 61, the molecular structure is shown in Fig. 15, p. 62. The 4-chlorophenyl ring is arranged almost perpendicularly to the $SN_2C_3$ ring, the coplanar atoms of the two rings forming an angle of 84° [1].

**References:**


[1] Gieren, A., Pertlik, F. (Angew. Chem. **88** [1976] 852/3; Angew. Chem. Intern. Ed. Engl. **15** [1976] 782/3).

[2] Kresze, G. (Org. Sulphur Chem. Proc. 6th Intern. Conf., Bangor, Wales, 1974 [1975], pp. 65/93, 72).

[3] Kresze, G., Grill, H. (Tetrahedron Letters **1969** 4117/20).

[4] Grill, H. (Diss. München 1970, pp. 1/91, 29/41, 46, 74/6, 81).

[5] Solouki, B., Bock, H., Glemser, O. (Z. Naturforsch. **33b** [1978] 284/90).

[6] Solouki, B. (Diss. Frankfurt a.M. 1974, pp. 1/153, 72/86).

[7] Lengyel, J., Kresze, G., Berger, M., Kosbahn, W., Schäfer, H. (Acta Chim. [Budapest] **96** [1978] 275/94, 282, 287, 292).

[8] Appel, R., Lundehn, J.-R., Lassmann, E. (Chem. Ber. **109** [1976] 2442/55, 2445/6, 2452/3).


### 9.4.1.2 4H-2,1,3-Benzothiadiazin-2-S$^{IV}$-4-one

**[=S=N–CO–{1,2-$C_6H_4$}–N=]**

The compound forms by addition of equimolar amounts of 4-$CH_3C_6H_4SO_2$–N=S=N–$SO_2C_6H_4CH_3$-4 to 2-aminobenzamide in absolute $CHCl_3$ under an $N_2$ atmosphere and stirring. Warming yields a clear red solution and starts the precipitation of 4-$CH_3C_6H_4SO_2NH_2$. After 8 d the yield is 70%. Recrystallization from ethyl acetate yields yellow flakes which melt at 119°C.

[=S=N–CO–{1,2-$C_6H_4$}–N=] is characterized by $^1H$ NMR chemical shifts: $\delta = 7.0$ to 8.45 ppm (in $(CD_3)_2SO$ with TMS standard), IR spectra showing $\nu$(NSN) vibrations at 1125 and 1030 cm$^{-1}$ and $\nu$(C=O) stretching vibrations at 1640 cm$^{-1}$, the UV absorption (in dioxane) at $\lambda_{max} = 374.5$ nm ($\varepsilon = 6166$ L·mol$^{-1}$·cm$^{-1}$), and by its mass spectrum with the molecular ion at m/e = 164 [1].

[=S=N–CO–{1,2-$C_6H_4$}–N=] shows thermal stability, but great moisture sensitivity. The compound hydrolyzes to form its sulfurous diamide [=SO–NH–CO–{1,2-$C_6H_4$}–NH=] [1] (see p. 78). With $(CH_3)_3O^+BF_4^-$ and $(C_2H_5)_3O^+BF_4^-$ in $CH_2Cl_2$ at room temperature, the 2,1,3-benzothiadiazinium salts [=S–N=COR–{1,2-$C_6H_2XY$}–N=]$^+BF_4^-$ (R = $CH_3$, $C_2H_5$) form (see below) [2].

Substituted compounds of this type can be obtained by reacting the appropriately substituted 2-aminobenzamides with 4-$CH_3C_6H_4SO_2$–N=S=N–$SO_2C_6H_4CH_3$-4 [2].

**References:**


[1] Grill, H., Kresze, G. (Liebigs Ann. Chem. **749** [1971] 171/7, 173, 176).

[2] Kosbahn, W., Schäfer, H. (Angew. Chem. **89** [1977] 826; Angew. Chem. Intern. Ed. Engl. **16** [1977] 780).

### 9.4.1.3 4-Alkoxy-2,1,3-benzothiadiazine Ion(1+), 4-Alkoxy-2,1,3-benzothiadiazin-2-ium Salts, and Derivatives

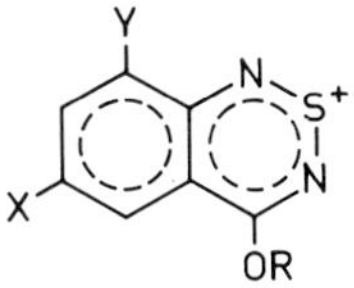

$R = CH_3, C_2H_5$
$X = H, Cl, NO_2;\ Y = H, Cl$

**[∸S∸N∸C(OR)∸{1,2-$C_6H_2XY$}∸N∸]$^+$**

Compounds with this type of cation were prepared by alkylating the appropriate benzothiadiazinone with oxonium salts in $CH_2Cl_2$:

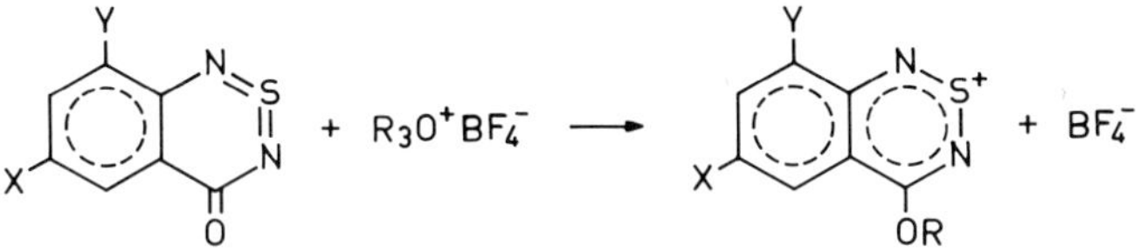

$$+ R_3O^+BF_4^- \longrightarrow \quad + BF_4^-$$

| compound | I | II | III | IV | V |
|---|---|---|---|---|---|
| X | H | H | Cl | Cl | $NO_2$ |
| Y | H | H | H | Cl | H |
| R | $CH_3$ | $C_2H_5$ | $C_2H_5$ | $C_2H_5$ | $C_2H_5$ |

The reaction occurs at room temperature within a few minutes (I, II) or a few days (IV). The products can be purified by dissolving in acetone and precipitating with diethyl ether. The yields are 13% (V) to 85% (I). No special salt is described.

The structure of the cations was established by IR, $^1H$, and $^{13}C$ NMR spectra (not given in the paper). The strong deshielding effect of the $N_2S^+$ group on the C-5 atom located in the position para to S is remarkable; the substituent increment of about +15.5 ppm has almost the same value as the diazonium group. The ring system can be regarded as being heteroaromatic because of its large resonance energy of 0.313 β calculated on an HMO basis.

The 2,1,3-benzothiadiazin-2-ium salts decompose in the 111 to 165°C range. With $H_2O$, they react slowly and $SO_2$ escapes. Compound II readily reacts at room temperature with 1,3-dienes like 2,3-dimethyl-1,3-butadiene, 2-methyl-1,3-butadiene (isoprene), or 1,3-pentadiene, the S=N1 double bond acting as a dienophile. The adducts can be isolated in 65% yield. The structure was established mainly by $^{13}C$ NMR spectra. The facility of addition and its marked regioselectivity may be explained by considering the frontier orbitals of the reactants. PPP and CNDO/S calculations show that the LUMO, a π orbital, is almost completely localized at N1 and S. This localization, a nodal plane between both atoms, and the low energy of the LUMO, enables an optimal interaction with the HOMO of the diene in the sense of a normal Diels-Alder reaction.

**Reference:**

Kosbahn, W., Schäfer, H. (Angew. Chem. **89** [1977] 826; Angew. Chem. Intern. Ed. Engl. **16** [1977] 780).

### 9.4.1.4 Naphtho[1,8-cd][1,2,6]thiadiazine

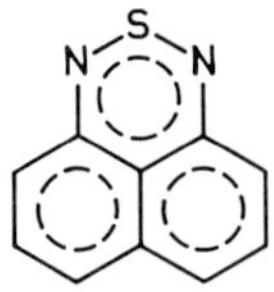

**[∸S∸N∸{1,8-$C_{10}H_6$}∸N∸]** (Reduced formula in the text $SN_2C_{10}H_6$)

The synthesis of the compound was first reported in 1965. A solution of 1,8-naphthalenediamine (I) and N-sulfinylaniline(II) in benzene was heated. The product (IV) was dehydrated by heating neat or in a solvent [1]. $SN_2C_{10}H_6$ formation is catalyzed by bases [2, 3]. It is prepared from IV in 88 to 95% yield by refluxing a suspension of IV in absolute benzene or toluene with an equimolar amount of II and $(C_2H_5)_3N$ (~10 wt%) [2]. $POCl_3$ also dehydrates IV [2].

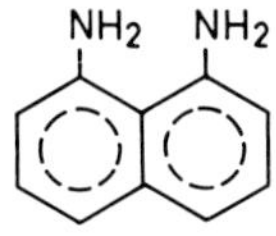

I 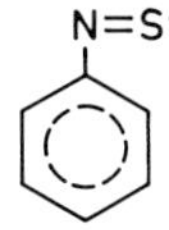 II 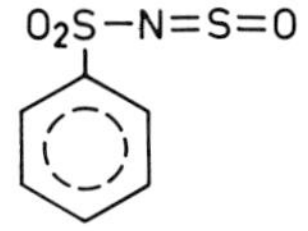 III 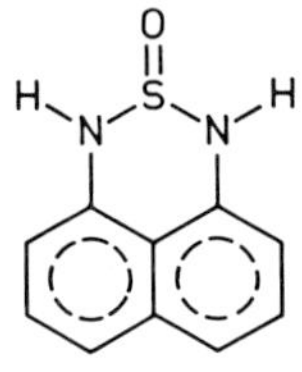 IV

The direct preparation of $SN_2C_{10}H_6$ from I and II is carried out in $(C_2H_5)_3N$ at room temperature and gives an 84% yield [3]. Pyridine or $(C_2H_5)_3N$ can be used as the catalyzing base if III is used instead of II [3]. A reaction scheme for the formation of $SN_2C_{10}H_6$ from I and II was proposed [4].

$SN_2C_{10}H_6$ is formed in 40 to 50% yield by passing $SO_2$ into a benzene solution of I containing some $(C_2H_5)_3N$ [2]. The preparation gives a 97% yield when a solution of I in $CH_2Cl_2$ and $(C_2H_5)_3N$ is saturated with $SO_2$, and the blue-green mixture is poured into a mixture of concentrated HCl, ice, and $CH_2Cl_2$ [3].

$SN_2C_{10}H_6$ crystallizes from acetone as thin brown-olive green platelets; m.p. 142.5 to 144°C [3]. Recrystallization from petroleum ether (80 to 100°C) yields bright blue-black scaly crystals; m.p. 142 to 143°C [2]. The reported decomposition on melting (m.p. 139 to 141°C) [1] was not confirmed [3].

The sulfur atom in the sulfur diimide $SN_2C_{10}H_6$ can be formally considered to be fourfold bound; but the ground state can also be described by mesomeric limiting structures [5]:

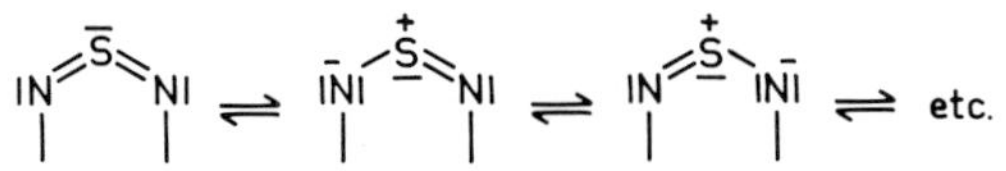

⇌ ⇌ ⇌ etc.

The participation of sulfur d orbitals in binding has been repeatedly discussed in connection with the tetravalency of sulfur. SCF MO-CNDO/2 calculations on compounds containing SN multiple bonds indicate that the SN π-bond order for $SN_2C_{10}H_6$ suggests the involvement of tetravalent sulfur. However, this must not be taken as evidence for d-orbital participation [6]. ESR spectra of the radical anion (see p. 68), as well as half-wave reduction potentials (see p. 67) can be satisfactorily interpreted with a p-orbital model [7]. Electronic spectra and photoelectron spectra (PES) and a comparison with related triazines data indicate that valency structures with tetravalent sulfur can be neglected, and that 3d orbitals are not important for the description of the ground state or excited states [5].

$^{15}N$ NMR (2 M in $(CH_3)_2SO$/external neat $CH_3NO_2$): $\delta = -89.3$ ppm. The conversion factor $-6.2$ is used for external 0.1 M $HNO_3$ [9].

UV spectrum ($\lambda_{max}$ in nm ($\varepsilon$ in $L \cdot mol^{-1} \cdot cm^{-1}$)) in stretched foils: $\lambda_{max} = 234$ (log $\varepsilon = 2.74$), 255 (3.93), 330 (3.92), 345 (4.07), and 643 (2.74) [5]. In solutions of $SN_2C_{10}H_6$ four UV absorptions are observed: in dioxane $\lambda_{max} = 234$ ($\varepsilon = 42200$), 331 (8100), 344 (8250), and 642 (540) [2]; in cyclohexane $\lambda_{max} = 235$ (log $\varepsilon = 475$), 332 (3.95), 347 (3.95), and 645 (2.70) [8]. The blue solution in hydrocarbons gave $\lambda_{max} = 658$ (log $\varepsilon = 2.76$) [1]. The absorption bands were tentatively assigned to $\pi \rightarrow \pi^*$ transitions. Calculated wavenumbers and directions of polarization are listed in the paper [5]. Configuration interaction calculations within the MIM (molecules in molecule) approximation indicate that the color determining electronic transition is governed mainly by symmetrical orbital interactions at the frontier NSN group [8].

The vertical ionization energies were determined from PE spectra: $E_i$ (in eV) = 7.46, 9.02, 9.90, and 10.14. They were tentatively assigned by comparison with calculated (EH and MINDO/3-method) orbital energies and assuming the Koopmans' theorem to be valid [5].

Polarographic investigations at a dropping mercury electrode in $CH_3CN$ (0.1 M in $(C_4H_9)_4NClO_4$) indicate a reversible one-electron reduction at $E_{1/2} = -0.96$ V vs. standard calomel electrode, and a further, irreversible reduction at $E_{1/2} = -1.8$ V. The approximate half-wave potential at a rotating platinum electrode in $CH_3CN$ (0.1 M in $(C_2H_5)_4NClO_4$) $E_{1/2} \approx +1.0$ V shows that $SN_2C_{10}H_6$ is a good electron donor [7].

Reduction with metallic potassium in $CH_3CH(OCH_3)_2$ yields the radical anion $SN_2C_{10}H_6^{\bullet -}$ (see p. 68) [7]. Reduction with zinc powder in ethyl acetate containing glacial acetic acid gives 1,2-dihydro-benz[cd]indazole (V) [3].

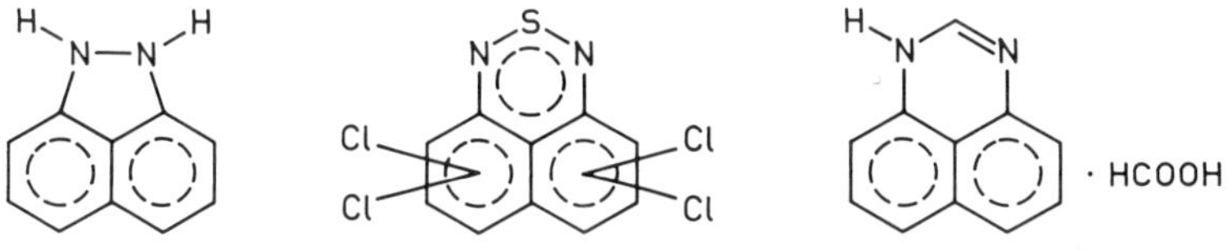

V VI VII

$SN_2C_{10}H_6$ is readily hydrolyzed by moist air [3], bases [1] (as $N(CH_3)_3$ in ethanol [2]), and acids (HCl in 2-propanol) [2] to form 1,8-naphthalenediamine (I) [1, 3]. Chromatography on active silica gel is not successful. When highly active $Al_2O_3$ is used, hydrolysis stops at the 1,2-dihydro-benz[cd]indazole (V) stage, presumably because there are not enough water molecules available [3].

A tetrachloro derivative $SN_2C_{10}H_6Cl_4$ (VI) forms from benzene solutions of $SOCl_2$, $S_2Cl_2$, or $PCl_5$. A suspension of $SN_2C_{10}H_6$ in refluxing 98% formic acid is solvolyzed to form the perimidine compound VII [2].

**References:**


[1] Dietz, R. (Chem. Commun. **1965** 57).
[2] Behringer, H., Leiritz, K. (Chem. Ber. **98** [1965] 3196/9).
[3] Beecken, H. (Chem. Ber. **100** [1967] 2164/9).
[4] Beecken, H. (Chem. Ber. **100** [1967] 2170/7, 2173/4).

### 9.4.1.5 Naphtho[1,8-cd][1,2,6]thiadiazine Radical Anion(1−)

**$[\dot{-}S\dot{-}N\dot{-}\{1,8\text{-}C_{10}H_6\}\dot{-}N\dot{-}]^{\dot{-}}$**

The radical anion forms when $SN_2C_{10}H_6$ is reduced with potassium metal in dimethoxyethane. It also forms in the polarographic reduction of $SN_2C_{10}H_6$ in $CH_3CN$ at a dropping mercury electrode, see p. 67. ESR spectra were measured in dimethoxyethane and hyperfine coupling constants (in G) were assigned on the basis of MO calculations: $a_H = 0.32(2)$ (position 5 and 8), 0.74(2) (6 and 7), 1.26(2) (4 and 9); $a_N = 4.69(2)$ (position 1 and 3). The ESR spectra and the polarographic studies agree with an MO model in which the p orbitals (but not d orbitals) of sulfur play a significant role in bonding.

**Reference:**

Atherton, N. M., Ockwell, J. N., Dietz, R. (J. Chem. Soc. A **1967** 771/7).

### 9.4.1.6 Naphtho[1,8-cd:4,5-c′d′]bis[1,2,6]thiadiazine

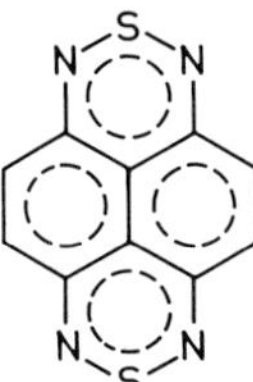


**$[\dot{-}N\dot{-}S\dot{-}N\dot{-}]_2\{1,8,4,5\text{-}C_{10}H_4\}$** (Reduced formula in the text $S_2N_4C_{10}H_4$)

The title compound is obtained in 35% yield by adding a slurry of the "tin double salt" of 1,4,5,8-naphthalenetetramine (i.e., the product of reduction of 1,4,5,8-tetranitronaphthalene with tin and hydrochloric acid [1]) in dry $CH_2Cl_2$ to $N(C_2H_5)_3$, and passing anhydrous $SO_2$ through the stirred mixture:

$$[C_{10}H_4(NH_2)_4]\ SnCl_4 \cdot 2HCl + (C_2H_5)_3N + SO_2 \longrightarrow$$

Recrystallization from 1,2-dichloroethane gives long, lustrous, metallic green needles which melt at 288 to 290°C (uncorrected; sealed tube) [2].

The X-ray structure determination shows the unit cell to be monoclinic, space group $P2_1/a\text{-}C^5_{2h}$ (No. 14); lattice parameters a = 14.935(8), b = 8.304(6), c = 3.794(4) Å, β = 91.53(7)°; V = 470.4 Å$^3$; $D_m = 1.72\ g/cm^3$, $D_x = 1.725\ g/cm^3$; Z = 2. The planar molecules form stacks in the c direction with a pseudohexagonal packing. The atoms are positioned similarly to those in the rhombohedral modification of graphite (see "Kohlenstoff" B 2, 1968, pp. 414, 422). The distance between the molecular planes along the stacks is 3.40 Å. The intermolecular distance between the stacks, S···N = 3.104 Å, is significantly less than the sum of van der Waals radii (3.35 Å). The result is a ribbon-like connection of molecules bound by two pairs of short intermolecular S···N contacts, see **Fig. 16**. The molecules are centrosymmetric with symmetry $D_{2h}$. The S–N bond length of 1.649 Å lies between that of a single and double bond. The C–N length, 1.314 Å, approaches the double bond length. The C–C bond lengths, 1.336 to 1.450 Å in the naphthalene skeleton also indicate intermediate bond orders. The geometry of the molecule agrees with the resonance formulas a to c, with structure a preferred [3]. For an HMO orbital energy scheme, see [4].

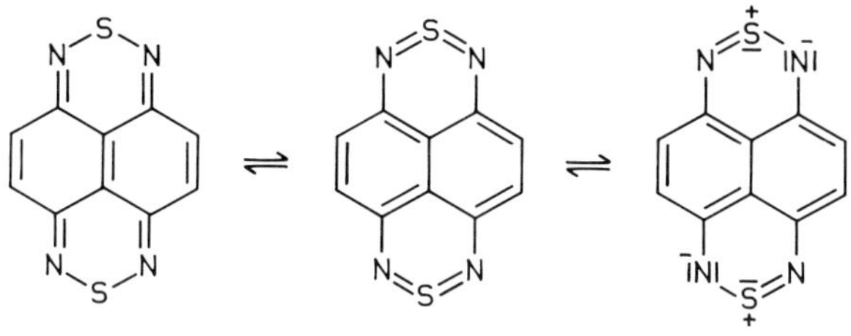

a ⇌ b ⇌ c

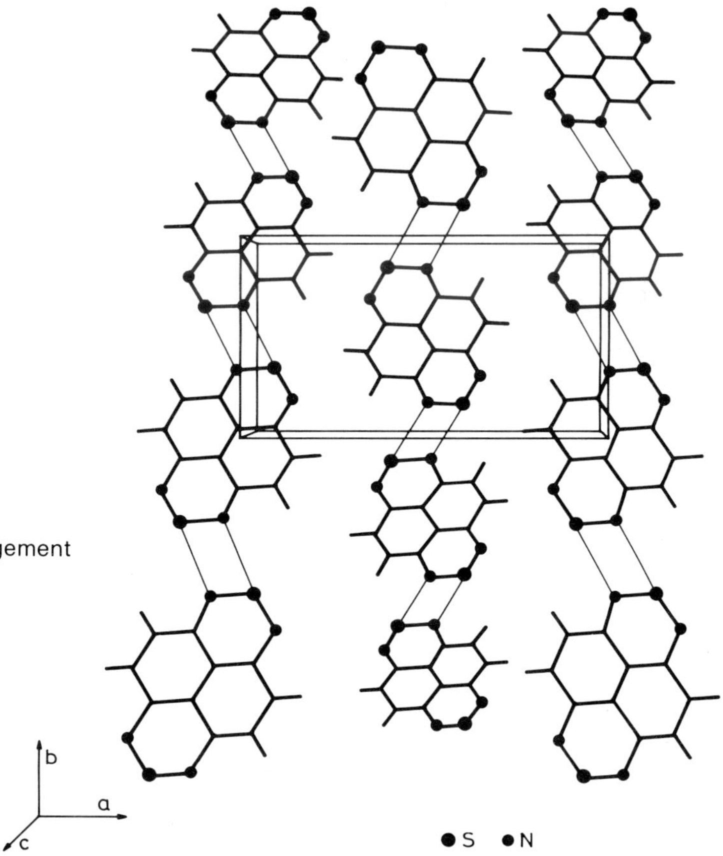

Fig. 16. Molecule arrangement in $S_2N_4C_{10}H_4$ crystals.

$^1H$ NMR spectrum ($CS_2$/TMS): $\delta = 4.45$ ppm [2], ($d_6$-DMSO/TMS): $\delta = 4.74$ ppm [4]. The high-field chemical shift is interpreted in terms of an induced paramagnetic ring current [2]. For hypothetical calculated ring currents, see [4].

UV spectrum ($\lambda_{max}$ in nm ($\varepsilon$ in $L \cdot mol^{-1} \cdot cm^{-1}$)) in $CH_3CN$: $\lambda_{max} = 255$ ($\varepsilon = 24000$), 273 (25500), 335 (sh) (4500), 352 (sh) (4000), 459 (14000), and 489 (19300) [2]; in dioxane: $\lambda_{max} = 261$ ($\varepsilon = 28900$) and 496 (25500) [4]. IR spectrum (in CsI): 18 absorption maxima between 335 and 3049 $cm^{-1}$ are listed without assignment [2].

Mass spectrum: molecular ion m/e = 244, seven strong peaks (not assigned) [2].

Cyclic voltammetry of $S_2N_4C_{10}H_4$ was carried out in $CH_3CN$ with 0.1 M $(n\text{-}C_4H_9)_4NBF_4$ as the supporting electrolyte with a Pt working electrode and an Ag|AgCl reference electrode. The half-wave potential for the oxidation is $E_{1/2} = +0.75$ V, and for the reduction $E_{1/2} = -0.77$ V (one electron) and $-1.38$ V (two electrons). With the exception of the second reduction the waves are reversible. ESR spectra of the radical ions (in nitrobenzene) give the g values and hyperfine coupling constants (a in G) for the anion g = 2.00253, $a_H = 2.235$, $a_N = 1.861$, for the cation g = 2.00560, $a_H = 0.30$, $a_N = 1.20$ [2].

$S_2N_4C_{10}H_4$ has poor solubility in the usual organic solvents, giving orange solutions. It is indefinitely stable to the atmosphere [2]. The electronic spectrum, $^1H$ NMR shift, and electrochemistry show that $S_2N_4C_{10}H_4$ can be easily reduced and has low-energy transitions. They also provide evidence for paramagnetic ring currents, and suggest that $S_2N_4C_{10}H_4$ is "antiaromatic" (i.e., it has negative resonance energy). Nevertheless, the compound has the chemical stability normally associated with an aromatic compound. Therefore the compound is said to possess "ambiguous aromatic character" [2].

**References:**

[1] Will, W. (Ber. Deut. Chem. Ges. **28** [1895] 2234/5).
[2] Haddon, R. C., Kaplan, M. L., Marshall, J. H. (J. Am. Chem. Soc. **100** [1978] 1235/9).
[3] Gieren, A., Lamm, V., Haddon, R. C., Kaplan, M. L. (J. Am. Chem. Soc. **101** [1979] 7277/81).
[4] Kaplan, M. L., Haddon, R. C., Schilling, F. C., Marshall, J. H., Bramwell, F. B. (J. Am. Chem. Soc. **101** [1979] 3306/8).
[5] Bartetzko, R., Gleiter, R. (Angew. Chem. **90** [1978] 481/2; Angew. Chem. Intern. Ed. Engl. **17** [1978] 468/9).
[6] Grunwell, J. R., Baker, H. S. (J. Chem. Soc. Perkin Trans. II **1973** 1542/4).
[7] Atherton, N. M., Ockwell, J. N., Dietz, R. (J. Chem. Soc. A **1967** 771/7).
[8] Fabian, J., Mayer, R., Bleisch, S. (Phosphorus Sulfur **7** [1979] 61/7).
[9] Yavari, I., Botto, R. E., Roberts, J. D. (J. Org. Chem. **43** [1978] 2542/4).

### 9.4.1.7 Derivatives of 4H-Furo[2,3-c][1,2,6]thiadiazin-2-S$^{IV}$-4-one

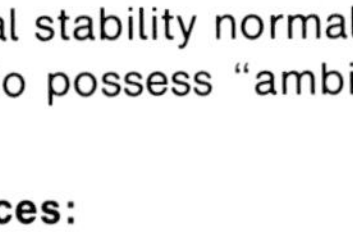

R = $CH_3$?, $C_6H_5$

**[=S=N–CO–{C(=)–CR=CR–O–C(=)}–N=]**

The ring system was isolated in the form of the phenyl ring stabilized derivative only. Attempts to prepare the ring system substituted with $CH_3$ instead of $C_6H_5$ had only limited success. The IR spectrum of the product mixture indicates the existence of the desired

[=S=N–CO–{C–C($CH_3$)=C($CH_3$)–O–C}–N=], and its hydrolysis product (N–H and S=O vibrations), but it was too unstable to be isolated.

**[=S=N–CO–{C–C($C_6H_5$)=C($C_6H_5$)–O–C}–N=]** is prepared in 87% yield by refluxing 10 mmol of 2-amino-4,5-diphenyl-3-furancarboxamide in a mixture of 5 mL $S_2Cl_2$ and 35 mL n-$C_6H_{14}$. The light red powder is recrystallized from ethanol. M.p. 220 to 223°C. IR spectrum (KBr): 1640 $cm^{-1}$ (C=O), 1120 and 1030 $cm^{-1}$ (–N=S=N–). UV spectrum ($CH_3OH$): $\lambda_{max}$ = 460 nm.

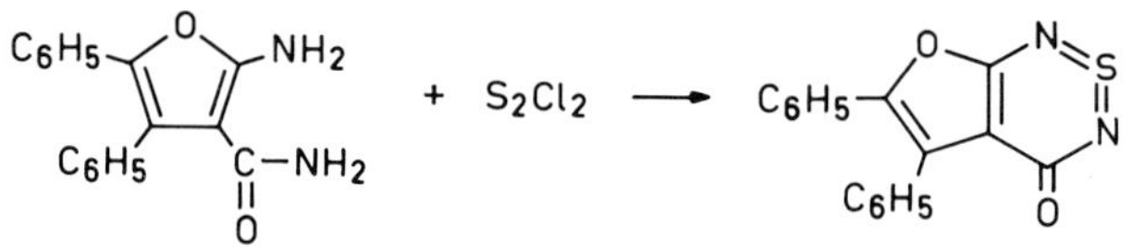

The compound is not sensitive to hydrolysis. It is stable for several days in neutral aqueous solutions and in HCl-containing solutions. In boiling HCl, the ring splits and the nitrile forms. The compound decomposes in dilute bases to form the starting compounds.

**Reference:**

Offermann, W., Eger, K., Roth, H. J. (Arch. Pharm. **314** [1981] 168/75, 169, 171, 174/5).

### 9.4.1.8 Derivatives of Pyrrolo[2,3-c][1,2,6]thiadiazin-2-$S^{IV}$-4(7H)-one

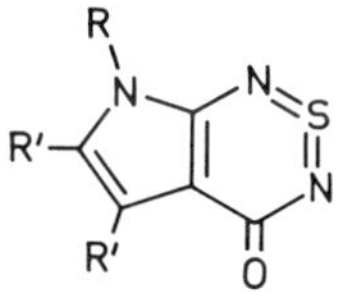

R is compiled in Table 30, p. 72.
R′ = $CH_3$?, $C_6H_5$

**[=S=N–CO–{C–CR′=CR′–NR–C}–N=]**

This ring system was isolated in the form of phenyl ring stabilized derivatives only (see below).

Attempts to prepare the $CH_3$-substituted compound (R = R′ = $CH_3$) resulted in a violent reaction. The products were very sensitive toward hydrolysis and could not be isolated in a pure state; IR spectra indicate a mixture of pyrrolo[2,3-c][1,2,6]thiadiazinones and hydrolysis products (N–H and S=O vibrations) [1].

**[=S=N–CO–{C–C($C_6H_5$)=C($C_6H_5$)–NR–C}–N=]**

The compounds form in good yields when the appropriate 1-substituted 2-amino-1H-pyrrole-3-carboxamides are reacted with $SOCl_2$, $S_2Cl_2$, or $SCl_2$:

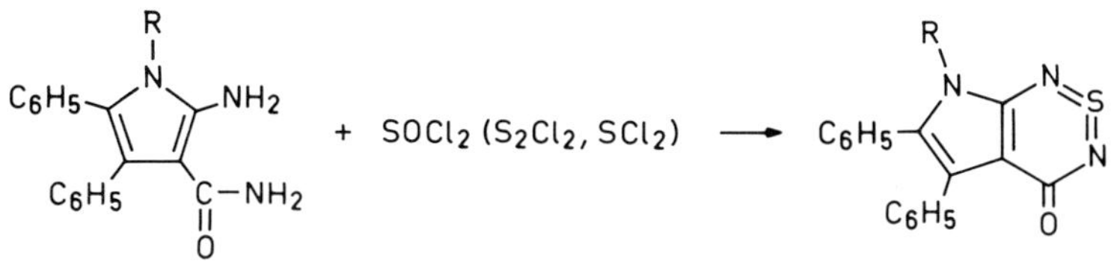

For preparation of compounds No. 1,2,4,5,6,8 in Table 30, the appropriate 1-substituted 2-amino-1H-pyrrole-3-carboxamide (20 mmol) is dissolved in hot dioxane or $CHCl_3$, and about 10 mL $SOCl_2$ is added dropwise. The precipitated crystals are washed with $CH_3OH$ [1].

Compound No. 6 was analogously prepared in THF [2]. For the preparation of compound No. 3, the appropriate amide (10 mmol) is dissolved in $CHCl_3$ and about 3 mL $SCl_2$ is added. The mixture is concentrated in vacuum. Compound No. 7 is prepared by refluxing the appropriate amide for 3 h in a mixture of $S_2Cl_2$ and n-hexane [1]. Yields and properties of the compounds are compiled in Table 30.

Table 30
Yields and Properties of Derivatives of Pyrrolo[2,3-c][1,2,6]thiadiazin-2-$S^{IV}$-4(7H)-one, [=S=N–CO–{C–CR′=CR′–NR–C}–N=] R′=$C_6H_5$

| No. | R | yield in % | properties | m.p. in °C (recryst. solvent) | Ref. |
|---|---|---|---|---|---|
| 1 | $CH_3$ | 92 | brick red crystals | 242 to 246 (ethyl acetate) | [1] |
| 2 | i-$C_3H_7$ | 64 | red crystals | 100 (dec.) (ethyl acetate) | [1] |
| 3 | n-$C_6H_{13}$ | 90 | brick red powder | 110 (ethyl acetate) | [1] |
| 4 | $(CH_2)_3N(CH_3)_2 \cdot HCl$ | 60 | light red crystals | 228 to 230 (ethanol) | [1] |
| 5 | $CH_2C_6H_5$ | 84 | brick red crystals | 184 to 186 (ethyl acetate) | [1] |
| 6 | $(CH_2)_2C_6H_5$ | 78 | brick red crystals | 161 to 163 (ethyl acetate) | [1] |
| 7 | c-$C_6H_{11}$ | 95 | dark red fine crystals | 177 to 180 (ethanol) | [1] |
| 8 | $C_6H_5$ | 60 | brick red fine crystals; | 265 (dioxane) | [2] |
| | | | soluble in ethanol, dioxane | 248 to 250 (ethanol) | [2, 3] |

The UV spectrum of the compounds typically shows an absorption maximum at $\lambda = 460$ nm (in $CH_3OH$). In the IR spectra (KBr), absorption bands at $\nu = 1640\ cm^{-1}$ (C=O), 1120 and 1030 $cm^{-1}$ (–N=S=N–) are characteristic [1].

The compounds are not sensitive to hydrolysis. They are stable for several days in aqueous and in dilute hydrochloric acid solutions. Boiling in HCl-containing solutions splits the ring and nitriles form. All of the compounds decompose in dilute bases to form the starting compounds [1].

**References:**

[1] Offermann, W., Eger, K., Roth, H. J. (Arch. Pharm. **314** [1981] 168/75, 169, 174/5).
[2] Kim-Su, M., Eger, K., Roth, H. J. (Arch. Pharm. **309** [1976] 721/4).
[3] Roth, H. J., Eger, K. (Arch. Pharm. **308** [1975] 186/9).

### 9.4.2 Derivatives of 1,2,6-Thiadiazine S (or 1)-Oxide

#### 9.4.2.1 2,6-Disubstituted Tetrahydro-2H-1,2,6-thiadiazine 1-Oxide

R = $CH_2CH(C_2H_5)C_4H_9$ (1) and $C_6H_5$ (2)

**[–SO–NR–(CH$_2$)$_3$–NR–]**

The first compound of this class, the 2-ethylhexyl derivative (1), is said to be formed in 73% yield as a "viscous concentrate", when 119 parts of $RNH(CH_2)_3NHR$ were allowed to react with 24 parts of $SOCl_2$. No exact data have been published [1]. Compound 2 (R = $C_6H_5$) forms in an 80% yield when a solution of $C_6H_5NH(CH_2)_3NHC_6H_5$ in pyridine is reacted with a slight excess of $SOCl_2$ at 0°C and the mixture subsequently stirred at room temperature for 2 h. Melting point 112 to 114°C [2].

Spectra of compound 2 (R = $C_6H_5$): IR spectrum (Nujol): $\nu = 1100\ cm^{-1}$ (S=O). $^1H$ NMR spectrum ($CDCl_3$/TMS): $\delta$ (in ppm) = 1.65 to 2.6 (2H, m, $CH_2$), 3.1 to 3.4 (2H, m, 2x$HC\underline{H}_{ax}$–N), 4.0 to 4.4 (2H, m, 2x$HC\underline{H}_{eq}$–N), and 6.95 to 7.6 (10H, m, arom.) [2].

Thermolysis experiments with compound 2 up to 200°C, with the goal of eliminating SO to give the pyrazole, were not successful even in the presence of phosphine derivatives [2]; (for comparison, see the thermolysis of the corresponding unsaturated compounds, p. 74).

**References:**

[1] Melamed, S., Croxall, W. L. (U.S. 2624729 [1950]; C.A. **1953** 11256).
[2] Barluenga, J., Lopez-Ortiz, J. F., Tomás, M., Gotor, V. (J. Chem. Soc. Perkin Trans. I **1981** 1891/5).

#### 9.4.2.2 Derivatives of 2H-1,2,6-Thiadiazine 1-Oxide

$R^1$, $R^2$, $R^3$, $R^4$ are compiled in Table 31, p. 74.

**[–SO–NR$^1$–CR$^2$=CR$^3$–CR$^4$=N–]**

The compounds No. 1 to 16, compiled in Table 31 are prepared by reaction of diimines with a slight excess of $SOCl_2$ in pyridine at 0°C. The reaction mixture is subsequently stirred at room temperature for 2 h [1, 2, 3].

$$R^4C(=NH)–C(R^3)=C(NHR^1)R^2 + SOCl_2 \xrightarrow[-C_5H_5N\cdot HCl]{+C_5H_5N} \text{2H-1,2,6-thiadiazine 1-oxide}$$

Table 31
Derivatives of 2H-1,2,6-Thiadiazine 1-Oxide, [–SO–NR¹–CR²=CR³–CR⁴=N–].

| No. | $R^1$ | $R^2$ | $R^3$ | $R^4$ | yield in % | m.p. in °C | Ref. |
|---|---|---|---|---|---|---|---|
| 1 | $c\text{-}C_6H_{11}$ | $C_6H_5$ | $CH_3$ | $C_6H_5$ | 85 | 146 to 147 | [1, 2] |
| 2 | $c\text{-}C_6H_{11}$ | $C_6H_5$ | $CH_3$ | $4\text{-}CH_3C_6H_4$ | 84 | 160 to 161 | [1, 2] |
| 3 | $C_6H_5$ | $C_6H_5$ | H | $C_6H_5$ | 86 | 153 to 154 | [1, 2] |
| 4 | $C_6H_5$ | $C_6H_5$ | H | $4\text{-}CH_3C_6H_4$ | 77 | 156 to 157 | [2] |
| 5 | $C_6H_5$ | $C_6H_5$ | Br | $C_6H_5$ | 82 | 165 to 167 | [3] |
| 6 | $C_6H_5$ | $C_6H_5$ | $CH_3$ | $C_6H_5$ | 80 | 133 to 135 | [1, 2] |
| 7 | $C_6H_5$ | $C_6H_5$ | $CH_3$ | $4\text{-}CH_3C_6H_4$ | 82 | 148 to 149 | [1, 2] |
| 8 | $C_6H_5$ | $4\text{-}ClC_6H_4$ | $CH_3$ | $4\text{-}CH_3C_6H_4$ | 80 | 156 to 158 | [2] |
| 9 | $4\text{-}CH_3C_6H_4$ | $C_6H_5$ | H | $c\text{-}C_6H_{11}$ | 84 | 122 to 124 | [1, 2] |
| 10 | $4\text{-}CH_3C_6H_4$ | $C_6H_5$ | H | $4\text{-}CH_3C_6H_4$ | 87 | 184 to 185 | [1, 2] |
| 11 | $4\text{-}CH_3C_6H_4$ | $C_6H_5$ | Br | $c\text{-}C_6H_{11}$ | 72 | 122 to 123 | [3] |
| 12 | $4\text{-}CH_3C_6H_4$ | $C_6H_5$ | Br | $C_6H_5$ | 78 | 141 to 142 | [3] |
| 13 | $4\text{-}CH_3C_6H_4$ | $C_6H_5$ | Br | $4\text{-}CH_3C_6H_4$ | 74 | 168 to 170 | [3] |
| 14 | $2\text{-}CH_3C_6H_4$ | $C_6H_5$ | $CH_3$ | $C_6H_5$ | 78 | 136 to 138 | [2] |
| 15 | $4\text{-}CH_3C_6H_4$ | $C_6H_5$ | $CH_3$ | $C_6H_5$ | 78 | 150 to 152 | [2] |
| 16 | $4\text{-}CH_3C_6H_4$ | $C_6H_5$ | $CH_3$ | $4\text{-}CH_3C_6H_4$ | 79 | 161 to 162 | [1, 2] |

The $^1H$ and $^{13}C$ NMR spectra of the compound with $R^1 = 4\text{-}CH_3C_6H_4$, $R^2 = C_6H_5$, $R^3 = Br$, and $R^4 = C_6H_5$ (Table 31, No. 12) in $CDCl_3$ solution with TMS as internal standard were recorded (δ in ppm). $^1H$ NMR: 2.2 (s, $CH_3$), 6.7 to 8.0 (m, arom. H). $^{13}C$ NMR: 20.6 (q), 98.2 (s), 126.3 (d), 127.7 (d), 128.0 (d), 128.7 (d), 129.5 (d), 129.9 (d), 130.6 (d), 133.8 (d), 136.5 (s), 137.7 (s), 138.0 (s), 148.5 (s), and 163.7 (s) [3]. The IR spectra of the compounds No. 1 to 4, 6 to 10, 14 to 16 in Table 31 show an absorption band at 1150 $cm^{-1}$ [2], and the compound No. 12 at 1130 $cm^{-1}$ [3], which are assigned to the ν(S=O) stretching vibration.

In the mass spectra the compounds No. 1 to 4, 6 to 10, 14 to 16 in Table 31 show the $M^+$–SO peak as base peak. The peak of the parent ion $M^+$ shows a very low intensity [2]. The molecular peak of compound 12 is observed at m/e = 436 [3]. On heating compounds No. 1 to 4, 6 to 10, 14 to 16 in toluene at 90°C [2] and 5, 11, 12 in toluene at 100°C [3] for 8 h, elimination of SO with formation of pyrazoles in high yields occurs [1, 2, 3].

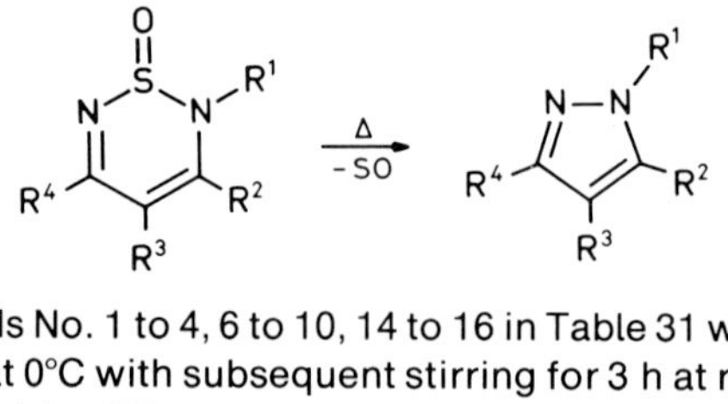

Reaction of compounds No. 1 to 4, 6 to 10, 14 to 16 in Table 31 with 3-chloroperbenzoic acid (1:2 mole ratio) in $CHCl_3$ at 0°C with subsequent stirring for 3 h at room temperature yields 2H-1,2,6-thiadiazine 1,1′-dioxides [2].

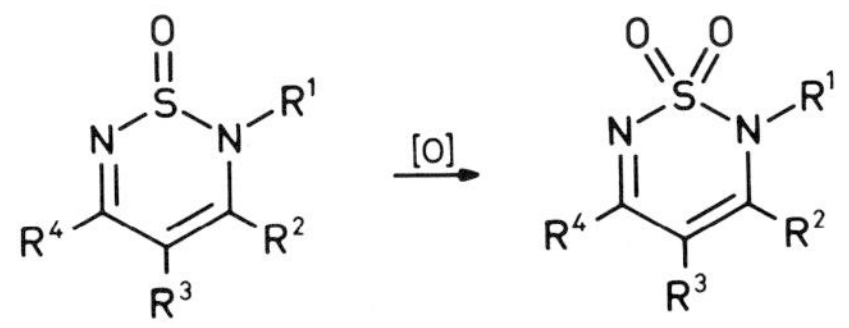

When solutions of the same compounds in THF are treated with 6N KOH at 60°C for 1 h, the starting diimines are regenerated quantitatively [2].

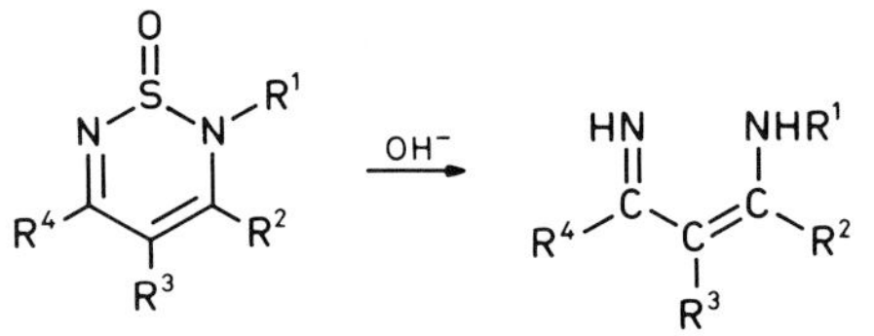

**References:**

[1] Barluenga, J., López-Ortiz, J. F., Gotor, V. (J. Chem. Soc. Chem. Commun. **1979** 891).
[2] Barluenga, J., López-Ortiz, J. F., Tomás, M., Gotor, V. (J. Chem. Soc. Perkin Trans. I **1981** 1891/5).
[3] Barluenga, J., Tomás, M., López-Ortiz, J. F., Gotor, V. (J. Chem. Soc. Perkin Trans. I **1983** 2273/6).

### 9.4.2.3 Derivatives of 4,5-Diphenyl-2H-1,2,6-thiadiazin-3(6H)-one 1-Oxide

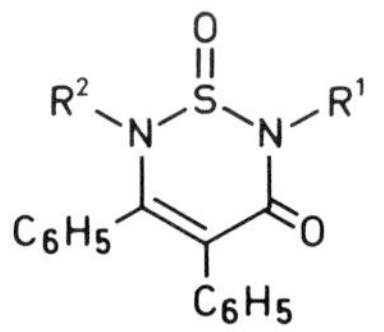

$R^1$, $R^2$ are compiled in Table 32, p. 76.

**[-SO-NR$^1$-CO-C($C_6H_5$)=C($C_6H_5$)-NR$^2$-]**

The compounds compiled in Table 32 are prepared by the reaction of α-aminocarboxamides with excess $SOCl_2$ in the presence of the HCl-trapping base, $(C_2H_5)_3N$ in toluene at 0°C followed by stirring at 50 to 60°C for 3 h. Use of pyridine instead of $(C_2H_5)_3N$ yields compounds free from sulfur, probably 4-amino-quinolines [1].

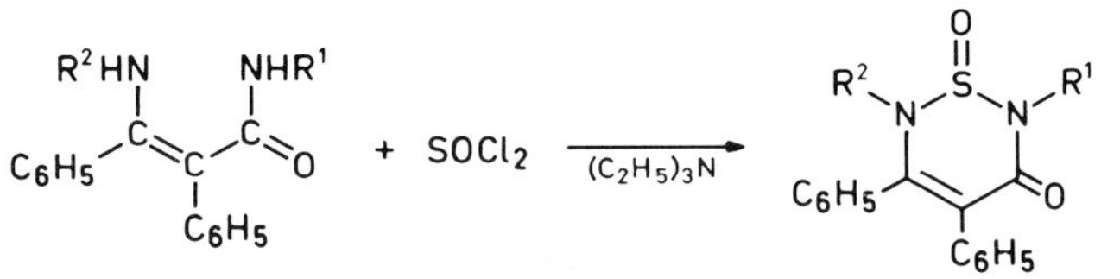

**1 to 5**

Table 32

Derivatives of 4,5-Diphenyl-2H-1,2,6-thiadiazin-3(6H)-one 1-Oxide, [–SO–NR$^1$–CO–C($C_6H_5$)=C($C_6H_5$)–NR$^2$–].

| No. | R$^1$ | R$^2$ | yield in % | m.p. in °C | $^1$H NMR spectra (standard TMS, δ in ppm); IR spectra (ν in cm$^{-1}$) |
|---|---|---|---|---|---|
| 1 | $C_6H_5$ | $C_6H_5$ | 36 | 231 | $^1$H NMR in $(CD_3)_2SO$: 7.00 to 7.67 (m, arom. H)<br>IR (KBr): ν(C=O) = 1650 to 1655, ν(S=O) = 1136 to 1155 |
| 2 | 4-$CH_3C_6H_4$ | 4-$CH_3C_6H_4$ | 37 | 198 | $^1$H NMR in $CDCl_3$: 2.22, 2.35 ($CH_3$), 6.93 to 7.55 (m, arom. H)<br>IR (KBr): ν(C=O) = 1650 to 1655, ν(S=O) = 1136 to 1155 |
| 3 | 4-$CH_3OC_6H_4$ | n-$C_3H_7$ | 50 | 168 | $^1$H NMR in $(CD_3)_2SO$: 0.78 ($CH_3$), 1.48, 3.52 ($CH_2$), 7.03 to 7.50 (m, arom. H) |
| 4 | 4-$CH_3OC_6H_4$ | c-$C_6H_{11}$ | 43 | 181 | $^1$H NMR in $(CD_3)_2SO$: 0.74 to 2.24 ($CH_2$), 3.25 (CH), 3.84 ($CH_3$), 7.00 to 7.50 (m, arom. H)<br>IR (KBr): ν(C=O) = 1650 to 1655, ν(S=O) = 1136 to 1155 |
| 5 | 4-$CH_3OC_6H_4$ | $C_6H_5$ | 47 | 173 | $^1$H NMR in $CDCl_3$: 3.82 (s, $CH_3$), 6.87 to 7.64 (m, arom. H)<br>IR (KBr): ν(C=O) = 1650 to 1655, ν(S=O) = 1136 to 1155 |

The mass spectra of the heterocycles show a weak molecular peak and the characteristic fragment $M^+ - R^1N{=}S{=}O$. In some cases a weak fragment $M^+ - SO_2$ is observed which probably forms from a 1,2,3-oxathiazine rearrangement product I.

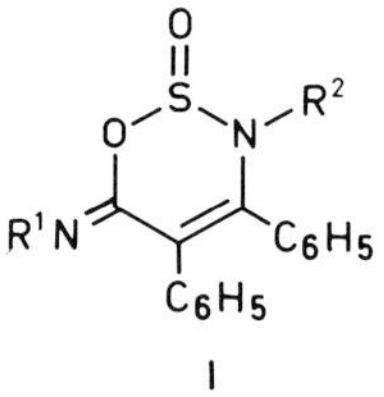

I

The compounds react with alkali under mild conditions to form 3-amino acrylamides. When the compound with R$^1$ = 4-$CH_3OC_6H_4$ and R$^2$ = $C_6H_5$ (Table 32, No. 5) is reacted in ethanol with 10% aqueous NaOH solution for 1 h, II forms quantitatively.

Acidic hydrolysis of compounds No. 1 to 5 with formation of aminoacylamides is more difficult. Thus, if an ethanolic solution of compound No. 5 is refluxed for 30 min with a 10% aqueous HCl solution, III is produced in 97% yield.

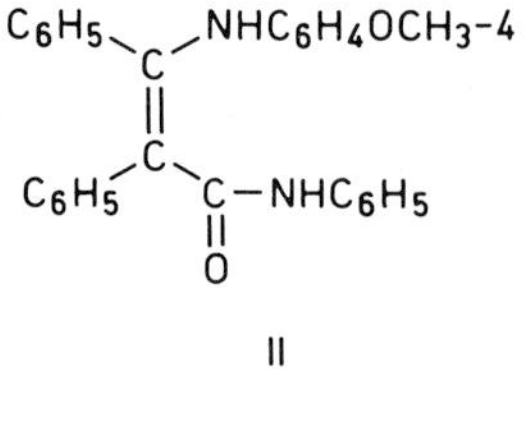

II

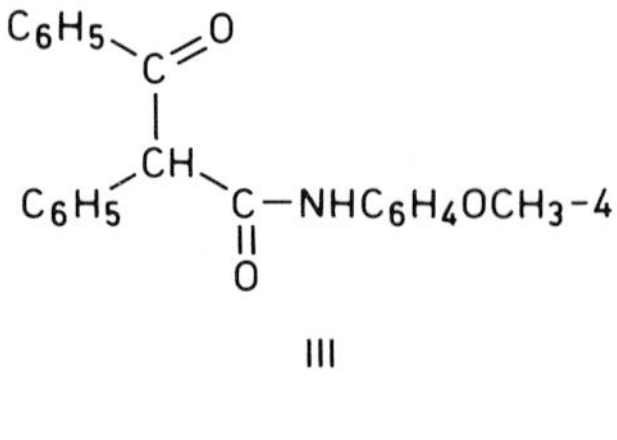

III

**Reference:**

Capuano, L., Urhahn, G., Willmes, A. (Chem. Ber. **112** [1979] 1012/22).

### 9.4.2.4 Derivatives of 2,6-Di-tert-butyl-4-methylene-2H-1,2,6-thiadiazine-3,5(4H,6H)-dione 1-Oxide

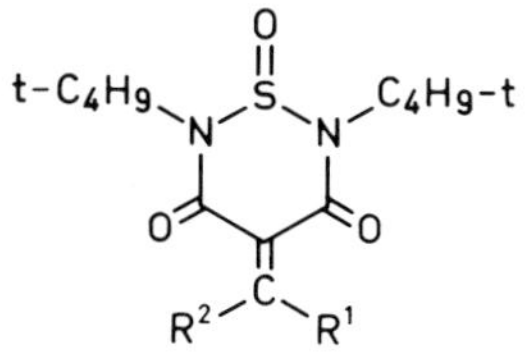

a) $R^1 = H$; $R^2 = C_6H_5$
b) $R^1 = R^2 = CH_3$

**[-SO-N($C_4H_9$-t)-CO-C(=C$R^1R^2$)-CO-N($C_4H_9$-t)-]**

Treatment of solutions of 1,2,6-thiadiazines Ia and Ib in glacial acetic acid at 60 to 80°C with $CrO_3$ (until green coloration disappears) followed by storing at 20°C for 1.5 h gives the 1,2,6-thiadiazine 1-oxides IIa and IIb in 38 and 40% yields, respectively. The colorless prisms of IIa melt at 153 to 154°C, the colorless needles of IIb at 188 to 189°C.

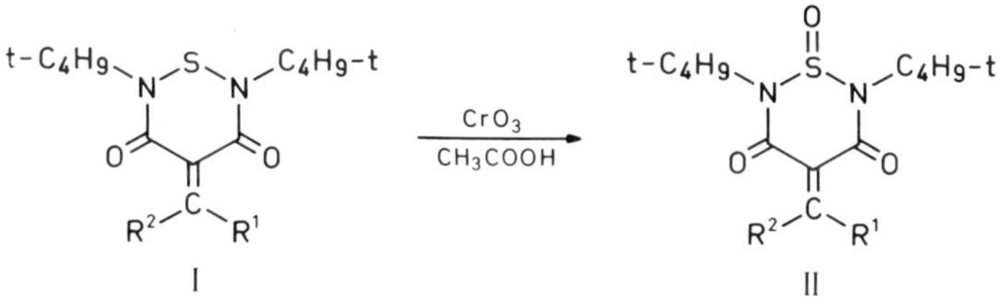

$CrO_3$ / $CH_3COOH$: I → II

**Reference:**

Wittmann, H., Ziegler, E., Sterk, H., Dworak, G. (Monatsh. Chem. **100** [1969] 959/67).

### 9.4.2.5 2,6-Di-tert-butyl-2H-1,2,6-thiadiazine-3,4,5(6H)-trione 1-Oxide

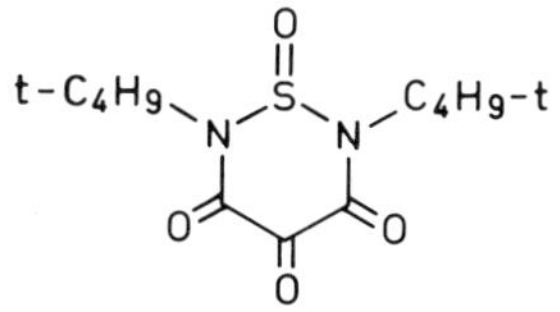

**[-SO-N($C_4H_9$-t)-$(CO)_3$-N($C_4H_9$-t)-]**

Treatment of a solution of di-tert-butyl-2H-1,2,6-thiadiazine-3,4,5(6H)-trione in glacial acetic acid at 60 to 80°C with $CrO_3$ (until green coloration disappears) results in oxidation of the sulfur atom. Completing the reaction by storing at 20°C for 1.5 h gives the title compound as light yellow prisms in a 40% yield. Melting point is 213°C with decomposition.

**Reference:**

Wittmann, H., Ziegler, E., Sterk, H., Dworak, G. (Monatsh. Chem. **100** [1969] 959/67).

### 9.4.2.6 3,4-Diphenyl-3,4-dihydro-1H-2,1,3-benzothiadiazine 2-Oxide

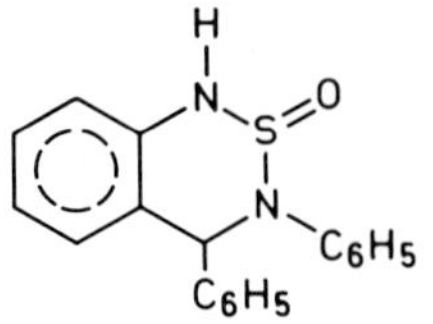

**[-SO-NC$_6$H$_5$-CH(C$_6$H$_5$)-{1,2-C$_6$H$_4$}-NH-]**

The title compound forms quantitatively as the dihydrate by mixing equimolar amounts of $C_6H_5NSO$ and $C_6H_5CH{=}NC_6H_5$, without a solvent, followed by storing the mixture at room temperature for 7 to 10 d. Neither six- nor four-membered ring products could be obtained when N-alkyl-substituted Schiff bases (e.g., $C_6H_5CH{=}NC_2H_5$) were used.

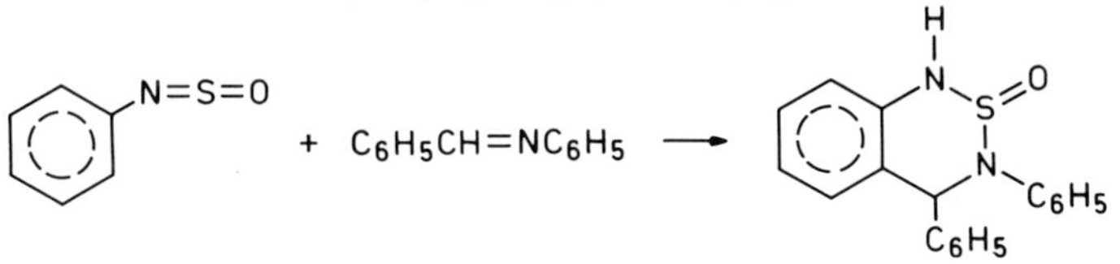

The compound melts at 122°C. IR spectrum (KBr): 3315 $cm^{-1}$ ($\nu$(N-H)), 1022 $cm^{-1}$ ($\nu$(S=O)). $^1$H NMR spectrum (DMSO-d$_6$/TMS): $\delta$ (in ppm) = 3.77 to 5.20 (s, broad, 5H, NH + 2H$_2$O), 6.37 to 6.87 (m, 4H, arom.), 6.93 (s, C-H), 6.97 to 7.33 (m, 5H, arom.), 7.24 to 7.59 (m, 3H, arom.), 7.64 to 8.03 (m, 2H, arom.).

Oxidation of the heterocycle with either $H_2O_2$ or 3-chloroperbenzoic acid or its reduction with $LiAlH_4$ afforded mixtures of unidentified oily products. The compound reacts with excess $(CH_3CO)_2O$ or $CH_3SO_2Cl$ in pyridine with cleavage of the thiadiazine ring at N3 to give the corresponding substituted anilides $C_6H_5NHCOCH_3$ and $C_6H_5NHSO_2CH_3$, respectively.

**Reference:**

Zoller, U., Rona, P. (Tetrahedron Letters **26** [1985] 2813/4).

### 9.4.2.7 Derivatives of 1H-2,1,3-Benzothiadiazin-4(3H)-one 2-Oxide

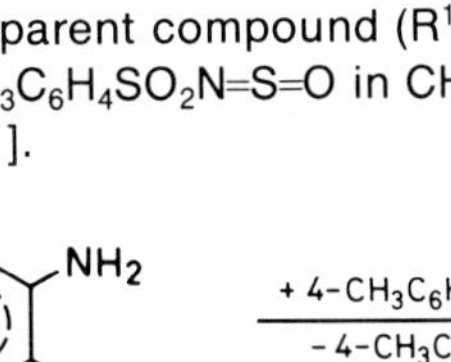

R$^1$, R$^2$, R$^3$ are compiled in Table 33.

**[-SO-NR$^2$-CO-{1,2-C$_6$H$_3$R$^3$}-NR$^1$-]**

The parent compound ($R^1 = R^2 = R^3 = H$, Table 33, No. 1) is prepared by reacting a solution of $4\text{-}CH_3C_6H_4SO_2N{=}S{=}O$ in $CHCl_3$ with a suspension of 2-aminobenzamide in $CHCl_3$ (1:1 mole ratio) [1].

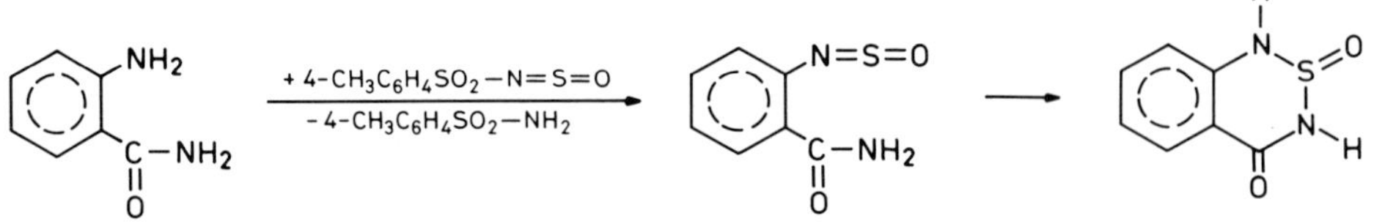

$$\xrightarrow[-\,4\text{-}CH_3C_6H_4SO_2-NH_2]{+\,4\text{-}CH_3C_6H_4SO_2-N=S=O}$$

It forms also by hydrolysis of [=S=N–CO–{1,2-$C_6H_4$}–N=] [1] (see p. 64).

Derivatives thereof are prepared by refluxing (1 to 3 h) α-aminocarboxamides with excess $SOCl_2$ [2, 3]. The compounds are summarized in Table 33.

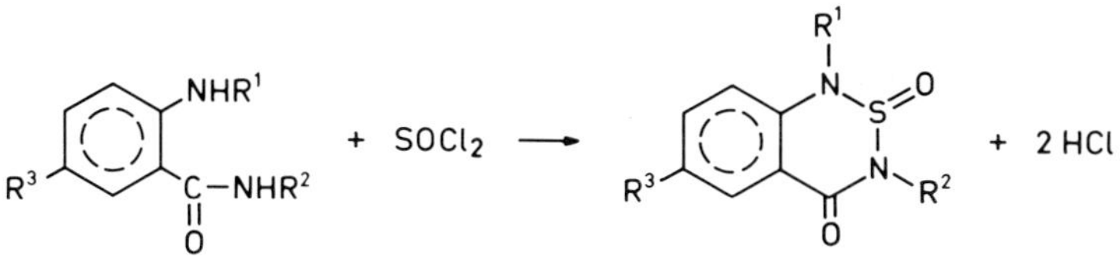

Table 33

Derivatives of 1H-2,1,3-Benzothiadiazin-4(3H)-one 2-Oxide, [–SO–$NR^2$–CO–{1,2-$C_6H_3R^3$}–$NR^1$–].

| No. | $R^1$ | $R^2$ | $R^3$ | yield in % | m.p. in °C | Ref. |
|---|---|---|---|---|---|---|
| 1 | H | H | H | 52 | 160 (red coloration) | [1] |
| 2 | $CH_3$ | $C_2H_5$ | Cl | 94 to 96 | 95 to 96.5 | [2, 3] |
| 3 | $CH_3$ | $C_2H_5$ | $SO_2NHC_2H_5$ | 62 to 83 | 150 to 152 | [2, 3] |
| 4 | $CH_3$ | $ClCH_2CH_2$ | Cl | 80 to 86 | 153 to 154 | [2, 3] |
| 5 | $CH_3$ | $ClCH_2CH_2$ | $NO_2$ | 67 to 72 | 161 to 162 | [2, 3] |
| 6 | $CH_3$ | $C_2H_5OCH_2CH_2$ | Cl | 62 to 100 | 73 to 74 | [2, 3] |
| 7 | $CH_3$ | $C_6H_5CH_2CH_2$ | Cl | 60 to 100 | 127 to 128.5 | [2, 3] |
| 8 | $CH_3$ | 2-$ClC_6H_4CH_2$ | H | — | — | [3] |
| 9 | $CH_3$ | 2-$ClC_6H_4CH_2$ | Cl | 85 to 100 | 148 to 149.5 | [2, 3] |
| 10 | $CH_3$ | c-$C_5H_9$ | Cl | 33 to 81 | 127 to 129 | [2, 3] |
| 11 | $C_2H_5$ | $C_2H_5$ | Cl | — | — | [3] |

Eleven other derivatives of 1H-2,1,3-benzothiadiazin-4(3H)-one 2-oxide have been claimed in a patent. But neither a description of preparative methods nor properties of the compounds are given therein [4].

$^1$H NMR spectrum of compound No. 1 (DMSO-$d_6$/TMS): δ (in ppm) = 7.0 to 8.2 (m, arom. H), 10.43 (s, NH), 11.08 (s, NH); IR spectrum (KBr; ν in $cm^{-1}$): 3 bands in the range 3300 to 2800 (NH and $CH_3$), 1080 (S=O) [1]. For compounds No. 2 to 4, 6, 7, 9, and 10 the IR spectrum (KBr) shows absorptions at 1645 to 1669 (C=O), 1121 to 1136 and 1318 to 1333 (S=O) [2].

Heating compound No. 1 above the melting point (160°C) produces a red melt, suggesting the formation of the starting compound 2-aminobenzamide. Dehydration experiments with the intention of generating a cyclic sulfur diimide I (see p. 64) by heating 1 in benzene or toluene [5] were not successful [1].

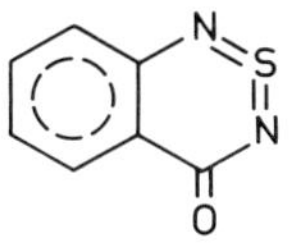

I

Oxidation of compound No. 1 with $H_2O_2$ at 20°C affords the dioxide, whereas at ~100°C the benzopyrazolinone forms preferentially [1]:

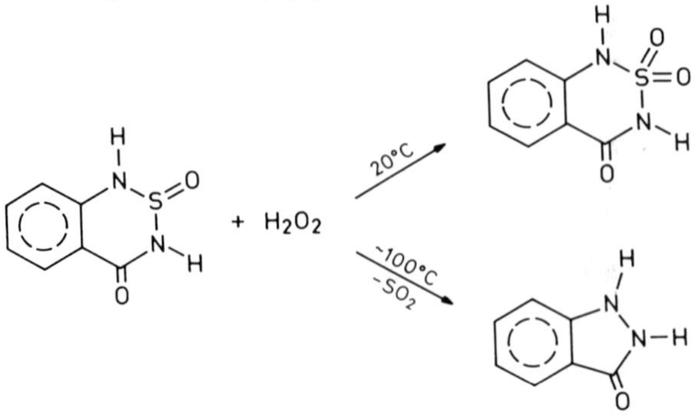

The derivatives are therapeutically active, affecting the central nervous system [3].

**References:**


[1] Grill, H., Kresze, G. (Liebigs Ann. Chem. **749** [1971] 171/7).

[2] Santilli, A. A., Osdene, T. S. (J. Org. Chem. **29** [1964] 2717/20).

[3] Santilli, A. A., Osdene, T. S., American Home Products Corporation, New York (U.S. 3217001 [1963/64] 1/4; C.A. **64** [1966] 3575).

[4] McKendry, L. H., Bland, W. P., Dow Chemical Company (U.S. 4155746 [1973/79] 1/7; C.A. **91** [1979] No. 103730).

[5] Behringer, H., Leiritz, K. (Chem. Ber. **98** [1965] 3196/9).


### 9.4.2.8 2-Benzyl-octahydro-pyrido[1,2-b][1,2,6]thiadiazine 1-Oxide

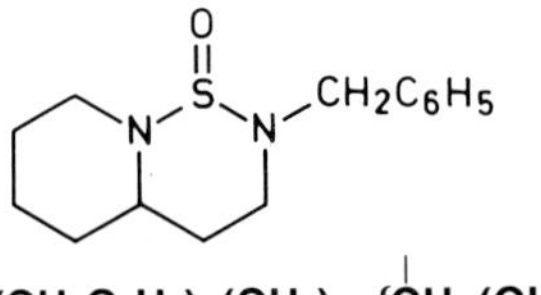

**[-SO-N($CH_2C_6H_5$)-($CH_2$)$_2$-{CH-($CH_2$)$_4$-N}-]**

The colorless crystalline compound forms in 80% yield by mixing equimolar amounts of benzyl[2-(piperidyl)ethyl]amine and $SOCl_2$ in the presence of triethylamine in $CHCl_3$. Melting point is 64 to 65.5°C.

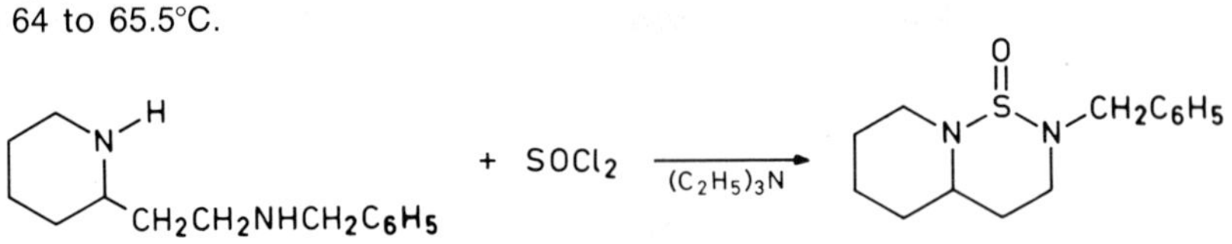

**Reference:**


Winterfeld, K., Häring, W. (Arch. Pharm. **295** [1962] 621/4).

### 9.4.2.9 1H-2,1,3-Benzothiadiazine-4(3H)-thione 2-Oxide

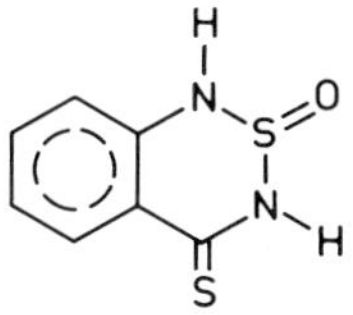

**[-SO-NH-CS-{1,2-$C_6H_4$}-NH-]**

The title compound forms in 56% yield by reacting 2-$NH_2C_6H_4$-$CSNH_2$ with equimolar amounts of 4-$CH_3C_6H_4SO_2$-N=S=O in $CHCl_3$. Melting point is 80 to 81°C (red coloration). $^1$H NMR spectrum (DMSO-$d_6$/TMS): δ (in ppm) = 6.9 to 7.7 (m, 4H), 8.36 (s, 1H), and 10.80 (s, 1H). The IR spectrum (KBr) shows for ν(NH) and ν(CH) three absorption bands in the range 3280 to 2850 $cm^{-1}$; ν(C=S) = 1090 $cm^{-1}$ [1]. Eight other compounds with various substituents on the benzo group have been claimed in a patent, but neither a description of preparation methods nor properties of the compounds are given therein [2].

**References:**


[1] Grill, H., Kresze, G. (Liebigs Ann. Chem. **749** [1971] 171/7).

[2] McKendry, L. H., Bland, W. P., Dow Chemical Company (U.S. 4155746 [1973/79]; C.A. **91** [1979] No. 103730).


### 9.4.2.10 1H,3H-Naphtho[1,8-cd][1,2,6]thiadiazine 2-Oxide

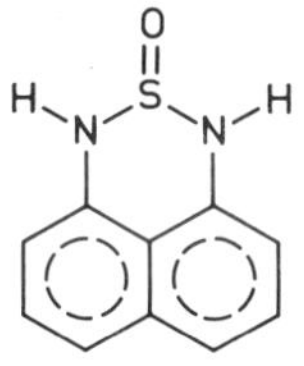

**[-SO-NH-{1,8-$C_{10}H_6$}-NH-]**

Treatment of 1,8-naphthalenediamine with an equimolar amount of $C_6H_5NSO$, in benzene at reflux [1, 2] or at ambient temperature [3] yields the title compound as colorless plates [1, 2, 3]. Yield 93% [1] and 98% [3]. It forms also by the reaction of 1,8-naphthalenediamine with $SO_2$ (~1:2 mole ratio) in a bomb tube at 80 to 90°C [1] or with $C_6H_5SO_2NSO$ in benzene at room temperature.

The UV spectrum (in $CH_3OH$, λ in nm) shows four absorption bands at $\lambda_{max}$ = 214 (log ε = 4.24), 235 (4.58), 330 (4.02), and 345 (4.01). IR spectrum (KBr): 3180 to 3190 $cm^{-1}$ (broad band, ν(N-H)), 1045 $cm^{-1}$ (strong, split band, ν(S=O)) [3].

The compound melts at 178 to 180°C with decomposition [2, 3] (dec. > 160°C [1]) and the product of the reaction of 1,8-naphthalenediamine with $C_6H_5SO_2NSO$ decomposes above 187°C [3]. Solutions of the compound in $C_6H_6$, $C_5H_5N$, $CH_3CN$, and alcohols show thermochromic behavior. Above 50°C the solutions become red, due to the formation of an 8-sulfinylamino-1-naphthalenamine intermediate. On cooling the color disappears, indicating a reversible process [3]. Thermal dehydration and formation of naphtho[1,8-cd][1,2,6]-

thiadiazine (see p. 66) was observed [2] when the compound was heated alone or in a solvent. Dehydration occurs also when mixtures of the compound with $C_6H_5NSO$ (1:1) and 10 wt% $(C_2H_5)_3N$ or $POCl_3$ were refluxed in benzene. The reaction with $PCl_5$ leads to the tetrachloro derivative of the thiadiazine; see VI, p. 66 [1].

**References:**

[1] Behringer, H., Leiritz, K. (Chem. Ber. **98** [1965] 3196/9).
[2] Dietz, R. (Chem. Commun. **1965** 57).
[3] Beecken, H. (Chem. Ber. **100** [1967] 2164/9).

### 9.4.2.11 2-Methyl-8,9-dihydro-7H-pyrido[3,2,1-ij][2,1,3]benzothiadiazin-3(2H)one 1-Oxide

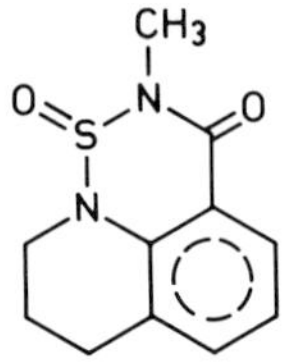

**[–SO–NCH₃–CO–{1,2-C₆H₃-2,3-{–(CH₂)₃–N}}–]**

The title compound forms in 97% yield on refluxing equimolar amounts of N-methyl-1,2,3,4-tetrahydro-8-quinoline-carboxamide and $SOCl_2$ in benzene for 1 h.

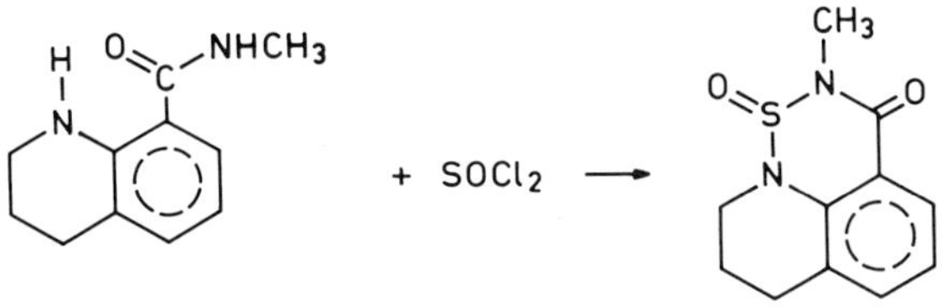

+ $SOCl_2$ ⟶

Melting point is 95 to 98°C. IR spectrum ($CHCl_3$): $\nu = 1665\ cm^{-1}$ (C=O) and $1140\ cm^{-1}$ (S=O); $^1H$ NMR spectrum ($CDCl_3$/TMS): $\delta$ (in ppm) = 2.1 (m, 2H), 2.85 (t, 2H), 3.4 (s, 3H), 3.75 (m, 2H), 7.1 (m, 2H), and 8.0 (m, 1H).

**Reference:**

Coppola, G. M. (J. Heterocycl. Chem. **15** [1978] 645/7).

### 9.4.2.12 Derivatives of 1H-Furo[2,3-c][1,2,6]thiadiazin-4(3H)-one 2-Oxide, 1,7-Dihydropyrrolo[2,3-c][1,2,6]thiadiazin-4(3H)-one 2-Oxide, and 1H-Thieno[2,3-c][1,2,6]thiadiazin-4(3H)-one 2-Oxide

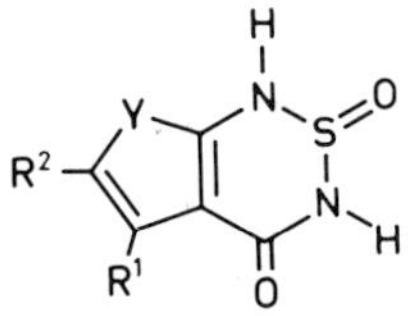

Y = O or N–R; $R^1 = R^2 = CH_3$
Y = S; $R^1 = R^2 = CH_3$ or $R^1R^2 = –(CH_2)_4–$

**[–SO–NH–CO–{C–CR¹=CR²–Y–C}–NH–]**

The formation of the compounds with $R^1=R^2=CH_3$, Y=O, and N–R ($R=C_6H_4Cl$-4, c-$C_6H_{11}$, and n-$C_6H_{13}$) was proposed to occur upon hydrolysis of unstable 4H-furo- and 4,7-dihydropyrrolo[2,3-c][1,2,6]thiadiazin-2-$S^{IV}$-4-ones (see pp. 70 and 71). They could not be isolated, but the appearance of the N–H band and the SO band ($\nu=1080$ $cm^{-1}$) in the IR spectra suggest a mixture of both starting compound and hydrolysis product [1].

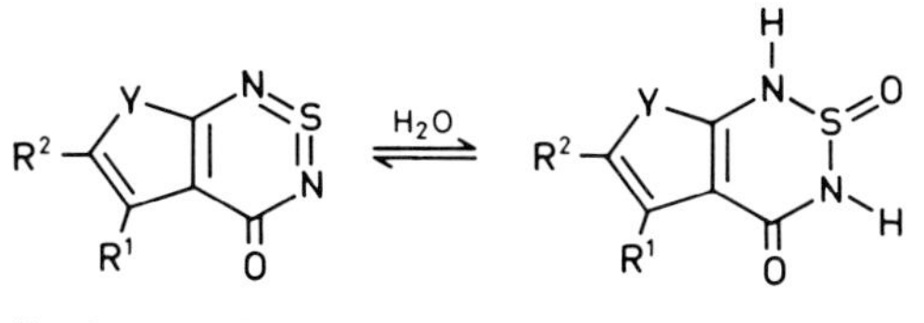

Y = O or N–R

The compounds with Y=S are prepared by the ring closure reaction of α-aminocarboxamides with excess $SOCl_2$, with elimination of HCl [1, 2].

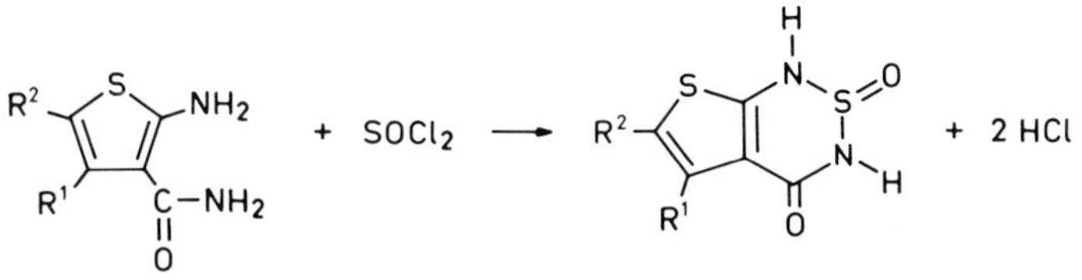

The compound with $R^1=R^2=CH_3$, Y=S was prepared in $SOCl_2$ at 25°C. The yellow-brown crystals melt at 206 to 208°C [1].

The compound with $R^1R^2=-(CH_2)_4-$ was prepared in 95% yield in THF. Melting point 187°C [2]. Previously it was prepared in $SOCl_2$ at 25°C; the lemon-colored crystals melt at 196°C [1]. $^1H$ NMR spectrum (DMSO-$d_6$/TMS): δ (in ppm) = 2.33 to 2.89 (m, $-\underline{CH_2}-CH_2-CH_2-\underline{CH_2}-$), 6.53 (s, 3NH), 7.73 (s, 1NH); IR spectrum (KBr): ν (in $cm^{-1}$) = 3380, 3180 (NH), 1640(CO) [2].

**References:**

[1] Offermann, W., Eger, K., Roth, H. J. (Arch. Pharm. **314** [1981] 168/75).
[2] Wamhoff, H., Ertas, M. (Synthesis **1985** 190/4).

### 9.4.2.13 6-Methyl-3,6-dihydro-2H-imidazo[1,2-b]pyrido[3,2-d][1,2,6]thiadiazine 5-Oxide

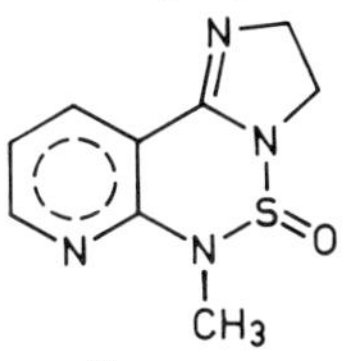

**[–SO–NCH₃–{C⁚N⁚CH⁚CH⁚CH⁚C}–{C=N–CH₂–CH₂–N}–]**

The title compound forms by refluxing a mixture of 2-(2-methylamino-3-pyridyl)-4,5-dihydro-1H-imidazole (I) and $SOCl_2$ (1:1.1 mole ratio) in $C_6H_6$ for 4 h. After cooling to room temperature and adding $H_2O$ and 10% aqueous $NaHCO_3$, the mixture was extracted with

$CH_3COOC_2H_5$. Evaporation to dryness and subsequent chromatographing of the residue gives the compound in 43% yield.

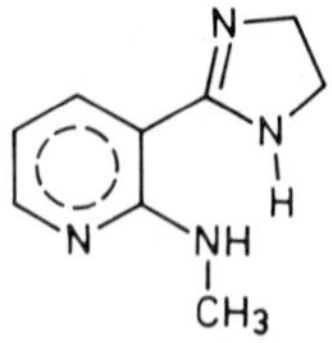

I

M.p. 116 to 119°C (from $CH_2Cl_2$–$(C_2H_5)_2O$); $^1H$ NMR spectrum ($CDCl_3$/TMS): δ (in ppm) = 3.47 (s, $CH_3$), 3.63 to 4.38 (m, $-CH_2CH_2-$), 7.0 (dd, 1 arom. H), and 8.2 to 8.52 (m, 2 arom. H); IR spectrum ($CHCl_3$): ν = 1638, 1605, 1480, 1450, 1414, 1220, and 1130 $cm^{-1}$.

**Reference:**

Coppola, G. M. (Syn. Commun. **15** [1985] 1013/7).

## 9.5 $S_2NCO_2$ Ring

### 9.5.1 Derivatives of 4,4-Bis(trifluoromethyl)-2,2,4,4-tetrahydro-1,3,2,4,5-dioxadithiazine

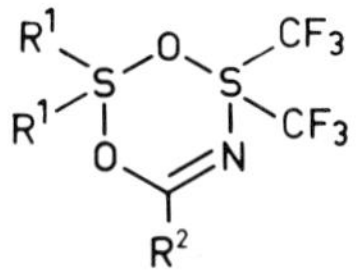

$R^1$ and $R^2$ are compiled in Table 34.

**[–S($CF_3$)$_2$–N=C$R^2$–O–S$R^1_2$–O–]**

1:1 mixtures of dimethyl or dimethoxy sulfoxide and an appropriate sulfimide react in the presence of $BF_3$ in dry ether at ambient temperature for 5 h to give the cyclic 1,3,2,4,5-dioxadithiazines compiled in Table 34.

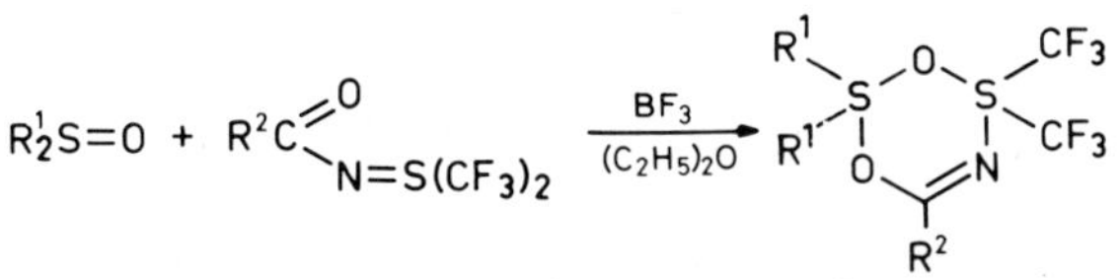

$$R^1_2S{=}O + R^2C({=}O)N{=}S(CF_3)_2 \xrightarrow[(C_2H_5)_2O]{BF_3} \text{cyclic product}$$

When the 1,3,2,4,5-dioxadithiazine with $R^1 = CH_3$ and $R^2 = C_6H_5$ is pyrolyzed at 150°C in a stainless steel Hoke vessel or hydrolyzed with water, $(CF_3)_2SO$, $(CH_3)_2SO$, $C_6H_5CN$, and $C_6H_5CONH_2$ are generated quantitatively. The reaction with primary amines results in the formation of $(CF_3)_2S{=}NR$.

**Reference:**

Kitazume, T., Shreeve, J. M. (J. Am. Chem. Soc. **100** [1978] 985/7).

Table 34

Derivatives of 4,4-Bis(trifluoromethyl)-2,2,4,4-tetrahydro-1,3,2,4,5-dioxadithiazine, [–$S(CF_3)_2$–N=$CR^2$–O–$SR^1_2$–O–] (Reference on p. 84).

| $R^1$ | $R^2$ | yield in % | m.p. in °C | $^1H$ NMR spectra ($CDCl_3$/TMS, δ in ppm, J in Hz) | $^{19}F$ NMR spectra (solvent?/$CFCl_3$, δ in ppm, J in Hz) | mass spectra m/e |
|---|---|---|---|---|---|---|
| $CH_3$ | $C_6H_5$ | 88 | 145 to 147 | δ = 3.79 (q, $CH_3$), 3.91 (q, $CH_3$), ($^4J$ = 0.6); 8.29 (m, $C_6H_5$) | δ = −55.2; −57.3; two quartets, ratio 1:1 ($^4J$ = 3.2) | 352 ($M^+ - CH_3$), 348 ($M^+ - F$), 298 ($M^+ - CF_3$), 103 ($C_7H_5N^+$), 69 ($CF_3^+$) |
| $CH_3$ | $C_6H_5CH_2$ | 79 | 161 to 162 | δ = 3.51 (m, $CH_2$); 3.87 (q, $CH_3$), 3.99 (q, $CH_3$), ($^4J$ = 0.7); 8.24 (m, $C_6H_5$) | δ = −56.8; −58.9; two quartets, ratio 1:1 ($^4J$ = 3.1) | 366 ($M^+ - CH_3$), 362 ($M^+ - F$), 312 ($M^+ - CF_3$) |
| $CH_3O$ | $C_6H_5$ | 81 | 171 to 172 | δ = 3.59 (s, $CH_3$), 3.77 (s, $CH_3$); 8.25 (m, $C_6H_5$) | δ = −55.9; −57.6; two quartets, ratio 1:1 ($^4J$ = 3.3) | 380 ($M^+ - F$), 353 ($M^+ - C_2H_6O$), 330 ($M^+ - CF_3$) |
| $CH_3O$ | $C_6H_5CH_2$ | 83 | 186 to 187 | δ = 3.49 (m, $CH_2$); 3.64 (s, $CH_3$), 3.71 (s, $CH_3$); 8.22 (m, $C_6H_5$) | δ = −56.1; −58.2; two quartets, ratio 1:1 ($^4J$ = 3.1) | 394 ($M^+ - F$), 367 ($M^+ - C_2H_6O$), 344 ($M^+ - CF_3$) |

## 9.6 $SNC_3O$ Ring

### 9.6.1 Derivatives of Tetrahydro-1,2,3-oxathiazine 2-Oxide and 3,4-Dihydro-1,2,3-benzoxathiazine 2-Oxide

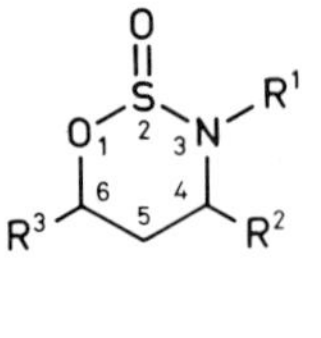

1 to 12 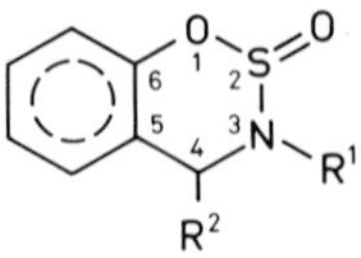 13 to 25

**[–SO–NR¹–CHR²–CH₂–CHR³–O–]** **[–SO–NR¹–CHR²–{1,2-C₆H₄}–O–]**

**Preparation**

The stereoisomeric compounds compiled in Table 35, are prepared by adding a solution of $SOCl_2$ with stirring to a solution of an amino alcohol or amino phenol in the presence of an HCl-trapping base B (triethylamine or pyridine). Oxathiazines with alkyl groups at the carbons in 4 and (or) 6 position (Table 35, No. 1, 4, 7 to 10) are unstable towards water, cannot be distilled, and were only characterized by $^1H$ NMR, IR, and mass spectra. The methods of preparation are described in detail by Deyrup and Moyer [1] and Morris and Collins [2], respectively.

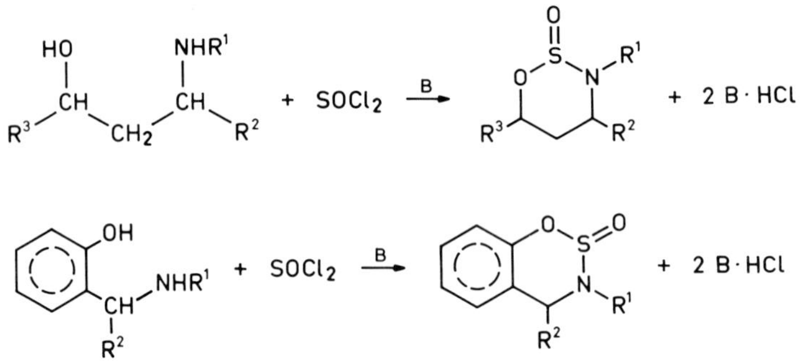

Table 35

Preparation, Boiling Point, and Melting Point of Derivatives of Tetrahydro-1,2,3-oxathiazine 2-Oxide, [–SO–NR$^1$–CHR$^2$–$CH_2$–CHR$^3$–O–], and 3,4-Dihydro-1,2,3-benzoxathiazine 2-Oxide, [–SO–NR$^1$–CHR$^2$–{1,2-$C_6H_4$}–O–]. The mole ratio of the reactants is given in the sequence amino alcohol or phenol, $SOCl_2$, and base (*c*= *cis*-, *t*= *trans*-).

| No. | R¹ | R² | R³ | reaction conditions (mole ratio, solvent, t in °C) | b.p. in °C/p in Torr | m.p. in °C | Ref. |
|---|---|---|---|---|---|---|---|
| 1 | $CH_3$ | $CH_3$ | H | 1:1:2, benzene, 5 to 10 | | | [4] |
| 2 | i-$C_3H_7$ | H | H | 1:1:2, benzene, 5 to 10 | 106/15 | | [3, 4] |
| 3 | t-$C_4H_9$ | H | H | 1:1:2, benzene, 5 to 10 | 92/3 [3, 4], or 60 to 65/0.02 [1] | | [1, 3, 4] |

Table 35 (continued)

| No. | $R^1$ | $R^2$ | $R^3$ | reaction conditions (mole ratio, solvent, t in °C) | b.p. in °C/p in Torr | m.p. in °C | Ref. |
|---|---|---|---|---|---|---|---|
| 4 | t-$C_4H_9$ | $CH_3$ | H | 1:1:2, benzene, 5 to 10 | 65/0.01, mixture of 4*t* and *c* | 36 4*c* only | [4] |
| 5 | $C_6H_5CH_2$ | H | H | 1:1:2, benzene, 5 to 10 | | 53 | [3, 4] |
| 6 | $C_6H_5$ | H | H | 1:1:2, benzene, 5 to 10 | 136/0.7 [3, 4], or 160/0.04 [5] | | [3 to 5] |
| 7 | $C_6H_5$ | $CH_3$ | H | 1:1:2, benzene, 5 to 10 | | | [4] |
| 8 | $C_6H_5$ | H | $CH_3$ | 1:1:2, benzene, 5 to 10 | | | [4] |
| 9 | $C_6H_5$ | $CH_3$ | $CH_3$ | 1:1:2, benzene, 5 to 10 | | 76 9*t* only | [3, 4] |
| 10 | $C_6H_5$ | t-$C_4H_9$ | t-$C_4H_9$ | 1:1:2, benzene, 5 to 10 | | | [4] |
| 11 | 4-$ClC_6H_4$ | H | H | 1:1:2, benzene, 5 to 10 | | 30 | [5] |
| 12 | 4-$CH_3OC_6H_4$ | H | H | 1:1:2, benzene, 5 to 10 | | 35 | [5] |
| 13 | $CH_3$ | H | — | 1:1:3, ether, 15 | 106/2 | | [6] |
| 14*) | $CH_3$ | $CH_3$ | — | 1:1:3, ether, 15 | 78/0.01, mixture of 14*t* and *c* | | [6] |
| 15 | i-$C_3H_7$ | H | — | 1:1:3, ether, 15 | 81/0.001 | | [6] |
| 16*) | i-$C_3H_7$ | $CH_3$ | — | 1:1:3, ether, 15 | 90/0.005, mixture of 16*t* and *c* | | [6] |
| 17 | t-$C_4H_9$ | H | — | 1:1:3, ether, 15 | | 60 | [6] |
| 18 | $C_6H_5CH(CH_3)$ | H | — | 1:3:5, toluene, 0, 4 h | | | [7] |
| 19 | $C_6H_5CH(CH_3)$ | $CH_3$ | — | 1:3:5, toluene, 0, 4 h | | | [7] |
| 20 | 1-$C_{10}H_7CH(CH_3)$ | H | — | 1:3:5, toluene, 0, 4 h | | | [7] |
| 21*) | 1-$C_{10}H_7CH(CH_3)$ | $CH_3$ | — | 1:1.5 to 3:3.6 to 5, toluene, THF, dimethylethylene, $CHCl_3$ or $CCl_4$, 0, 4 h | | 121 for 21*t* and 98 to 99 for 21*c* | [7, 8] |
| 22 | 1-$C_{10}H_7CH(CH_3)$ | $C_2H_5$ | — | 1:3:5, toluene, 0, 4 h | | | [7] |

Table 35 (continued)

| No. | $R^1$ | $R^2$ | $R^3$ | reaction conditions (mole ratio, solvent, t in °C) | b.p. in °C/p in Torr | m.p. in °C | Ref. |
|---|---|---|---|---|---|---|---|
| 23 | $1\text{-}C_{10}H_7CH(CH_3)$ | $C_4H_9$ | — | 1:3:5, toluene, 0, 4 h | | | [7] |
| 24 | $C_6H_5$ | H | — | 1:1:3, ether, 15 | | 62 | [6] |
| 25*) | $C_6H_5$ | $CH_3$ | — | 1:1:3, ether, 15 | | 71 | [4] |

*) Depending on the base used in the reaction the two diastereoisomers (4-*cis* and 4-*trans*) of No. 14, 16, and 25 form in different mole ratios. $(NC_2H_5)_3$: 14*t*:14*c*=0.45:0.55, 16*t*:16*c*=0.30:0.70, 25*t*:25*c*=0.70:0.30; pyridine: 14*t*:14*c*=0.80:0.20, 16*t*:16*c*=0.35:0.65, 25*t*:25*c*=0.65:0.35 [6]. No. 21 forms in 93% yield on reacting the 2-aminophenol with $SOCl_2$ in the presence of $N(C_2H_5)_3$ in the 1:1.5:5 mole ratio. The ratio of the isomers is 21*t*:21*c*=2:1 [7].

**Spectra**

**UV spectra** are recorded for the compounds $[-SO-N(C_3H_7\text{-}i)-CH_2-CH_2-CH_2-O-]$ (Table 35, No. 2), $[-SO-N(C_4H_9\text{-}t)-CH_2-CH_2-CH_2-O-]$ (Table 35, No. 3), and the diastereomer $[-SO-NC_6H_5-CH(CH_3)-CH_2-CH(CH_3)-O-]$ (Table 35, No. 9) [3]:

| No. | 2 | | 3 | 9*t* | 9*c* |
|---|---|---|---|---|---|
| solvent | $C_2H_5OH$ | | $c\text{-}C_6H_{12}$ | $C_2H_5OH$ | $C_2H_5OH$ |
| $\lambda_{max}$ in nm | 242 | 292 | 199 | 257 | 240 |
| $\varepsilon$ in $L\cdot mol^{-1}\cdot cm^{-1}$ | 8200 | 850 | 2150 | 680 | 1440 |

**$^1$H NMR spectra** measured for compounds No. 1 to 10 [3, 4] and 13 to 17, 24, and 25 [6] are given in Table 36, pp. 90/1, and Table 37, p. 89. Compound numbers taken from Table 35, pp. 86/8. The $^1$H NMR was measured with use of the paramagnetic lanthanide shift reagent $Eu(fod)_3$, with fod=1,1,1,2,2,3,3-heptafluoro-7,7-dimethyl-4,6-octanedionate and a static 1:1 model of complex formation.

**Conformation** (compound numbers taken from Table 35, pp. 86/8)

The conformations are presented in Fig. 17. In the solid state, the conformation of compound No. 4 is a chair with axial S=O and methyl groups (conformation b); see p. 92.

Conformational analysis of compounds No. 1 to 7, 9, 10, 13, 17, and 24 in solution relies only on $^1$H NMR spectra; see Table 36 and 37. For the oxathiazines different conformations are found in $CDCl_3$ solution: a) the most stable conformation is the standard chair with an axial S=O group when the molecule is not substituted in the 4 and 6 positions or when the substituents are equatorial; the substituents at the N atom ($R^1=CH_3$, $i\text{-}C_3H_7$, or $t\text{-}C_4H_9$) are preferentially axial (No. 1a, 2, 3, 5, 6, 7a, 9a, and 10a); b) strained chairs with axial $R^2$ and S=O groups; in this conformation $R^1=C_6H_5$ may be partially conjugated and $R^1=CH_3$ or $t\text{-}C_4H_9$ may adopt the more favorable axial orientation (No. 1b, 4b, and 9b); c) twist conformations with a 1,4 axis and an axial S=O group (No. 10c); d) the twist conformation with a 3,6 axis and an axial S=O group for No. 4d, because of the 4-methyl···3-tert-butyl-1,2 interaction.

For the benzoxathiazines (No. 13, 17, and 24) half chair conformations with an axial S=O group are proposed [4, 9].

Table 37

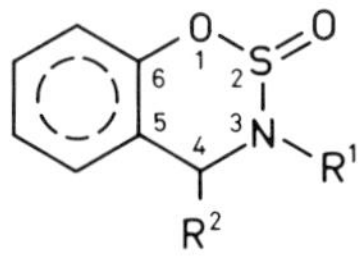

$^1$H NMR Spectra of Derivatives of 3,4-Dihydro-1,2,3-benzoxathiazine 2-Oxide ($CDCl_3$/internal TMS, δ in ppm; *c* = *cis*-, *t* = *trans*-position to the S=O group). If not stated otherwise δ is assigned to protons bound directly to the six-membered ring. The conformations are given in Fig. 17. Numbers of compounds correspond to those of Table 35, pp.86/8 [6].

| No. | $R^1$ | $R^2$ | δ for position 4 4*c* | 4*t* |
|---|---|---|---|---|
| 13 | $CH_3$ | H | 4.76 | 3.83 |
| 14*t* | $CH_3$ | $CH_3$ | 4.83 | 1.70 ($CH_3$) |
| 14*c* | | | 1.62 ($CH_3$) | 4.48 |
| 15 | i-$C_3H_7$ | H | 4.73 | 3.90 |
| 16*t* | i-$C_3H_7$ | $CH_3$ | 4.75 | 1.73 ($CH_3$) |
| 16*c* | | | 1.67 ($CH_3$) | 4.28 |
| 17 | t-$C_4H_9$ | H | 4.73 | 4.00 |
| 24 | $C_6H_5$ | H | 5.20 | 4.33 |
| 25*t* | $C_6H_5$ | $CH_3$ | 5.28 | 1.87 ($CH_3$) |
| 25*c* | | | 1.51 ($CH_3$) | 4.81 |

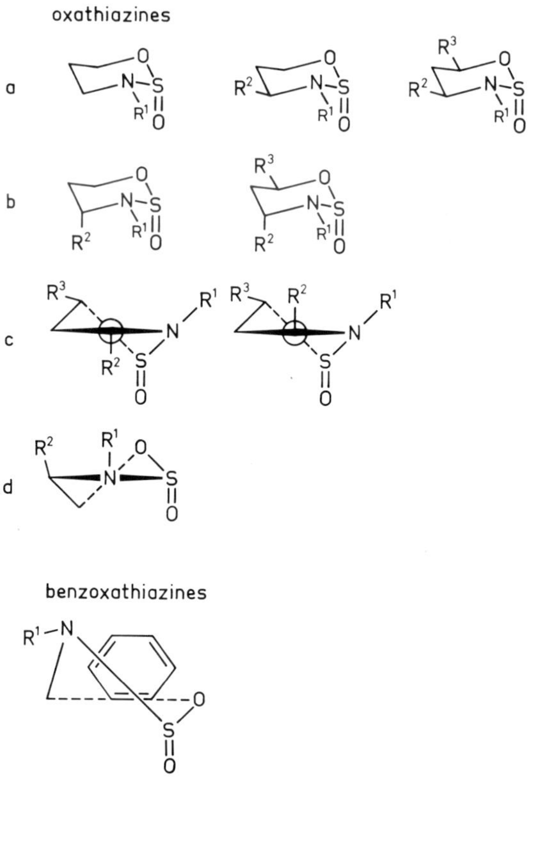

oxathiazines

a

b

c

d

benzoxathiazines

Fig. 17. Conformations of tetrahydro-1,2,3-oxathiazine 2-oxides (compound numbers taken from Tables 36/7, pp. 89/91): a) standard chairs with an axial S=O group (No. 1a, 2, 3, 5, 6, 7a, 9a, and 10a); b) strained chairs with axial $CH_3$-4 and S=O groups (No. 1b, 4b, and 9b); c) twist conformation with a 1,4 axis and axial S=O group (No. 10c); d) twist conformation with a 3,6 axis and axial S=O group (No. 4d). Conformations of 3,4-dihydro-1,2,3-benzoxathiazine 2-oxides: half chairs with axial S=O group (No. 13, 17, 24).

Table 36

[1]H NMR Spectra of Derivatives of Tetrahydro-1,2,3-oxathiazine 2-Oxide
(standard internal TMS, δ in ppm, J in Hz; c = *cis*-, t = *trans*-position to the S=O group). If not stated otherwise δ is assigned to protons bound directly to the six-membered ring. The conformations are given in Fig. 17, p. 89. Numbers of compounds correspond to those of Table 35, p. 86/7.

| No. | $R^1$ | $R^2$ | $R^3$ | solvent | δ for positions 4, 5, and 6 | | | | | | confor-mation | Ref. |
|---|---|---|---|---|---|---|---|---|---|---|---|---|
| | | | | | 4 *c* | 4 *t* | 5 *c* | 5 *t* | 6 *c* | 6 *t* | | |
| 1a | $CH_3$ | $CH_3$ | H | $CDCl_3$ | 3.92 | 1.16 ($CH_3$) | 1.49 | 2.04 | 4.83 | 3.83 | a | [4] |
| 1b | | | | $CDCl_3$ | 1.39 ($CH_3$) | 3.58 | 1.84 | 1.67 | 4.45 | 4.12 | b | [4] |
| 2 | i-$C_3H_7$ | H | H | $CCl_4$ | 3.59 | 2.67 | 1.64 | 2.20 | 4.70 | 3.72 | a | [3, 4] |
| 3 | t-$C_4H_9$ | H | H | $CCl_4$ | 3.61 | 2.79 | 1.67 | 2.18 | 4.66 | 3.68 | a | [3, 4] |
| 4d | t-$C_4H_9$ | $CH_3$ | H | $CDCl_3$ | 3.64 | 1.34 ($CH_3$) | 2.92 | 1.85 | 4.39 | 4.16 | d | [4] |
| 4b | | | | $CDCl_3$ | 1.61 ($CH_3$) | 3.53 | 1.82 | 2.27 | 4.89 | 3.82 | b | [4] |
| 5 | $C_6H_5CH_2$ | H | H | $CCl_4$ | 3.53 | 2.57 | 1.48 | 2.21 | 4.72 | 3.73 | a | [3, 4] |
| 6 | $C_6H_5$ | H | H | $CCl_4$ | 4.13 | 3.07 | 1.61 | 2.32 | 4.78 | 3.76 | a | [3, 4] |
| 7a | $C_6H_5$ | $CH_3$ | H | $CCl_4$ | 4.25 | 0.88 ($CH_3$) | 1.70 | 2.01 | 4.92 | 3.85 | a | [4] |
| 7b | | | | $CCl_4$ | 1.46 ($CH_3$) | | | 2.44 | | | | [4] |
| 8 | $C_6H_5$ | H | $CH_3$ | $CDCl_3$ | 4.20 | 3.15 | 1.70 | 2.02 | 4.98 | 1.27 ($CH_3$) | | [4] |
| 9a[a)] | $C_6H_5$ | $CH_3$ | $CH_3$ | $CCl_4$ | 4.26 | 0.88 ($CH_3$) | 1.75 | 1.70 | 5.09 | 1.30 ($CH_3$) | a | [3, 4] |
| 9b[a)] | | | | $CCl_4$ | 1.51 ($CH_3$) | 3.86 | 1.81 | 2.20 | 5.15 | 1.35 ($CH_3$) | b | [3, 4] |

| | | | | | | | | | | | |
|---|---|---|---|---|---|---|---|---|---|---|---|
| 10a | $C_6H_5$ | t-$C_4H_9$ | t-$C_4H_9$ | $CCl_4$ | 4.25 | 0.79 (t-$C_4H_9$) | 1.69 | 2.14 | 4.63 | 1.01 (t-$C_4H_9$) | a | [4] |
| 10c | | | | $CCl_4$ | 0.88 (t-$C_4H_9$) | 3.89 or 3.92 | 2.83 | 1.99 | 0.98 (t-$C_4H_9$) | 3.92 or 3.89 | c | [4] |
| 10c | | | | $CCl_4$ | 3.71 | 0.92 or 0.98 (t-$C_4H_9$) | 2.50 | 2.03 | 0.98 or 0.92 (t-$C_4H_9$) | 3.98 | c | [4] |

| No. | J for positions 4, 5, and 6 (*cis* and *trans*) | | | | | | | | Ref. |
|---|---|---|---|---|---|---|---|---|---|
| | 4*c*, 5*c* | 4*c*, 5*t* | 4*t*, 5*c* | 4*t*, 5*t* | 6*c*, 5*c* | 6*c*, 5*t* | 6*t*, 5*c* | 6*t*, 5*t* | |
| 1a | 2.8 | 11.6 | — | — | 2.6 | 12.9 | 1.8 | 4.9 | [4] |
| 1b | — | — | 8.5 | 3.8 | 4.0 | 5.4 | 8.7 | 3.6 | [4] |
| 2 | 2.8 | 12.6 | 2.9 | 4.6 | 2.6 | 12.9 | 1.7 | 4.8 | [3, 4] |
| 3 | 2.9 | 12.9 | 3.1 | 4.5 | 2.7 | 13.1 | 1.7 | 4.6 | [3, 4] |
| 4d | 4.0 | 3.6 | — | — | 8.3 | 1.4 | 11.6 | 7.6 | [4] |
| 4b | — | — | 4.0 | 6.2 | 2.8 | 11.0 | 4.4 | 4.6 | [4] |
| 5 | 2.8 | 12.7 | 2.9 | 4.5 | 2.6 | 12.9 | 1.8 | 4.7 | [3, 4] |
| 6 | 2.8 | 12.8 | 2.9 | 4.5 | 2.7 | 12.9 | 1.7 | 4.8 | [3, 4] |
| 7a | 2.8 | 11.4 | — | — | 2.4 | 13.6 | 1.6 | 4.8 | [4] |
| 7b | — | — | — | 5.0 | — | 10.2 | — | 5.0 | [4] |
| 8 | 2.9 | 12.6 | 2.8 | 4.1 | 2.2 | 11.7 | — | — | [4] |
| 9a[a)] | 2.8 | 11.2 | — | — | 2.1 | 11.6 | — | — | [3, 4] |
| 9b[a)] | — | — | 2.8 | 5.4 | 2.6 | 10.7 | — | — | [3, 4] |
| 10a | 3.3 | 12.0 | — | — | 3.0 | 12.0 | — | — | [4] |
| 10c | — | — | 12.0 | 6.0 | — | — | 12.0 | 2.4 | [4] |
| 10c | 6.7 | 4.5 | — | — | — | — | 12.2 | 4.0 | [4] |

a) Use of a polar solvent, $CD_3CN$, instead of $CCl_4$ has no effect on chemical shifts and coupling constants.

**Crystal Structure**

X-ray diffraction measurements at 20°C show the crystals of the compound [–SO–N(t-$C_4H_9$)–CH($CH_3$)–$CH_2$–$CH_2$–O–] (Table 36, No. 4b) to be monoclinic, space group $P2_1$-$C_2^2$ (No. 4) with a = 6.57(1), b = 10.68(1), c = 7.72(1) Å, and β = 109.14(13)°; Z = 2, $D_m$ = 1.23 and $D_x$ = 1.24 g/cm³; R = 0.065, $R_w$ = 0.079. Bond lengths and bond angles are given in Table 38. For atomic coordinates and temperature factors, see the paper. **Fig. 18** shows that this compound is in the chair conformation with axial S=O and methyl groups. O(1), N(3), C(4), and C(6) atoms are close to the mean plane with deviations of, respectively, +0.04, −0.05, +0.05, and −0.06 Å from this plane. C(5) and S(2) are on opposite sides of this plane with deviations of, respectively, +0.64 and −0.74 Å. The S(2)–O(2) bond has an almost perfect axial orientation, whereas the methyl group in the 4-position is displaced from this orientation by 19.6° toward the geminal H(3). This deviation may be divided into a rotation of about 12° around the N(3)–C(4) bond and a decrease in 7° of the H(3)–C(4)–$CH_3$ angle. The rotation is indicative of a slight deformation of the ring toward the 3,6-axis twist form as it was demonstrated for the diastereomer compound (see Fig. 17d on p. 89; Table 36, No. 4d) by $^1$H NMR spectroscopy in $CDCl_3$ solution [10].

Table 38
Bond Lengths and Bond Angles of [–SO–N($C_4H_9$-t)–CH($CH_3$)–$CH_2$–$CH_2$–O–] (Table 36, p. 90, No. 4b).

| bond length in Å | | | | | |
|---|---|---|---|---|---|
| S(2)–O(2) | 1.468(5) | C(4)–C(5) | 1.541(13) | N(3)–C(4) | 1.486(9) |
| S(2)–O(1) | 1.677(7) | C(6)–O(1) | 1.483(12) | C(4)–C(7) | 1.509(11) |
| N(3)–C(8) | 1.519(8) | S(2)–N(3) | 1.653(7) | C(5)–C(6) | 1.503(13) |

| bond angle in ° | | | | | |
|---|---|---|---|---|---|
| O(2)–S(2)–N(3) | 105.9(3) | C(6)–C(5)–C(4) | 114.1(8) | C(4)–N(3)–S(2) | 119.2(5) |
| N(3)–S(2)–O(1) | 101.4(3) | C(10)–C(8)–N(3) | 109.8(6) | N(3)–C(4)–C(7) | 112.5(7) |
| C(4)–N(3)–C(8) | 117.8(6) | N(3)–C(8)–C(11) | 111.1(9) | C(7)–C(4)–C(5) | 112.6(7) |
| C(8)–N(3)–S(2) | 116.1(5) | O(2)–S(2)–O(1) | 104.6(3) | O(1)–C(6)–C(5) | 109.7(7) |
| N(3)–C(4)–C(5) | 111.6(6) | C(6)–O(1)–S(2) | 113.4(5) | N(3)–C(8)–C(9) | 109.9(8) |

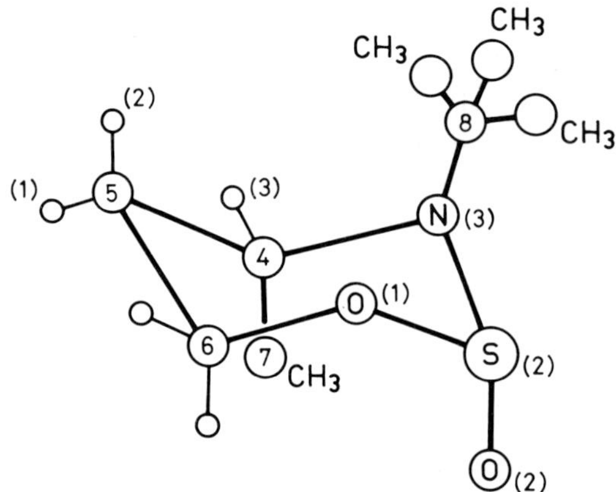


Fig. 18. Structure of [–SO–N($C_4H_9$-t)–CH($CH_3$)–$CH_2$–$CH_2$–O–] (Table 36, No. 4b).

**Epimerization**

Epimerization of the optically active *trans*-isomers (*t*) of some 3,4-dihydro-benzoxathiazine 2-oxides (Table 35, pp. 86/8, No. 18 to 23), was observed by reaction with hydrogen chloride. Treating these compounds in toluene at 25°C with HCl induces epimerization, producing the *cis*-isomers 18*c* to 23*c* [7].

HCl

**18t** to 23t

**18c** to 23 c

| No. | Y | $R^2$ | No. | Y | $R^2$ |
|---|---|---|---|---|---|
| 18*c, t* | $C_6H_5$ | H | 21*c, t* | 1-$C_{10}H_7$ | $CH_3$ |
| 19*c, t* | $C_6H_5$ | $CH_3$ | 22*c, t* | 1-$C_{10}H_7$ | $C_2H_5$ |
| 20*c, t* | 1-$C_{10}H_7$ | H | 23*c, t* | 1-$C_{10}H_7$ | $C_4H_9$ |

The epimerization of compound 21*t* was studied in detail. The rate of epimerization 21*t*→21*c* in toluene at 0°C depends on the concentration of hydrogen chloride and on the reaction time. With 0.016N HCl the ratio 21*t*:21*c*=33:67 in 2 h. Using 0.032N HCl produces 21*t*:21*c* ratios of 87:13 in 0.5 h, 39:61 in 1 h, 4:96 in 2 h, and 3:97 in 3 h. The results indicate that the isomer 21*c* is thermodynamically more stable than 21*t* [8].

The results of the epimerization of compound 21*t* with $BF_3 \cdot O(C_2H_5)_2$, $CF_3COOH$, $CH_3COOH$, and $AlCl_3$ in toluene at 0°C are listed in Table 39 [8].

Table 39

Studies on the Epimerization of Compound No. 21*t* with Various Acidic Catalysts in Toluene [8].

| catalyst | (equiv.) | reaction time in h | ratio of 21*t* to 21*c* 21*t* | 21*c* |
|---|---|---|---|---|
| $BF_3 \cdot O(C_2H_5)_2$ | 0.04 | 3.0 | 46 | 54 |
| $BF_3 \cdot O(C_2H_5)_2$ | 0.07 | 1.0 | — | 100 |
| $CF_3COOH$ | 0.07 | 3.0 | 33 | 67 |
| $CF_3COOH$ | 0.12 | 1.5 | — | 100 |
| $CH_3COOH$ | 0.20 | 4.0 | 25 | 75 |
| $CH_3COOH$ | 0.60 | 14.0 | — | 100 |
| $AlCl_3$ | 0.07 | 1.0 | 50 | 50 |
| $AlCl_3$ | 0.20 | 14.0 | — | 100 |

**Chemical Reactions**

**Hydrolysis**

The kinetics of hydrolysis of 3-phenyl-tetrahydro-1,2,3-oxathiazine 2-oxide (Table 35, pp. 86/8, No. 6) and 3-phenyl-3,4-dihydro-1,2,3-benzoxathiazine 2-oxide (Table 35, No. 24) were studied spectrophotometrically over a wide pH range. A reaction mechanism was proposed, as shown in the scheme on p. 94. Hydrolysis occurs in two steps: (1) cleavage of the

$SNC_2O$ ring and (2) splitting of the OSO group with formation of an amino alcohol. The intermediates depend on the pH of the medium [5, 11].

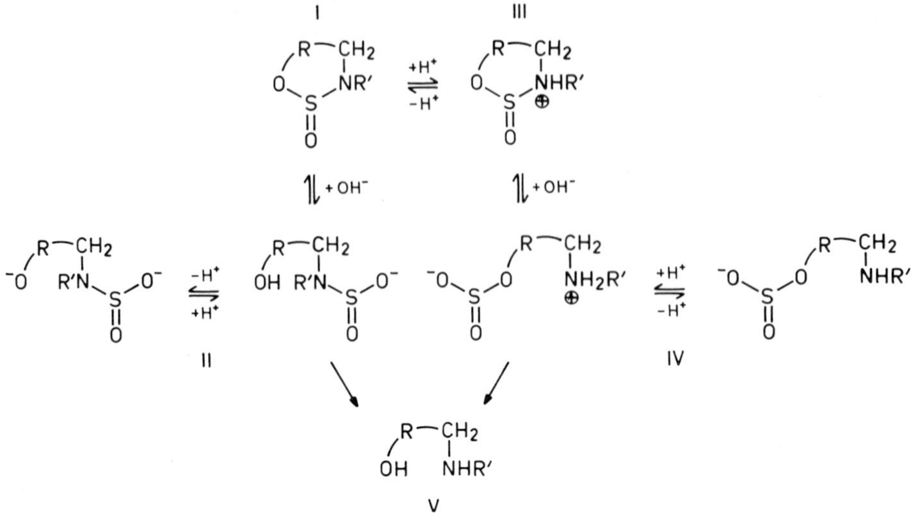

$R' = C_6H_5$
a: $R = [CH_2]_2$ (Ia = No. 6 in Table 35)
b: $R = -\{C_6H_4\}-$ (Ib = No. 24 in Table 35)

For [–SO–$NC_6H_5$–$CH_2$–$CH_2$–$CH_2$–O–] (Table 35, No. 6) the first reaction step is slow. The opening of the amidosulfite ring in acidic and slightly alkaline media by base-assisted water attack on the N-protonated substrate (IIIa) results in the formation of the ionic intermediate (IVa). This is subsequently hydrolyzed to an amino alcohol (Va) in a very slow step. In strongly alkaline media the measured data suggest hydroxide ion attack on the neutral substrate (Ia) and cleavage of the ring S–O bond resulting in the formation of the ionic intermediate (IIa), which rapidly decomposes into an amino alcohol (Va). Both reaction processes may involve either a concerted or a stepwise mechanism [5].

The first step in the hydrolysis of [–SO–$NC_6H_5$–$CH_2$–{1,2-$C_6H_4$}–O–] (Table 35, No. 24) is very fast. Splitting of the benzoxathiazine ring can give two ionic open-chain products, depending on the pH of the solutions as with the hydrolysis of compound No. 6. Between pH 11 and 14 the ion, IIb, was found to be the major product in the reaction medium. In the pH range 2 to 9 the ion, IVb, is obtained. Between pH 9 and 11 both ionic species, IIb and IVb, are present and the two hydrolysis reactions compete. The hydrolysis of the ion, IVb, occurs by attack at the oxygen atom bound to the aromatic ring, concerted with the opening of its bond to the sulfur atom. In slightly alkaline media ($7 < pH < 9$) hydroxide ions cleave the species possessing a nonprotonated nitrogen. In acidic media ($2 < pH < 7$) acid attack occurs. The hydrolysis of the ion, IIb, studied in the pH range 11 to 14 occurs via hydroxide ion attack on the phenol form, which is much more reactive than the phenolate ion [11].

#### Reactions with $C_6H_5MgBr$

Reacting the crude product of No. 21, [–SO–$NR^1$–$CHR^2$–{1,2-$C_6H_4$}–O–] with $R^1 = 1\text{-}C_{10}H_7CH(CH_3)$ and $R^2 = CH_3$, containing a 2:1 mixture of 21*t* and 21*c*, with $C_6H_5MgBr$ at −78°C for 2 h followed by addition of methyl- or butyllithium at −78°C for 2 h in THF affords the (R)-

(+)-methyl phenyl sulfoxide, or the (R)-(+)-butyl phenyl sulfoxide in 41 or 42% yield with 32 or 33% enantiomeric excess, respectively. If the crude product of No. 21 is treated with 0.032 N HCl in toluene at 0°C for 2 h (see p. 93) accompanied by the same sequence of reactions ($C_6H_5MgBr$, −78°C, 2 h, and $LiCH_3$ or $LiC_4H_9$, −78°C, 2 h), (S)-(−)-methyl phenyl sulfoxide or (S)-(−)-butyl phenyl sulfoxide form in 40 or 42% yield with extremely high enantiomeric excess (75 to 81%), respectively. Toluene is the most effective solvent for this asymmetric synthesis. With similar compounds (Table 35, No. 18 to 20, 22, and 23) the reaction sequence described above yields (S)-(−)-methyl phenyl sulfoxide and (S)-(−)-butyl phenyl sulfoxide in much lower enantiomeric excess (15 to 45%) [7, 8].

18 to 23 $\xrightarrow{C_6H_5MgBr}$ [intermediate: OH, $C_6H_5$, O←S, H, C—N---C---Y, H, $R^2$, $CH_3$] $\xrightarrow{R^3Li}$ $C_6H_5$–S→O–$R^3$

$R^3 = CH_3, C_4H_9$

**References:**


[1] Deyrup, J. A., Moyer, C. L. (J. Org. Chem. **34** [1969] 175/9).
[2] Morris, L. R., Collins, L. R. (J. Heterocycl. Chem. **12** [1975] 309/10).
[3] Cazaux, L., Tisnès, P., Roca, C. (Compt. Rend. C **280** [1975] 589/92).
[4] Tisnès, P., Maroni, P., Cazaux, L. (Org. Magn. Resonance **12** [1979] 481/9).
[5] Maroni, P., Calmon, M., Cazaux, L., Tisnès, P., Satoré, G., Aknin, M. (J. Chem. Soc. Perkin Trans. II **1978** 1207/10).
[6] Cazaux, L., Tisnès, P. (J. Heterocycl. Chem. **13** [1976] 665/8).
[7] Hiroi, K., Sato, S., Kitayama, R. (Chem. Letters **1980** 1595/8).
[8] Hiroi, K., Kitayama, R., Sato, S. (Heterocycles **15** [1981] 879/82).
[9] Tisnès, P., Maroni, P., Cazaux, L. (Org. Magn. Resonance **12** [1979] 490/8).
[10] Cazaux, L., Tisnès, P., Jaud, J. (J. Chem. Res. S **1980** 10/1; J. Chem. Res. M **1980** 156/73).

[11] Maroni, P., Cazaux, L., Tisnès, P., Aknin, M., Sartoré, G. (J. Chem. Soc. Perkin Trans. II **1978** 1211/4).


### 9.6.2 Derivatives of 1,2,3-Oxathiazin-4(3H)-one 2-Oxide and 1,2,3-Benzoxathiazin-4(3H)-one 2-Oxide

$R^1$, $R^2$, $R^3$, and $R^4$ are compiled in Table 40, pp. 97/9.

1 to 14 | 15 to 18

**[–SO–NR¹–CO–CR²=CR³–O–]** | **[–SO–NR¹–CO–{1,2-C₆HR²R³R⁴}–O–]**

**Preparation. Properties**

The compounds, compiled in Table 40, are prepared by four general methods.

Method I: The 1,2,3-oxathiazine 2-oxides No. 2, 5, 6, 8, and 12 are prepared by the reactions of N-sulfinylamines with benzoylketene (1:1 mole ratio) in benzene at 70°C [1].

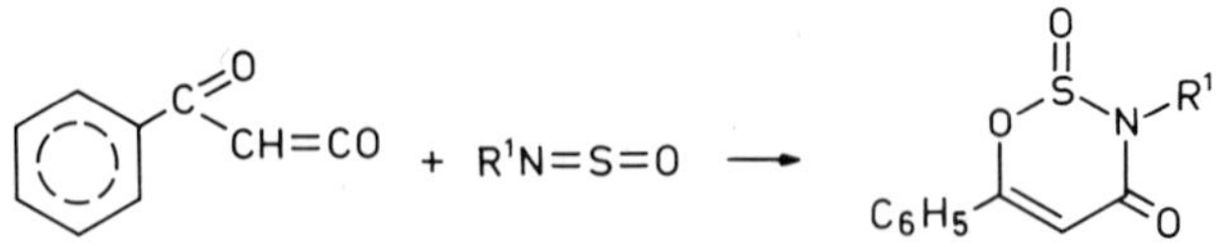

Method II: The compounds No. 1, 3, 7, 9, 10, 11, 13, and 14 are prepared by thermolysis of diazoketones with N-sulfinylamines (~1:1 mole ratio) in refluxing toluene for 25 min [2].

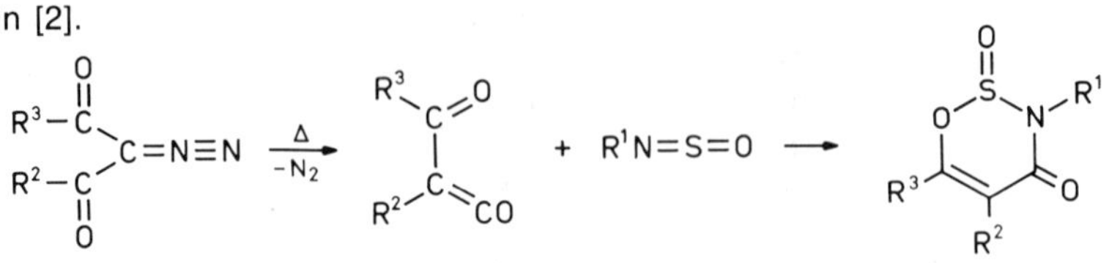

Method III: The compounds No. 4, 9, 13, and 14 form in better yields by the base-catalyzed condensation of benzoylphenylacetamides with thionyl chloride. Solutions of benzoylphenylacetamides in benzene are treated with 2 to 3 equivalents of $SOCl_2$ in the presence of pyridine. Removal of the solvent leaves crystalline residues [2].

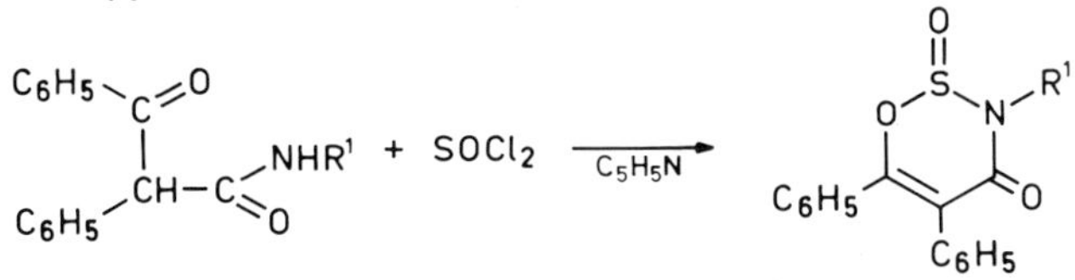

Method IV: The 1,2,3-benzoxathiazine 2-oxide, No. 15 (with $R^1 = R^2 = R^3 = R^4 = H$), forms from the reaction of equimolar amounts of 2-$HOC_6H_4CONH_2$ and $SOCl_2$ in refluxing benzene for 2 h. Otherwise substituted compounds, No. 16 to 18, form when equimolar amounts of the appropriate amide and $SOCl_2$ are heated with two equivalents of triethylamine in refluxing $CH_2Cl_2$ for 1 h, giving a slurry. Removal of the solvent leaves brown liquids which become solid on recrystallization [3].

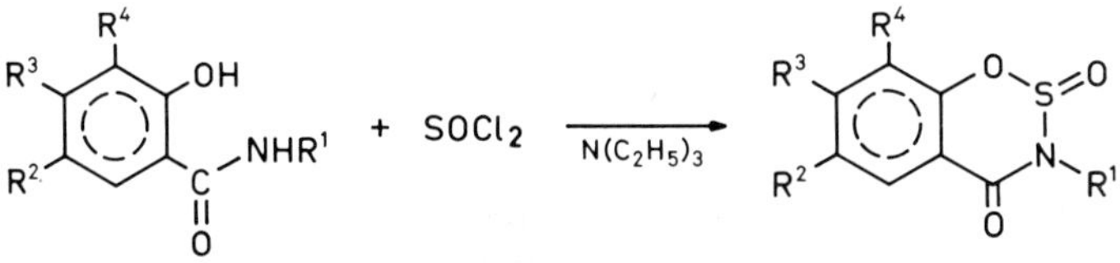

## Chemical Reactions

Thermolysis occurs when [–SO–N($C_6H_4$Cl-4)–CO–CH=C($C_6H_5$)–O–] (Table 40, p. 98, No. 8) is refluxed in toluene for 3 h. The S–N part of the ring splits and sulfinylamine forms along with the pyran ring [1].

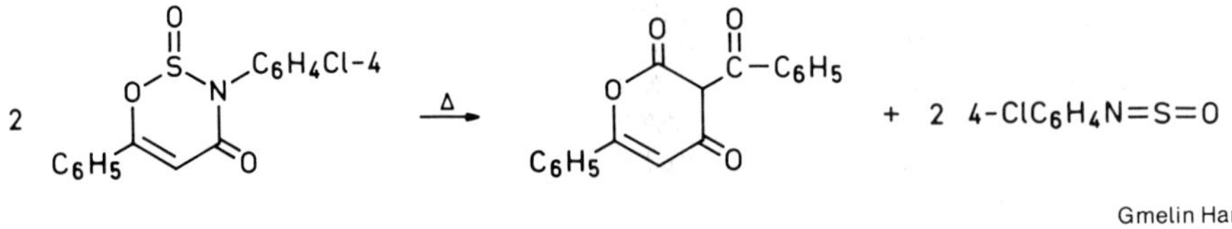

Table 40

Preparation and Properties of Derivatives of 1,2,3-Oxathiazin-4(3H)-one 2-Oxide, [–SO–NR$^1$–CO–CR$^2$=CR$^3$–O–], and 1,2,3-Benzoxathiazin-4(3H)-one 2-Oxide, [–SO–NR$^1$–CO–{1,2-$C_6HR^2R^3R^4$}–O–].

| No. | $R^1$ | $R^2$ | $R^3$ | $R^4$ | method of preparation | reaction time | yield in % | m.p. in °C | $^1$H NMR spectra (standard TMS, δ in ppm); IR spectra (in Nujol, ν in $cm^{-1}$) | Ref. |
|---|---|---|---|---|---|---|---|---|---|---|
| 1 | n-$C_3H_7$ | $C_6H_5$ | $C_6H_5$ | — | II | 4 h | 20 | 67 | $^1$H NMR (DMSO-$d_6$): 0.94 ($CH_3$), 1.73, 3.75 ($[CH_2]_2$), 7.35 (m, arom. H)<br>IR: 1660 to 1684 (C=O), 1155 to 1164, 1171 to 1175 (S=O) | [2] |
| 2 | c-$C_6H_{11}$ | H | $C_6H_5$ | — | I | 13 h | 67 | 90 to 91 | $^1$H NMR ($CDCl_3$): 1.0 to 2.4 (m, $[CH_2]_5$), 4.0 to 4.6 (m, N–CH), 6.35 (s, CH), 7.20 to 7.80 (m, arom. H)<br>IR: 1650 (C=O), 1620 (C=C), 990 (S=O) | [1] |
| 3 | c-$C_6H_{11}$ | $C_6H_5$ | $C_6H_5$ | — | II | 12 h | 40 | 141 | $^1$H NMR (DMSO-$d_6$): 1.10 to 2.20 (m, $[CH_2]_5$), 4.26 (m, CH), 7.35 (m, arom. H)<br>IR: analogous to No. 1 | [2] |
| 4 | $C_6H_5(CH_3)N$ | $C_6H_5$ | $C_6H_5$ | — | III | 1 d | 65 | 163.5 | $^1$H NMR ($CDCl_3$): 3.50 (s, $CH_3$), 7.35 (m, arom. H)<br>IR: analogous to No. 1 | [2] |
| 5 | 4-$CH_3C_6H_4SO_2$ | H | $C_6H_5$ | — | I | 7 h | 51 | 118 to 121 | $^1$H NMR ($CDCl_3$): 2.4 (s, $CH_3$), 6.18 (s, CH), 7.15 to 8.15 (m, arom. H)<br>IR: 1670 (C=O), 1615 (C=C), 990 (S=O) | [1] |

Table 40 (continued)

| No. | $R^1$ | $R^2$ | $R^3$ | $R^4$ | method of preparation | reaction time | yield in % | m.p. in °C | $^1H$ NMR spectra (standard TMS, $\delta$ in ppm); IR spectra (in Nujol, $\nu$ in $cm^{-1}$) | Ref. |
|---|---|---|---|---|---|---|---|---|---|---|
| 6 | $C_6H_5$ | H | $C_6H_5$ | — | I | 10.5 h | 49 | 121 to 122 | $^1H$ NMR ($CDCl_3$): 6.40 (s, CH), 7.15 to 7.85 (m, arom. H)<br>IR: 1655 (C=O), 1640 (C=C), 1010 (S=O) | [1] |
| 7 | $C_6H_5$ | $C_6H_5$ | $C_6H_5$ | — | II | 3 d | 38 | 152 | $^1H$ NMR (DMSO-$d_6$): 7.20 to 7.70 (m, arom. H)<br>IR: analogous to No. 1 | [2] |
| 8 | 4-$ClC_6H_4$ | H | $C_6H_5$ | — | I | 15 h | 37 | 129 to 130 | $^1H$ NMR ($CDCl_3$): 6.40 (s, CH), 7.15 to 7.85 (m, arom. H)<br>IR: 1680 (C=O), 1620 (C=C), 1010 (S=O) | [1] |
| 9 | 4-$CH_3OC_6H_4$ | $C_6H_5$ | $C_6H_5$ | — | II<br>III | 12 h<br>2 d | 29<br>58 | 143 | $^1H$ NMR (DMSO-$d_6$): 3.80 (s, $CH_3$), 7.10 to 7.66 (m, arom. H)<br>IR: analogous to No. 1 | [2] |
| 10 | 4-$CH_3OC_6H_4$ | 4-$CH_3$-$OC_6H_4$ | 4-$CH_3$-$OC_6H_4$ | — | II | several hours | 16 | 139 | $^1H$ NMR (DMSO-$d_6$): 3.72 and 3.82 ($CH_3$), 6.80 to 7.60 (m, arom. H)<br>IR: analogous to No. 1 | [2] |
| 11 | 4-$NO_2C_6H_4$ | $C_6H_5$ | $C_6H_5$ | — | II | 12 h | 31 | 154 | $^1H$ NMR (DMSO-$d_6$): 7.25 to 8.66 (m, arom. H)<br>IR: analogous to No. 1 | [2] |

| | | | | | | | | | | |
|---|---|---|---|---|---|---|---|---|---|---|
| 12 | 4-$CH_3C_6H_4$ | H | $C_6H_5$ | — | I | 6 h | 47 | 152 to 153 | $^1H$ NMR ($CDCl_3$): 2.4 (s, $CH_3$), 6.40 (s, CH), 7.15 to 8.15 (m, arom. H)<br>IR: 1660 (C=O), 1615 (C=C), 1020 (S=O) | [1] |
| 13 | 4-$CH_3C_6H_4$ | $C_6H_5$ | $C_6H_5$ | — | II<br>III | 12 h<br>1 d | 30<br>68 | 151 | $^1H$ NMR (DMSO-$d_6$): 2.40 (s, $CH_3$), 7.24 to 7.76 (m, arom. H)<br>IR: analogous to No. 1 | [2] |
| 14 | 2,6-$(CH_3)_2C_6H_3$ | $C_6H_5$ | $C_6H_5$ | — | II<br>III | 12 h<br>2 d | 18<br>30 | 150 | $^1H$ NMR ($CDCl_3$): 2.32 (s, $CH_3$), 2.38 (s, $CH_3$), 7.16 to 7.45 (m, arom. H)<br>IR: analogous to No. 1 | [2] |
| 15 | H | H | H | H | IV | | 80 | 183 to 185 | IR: 1685, 1330, 1180, 1125 | [3] |
| 16 | $C_6H_5$ | H | H | H | IV | | 90 | 114 to 115.5 | IR: similar to No. 15 | [3] |
| 17 | 4-$BrC_6H_4$ | Br | H | Br | IV | | 20 | 193.5 to 194.5 | IR: similar to No. 15 | [3] |
| 18 | 2-Cl-5-$NO_2C_6H_3$ | Cl | H | H | IV | | 63 | 209.5 to 211 | IR: similar to No. 15 | [3] |

Alkaline hydrolysis of the compounds [–SO–NR[1]–CO–CH=C($C_6H_5$)–O–] with $R^1 = C_6H_5$ (Table 40, No. 6) and $R^1 = 4\text{-}ClC_6H_4$ (Table 40, No. 8) in 95% ethanol containing NaOH at reflux for 5 h results in the loss of $SO_2$ [1].

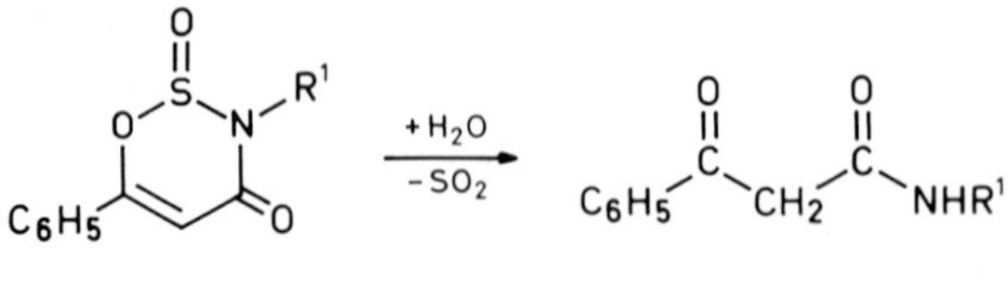

$$\text{6 } (R^1 = C_6H_5),\ \text{8 } (R^1 = 4\text{-}ClC_6H_4) \xrightarrow[-SO_2]{+H_2O} C_6H_5\text{–CO–}CH_2\text{–CO–}NHR^1$$

6 ($R^1$ = $C_6H_5$)
8 ($R^1$ = 4-$ClC_6H_4$)

Similarly the alkaline hydrolysis with 10% aqueous NaOH as well as the acidic hydrolysis with concentrated HCl of ethanolic solutions of [–SO–NH–CO–{1,2-$C_6H_4$}–O–] (Table 40, No. 15) under reflux for 15 min produces salicylamide [3]. When [–SO–N($C_6H_4CH_3$-4)–CO–C($C_6H_5$)= C($C_6H_5$)–O–] (Table 40, No. 13) is either hydrolyzed with concentrated $NH_3$ for 1 d or with refluxing HCl for 4 h, $C_6H_5CH(COC_6H_5)CONHC_6H_4CH_3$-4 is obtained [2].


**References:**

[1] Minami, T., Yamauchi, Y., Ohshiro, Y., Agawa, T., Murai, S., Sonoda, N. (J. Chem. Soc. Perkin Trans. I **1977** 904/8).
[2] Capuano, L., Urhahn, G., Willmes, A. (Chem. Ber. **112** [1979] 1012/22).
[3] Morris, L. R., Collins, L. R. (J. Heterocycl. Chem. **12** [1975] 309/10).


### 9.6.3 Derivatives of 3,2,1-Benzoxathiazin-4(1H)-one 2-Oxide

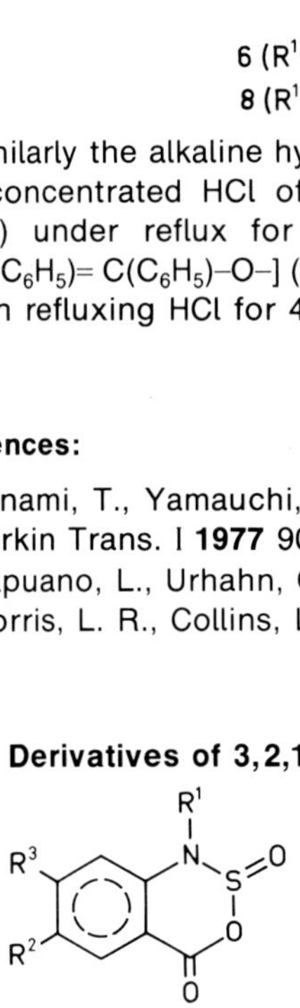

$R^1$, $R^2$, and $R^3$ are compiled on p. 101.

**[–SO–O–CO–{1,2-$C_6H_2R^2R^3$}–NR[1]–]**

The first attempted synthesis of compound No. 1 ($R^1 = R^2 = R^3 = H$) was reported by Graf and Langer in 1937 [1]. Recently, this parent compound and derivatives thereof were prepared but were not, except for compound No. 1, isolated or characterized. They are formed as unstable intermediates in the synthesis of quinazolones from anthranilic acid and $SOCl_2$. Compounds No. 1 to 5 form on reacting 2-aminobenzoic acid (anthranilic acid) or its derivatives N-methylanthranilic acid [2 to 6], 2-amino-4,5-dimethoxybenzoic acid [6, 7], 2-amino-4,5-methylenedioxybenzoic acid (6-amino-1,3-benzodioxole-5-carboxylic acid) [6], and 2-amino-5-methylbenzoic acid [8] with an excess of thionyl chloride in dry refluxing benzene, according to the following reaction:

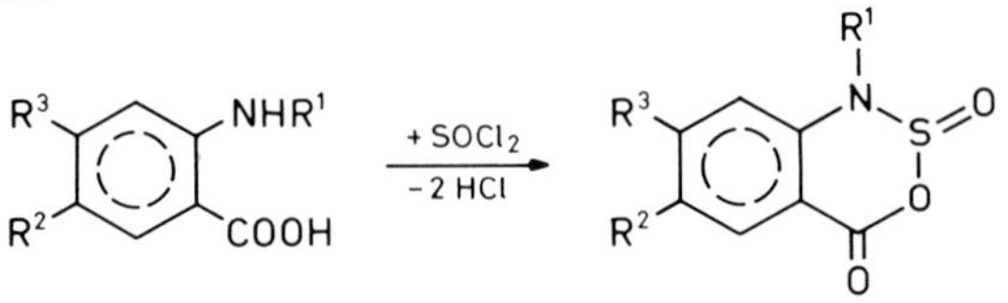

$$\text{2-}R^1NH\text{-4-}R^2\text{-5-}R^3C_6H_2COOH \xrightarrow[-2\,HCl]{+SOCl_2} \textbf{1 to 5}$$

1 to 5

| No. | $R^1$ | $R^2$ | $R^3$ | Ref. |
|---|---|---|---|---|
| 1 | H | H | H | [2 to 6] |
| 2 | $CH_3$ | H | H | [2 to 6] |
| 3 | H | $CH_3O$ | $CH_3O$ | [6, 7] |
| 4 | H | $-OCH_2O-$ | | [6] |
| 5 | H | H | $CH_3$ | [8] |

Compounds No. 1 and 2 are described as oily yellow liquids [4, 5, 6] and No. 3 as a brown solid [6]. In the mass spectrum of No. 1 the peak of the molecular ion is observed at m/e = 183. For the IR spectrum (in $CHCl_3$) of No. 1 two characteristic bands at $\nu = 1760$ (–CO–O–SO–) and 1135 $cm^{-1}$ (–SO–NH–) are reported [4].

When the 1-methyl derivative (No. 2) is kept at room temperature for several hours under moist conditions, compound I forms quantitatively [6]:

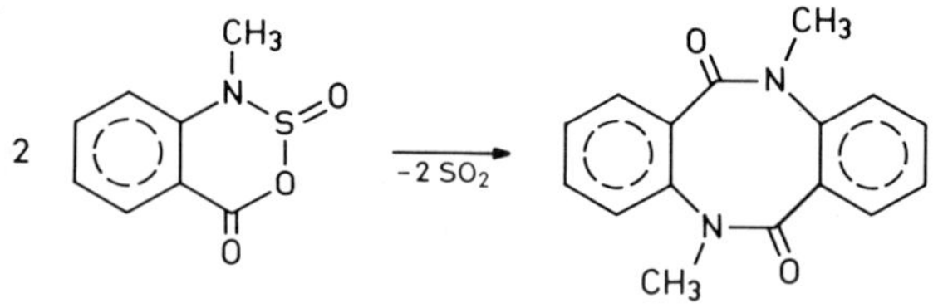

$$2\ \text{(1-methyl-benzoxathiazinone S-oxide)} \xrightarrow{-2\,SO_2} \text{I}$$

Cycloaddition of primary and secondary amines or cyclic imines in benzene at room temperature affords the quinazolone system according to the general equation:

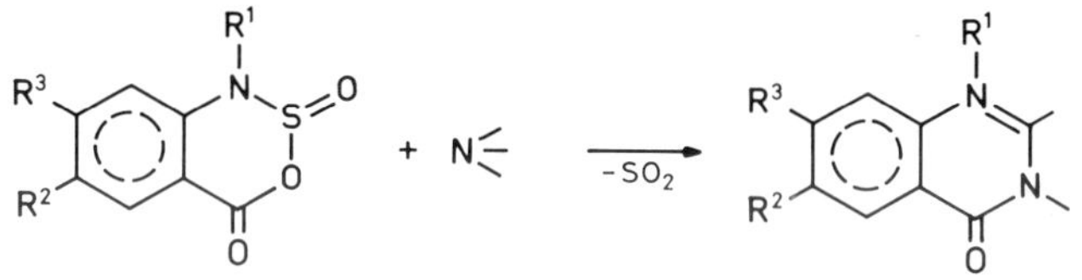

$$\text{1 to 5} + N\!\lessgtr \xrightarrow{-SO_2} \text{quinazolone}$$

1 to 5

The reactions are summarized in Table 41, pp. 102/3.

**References:**


[1] Graf, R., Langer, W. (J. Prakt. Chem. **148** [1937] 161/9).
[2] Kametani, T., Higa, T., Fukumoto, K., Koizumi, M. (Heterocycles **4** [1976] 23/8).
[3] Kametani, T., Loc, C. V., Higa, T., Koizumi, M., Ihara, M., Fukumoto, K. (Heterocycles **4** [1976] 1487/92).
[4] Kametani, T., Higa, T., Loc, C. V., Ihara, M., Koizumi, M., Fukumoto, K. (J. Am. Chem. Soc. **98** [1976] 6186/8).
[5] Kametani, T., Loc, C. V., Higa, T., Koizumi, M., Ihara, M., Fukumoto, K. (J. Am. Chem. Soc. **99** [1977] 2306/9).
[6] Kametani, T., Loc, C. V., Higa, T., Ihara, M., Fukumoto, K. (J. Chem. Soc. Perkin Trans. I **1977** 2347/9).
[7] Kametani, T. (Japan. Kokai Tokkyo Koho 78-147096 [1977/78]; C.A. **91** [1979] No. 20869).
[8] Parker, K. A., Fedynyshyn, T. H. (Tetrahedron Letters **1979** 1657/60).
[9] Kametani, T., Fukumoto, K. (Accounts Chem. Res. **9** [1976] 319/25).

Table 41

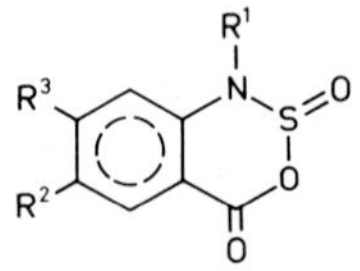

Reactions of 3,2,1-Benzoxathiazin-4(1H)-one 2-Oxides (compounds No. 1 to 5) with Amines and Cyclic Imines (References on p. 101)

[1: $R^1=R^2=R^3=H$; 2: $R^1=CH_3$, $R^2=R^3=H$; 3: $R^1=H$, $R^2=R^3=OCH_3$; 4: $R^1=H$, $R^2R^3=OCH_2O$; 5: $R^1=R^2=H$, $R^3=CH_3$].

| compound | reagents | products | Ref. |
|---|---|---|---|
| 1 | $R'CONH_2$ with $R'=C_6H_5CH_2$ [5] or $R'=C_6H_5CH_2OCH(CH_3)$ [6] | 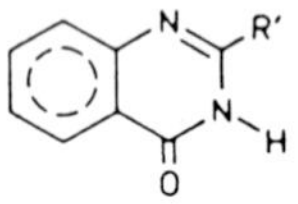 | [5, 6] |
| 2 | $R'CONH_2$ with $R'=H$, $CH_3$, and $C_2H_5$ [6] or $R'=C_6H_5CH_2$ [3, 5] | 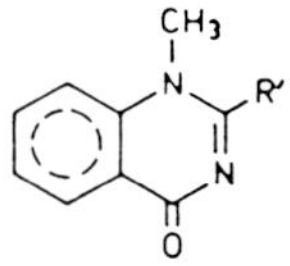 | [3, 5, 6] |
| 1 | 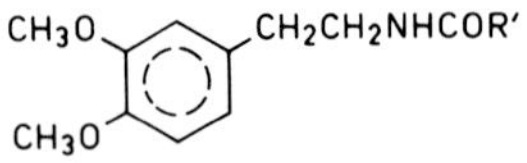 $R'=H$ or $CH_3$ | 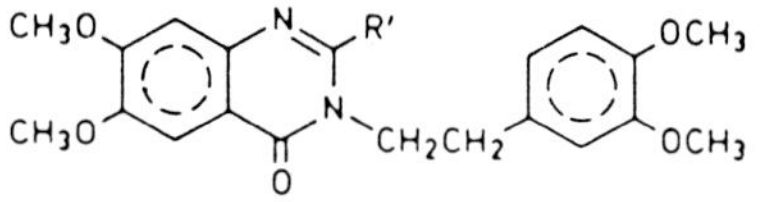 | [5] |
| 1 | 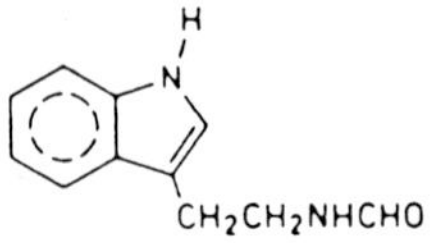 | 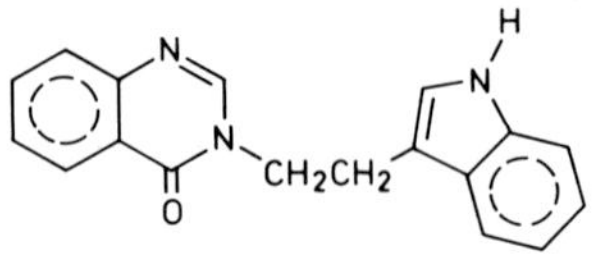 | [3, 5] |
| 1 | 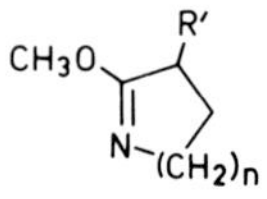 $R'=H$, $n=2$ [3, 4, 5]<br>$R'=H$, $n=2$ and<br>$R'=CH_3O$, $n=2$ [3] | 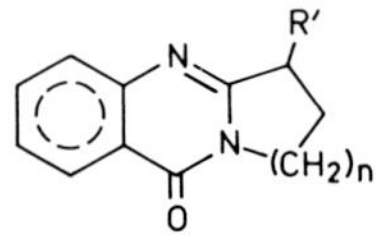 | [3, 4, 5] |
| 1 | 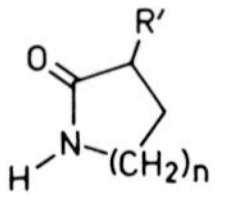 $R'=H$, $n=1$ [3, 5]<br>$R'=H$, $n=2$ [3, 5]<br>$R'=CH_3O$, $n=2$ [3, 5]<br>$R'=C_2H_5OOC$, $n=2$ [5] | 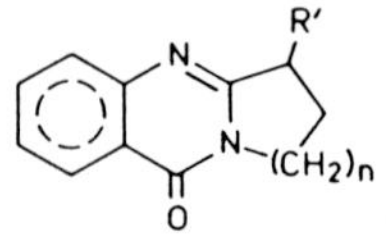 | [3, 5] |
| 5 | 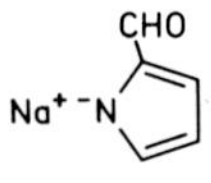 | 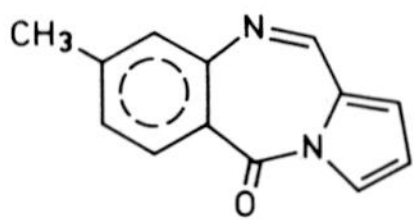 | [8] |

Table 41 (continued)

| com-pound | reagents | products | Ref. |
|---|---|---|---|
| 1 | 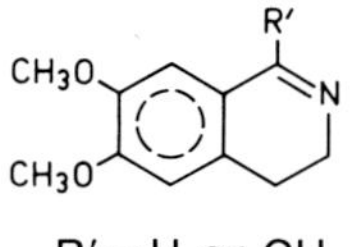 R′ = H or $CH_3$ | 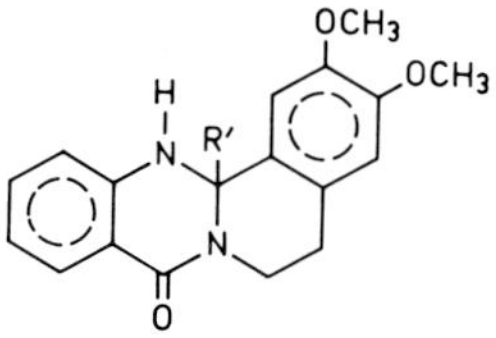 | [4] |
| 1 | 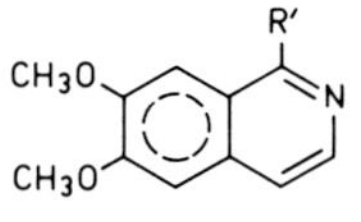 R′ = H or 4,5-$(CH_3O)_2C_6H_3CH_2$ | 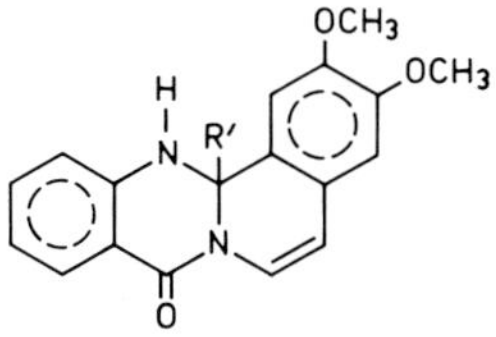 | [4] |
| 1 to 4 | 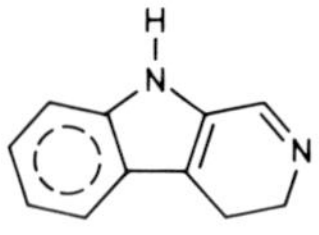 | 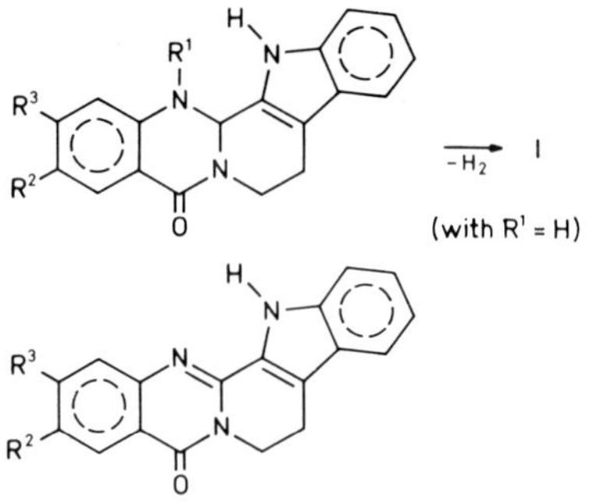 $\xrightarrow{-H_2}$ I (with R¹ = H) I | [2, 4, 5, 7, 9] |

## 9.7 $SN_2C_2O$ Ring

### 9.7.1 Derivatives of 5,6-Dihydro-1,2,3,5-oxathiadiazin-4(3H)-one 2-Oxide

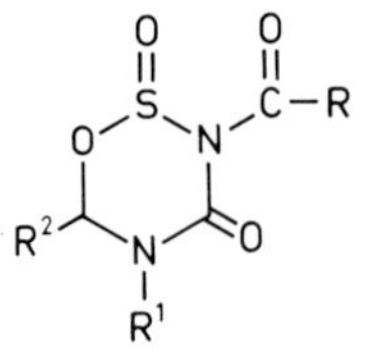

R, $R^1$, and $R^2$ are compiled in Table 42.

**[–SO–N(COR)–CO–NR$^1$–CHR$^2$–O–]**

The compounds, compiled in Table 42, are prepared by the following method: Benzoyl or trichloroacetyl isocyanates RCONCO in ether are treated with an ethereal solution of the respective azomethine $R^2CH{=}NR^1$ in the presence of $SO_2$ (see the following scheme). A mixture of compounds of type I and type II forms as orange or yellow-orange precipitates. The compounds of type I readily isomerize to the compounds of type II. Compounds No. 4 and 5, in Table 42 are obtained as the only products of the reaction of trichloroacetyl isocyanate with N-(4-nitrobenzylidene)aniline and 4-methoxy-N-(benzylidene)aniline in $CCl_4$ solution with subsequent treatment with $SO_2$ for 2 h [1, 2, 3].

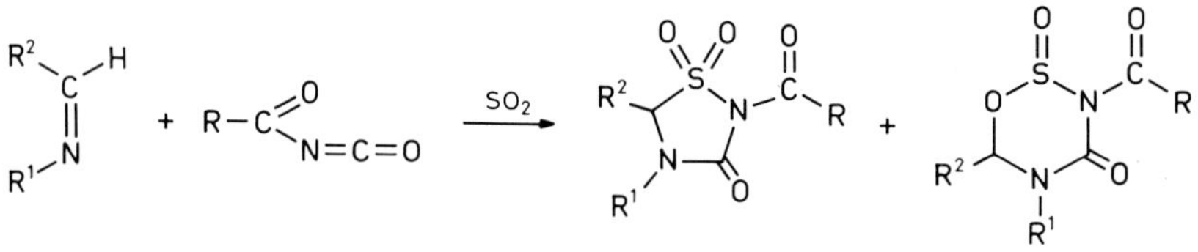

I II

**References:**


[1] Arbuzov, B. A., Zobova, N. N., Rubinova, N. R. (Izv. Akad. Nauk SSSR Ser. Khim. **1975** 1438/40; Bull. Acad. Sci. USSR Div. Chem. Sci. **24** [1975] 1333/5).

[2] Arbuzov, B. A., Zobova, N. N., Rubinova, N. R. (Izv. Akad. Nauk SSSR Ser. Khim. **1980** 1164/6; C.A. **93** [1980] No. 95250).

[3] Arbuzov, B. A., Zobova, N. N. (Synthesis **1982** 433/50).


### 9.7.2 Derivatives of 2-Imino-2,3-dihydro-1,4,3,5-oxathiadiazine 4-Oxide

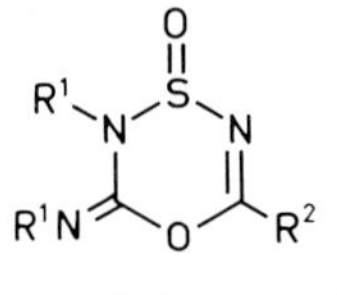

$R^1 = c\text{-}C_6H_{11}$; $R^2 = C_6H_5$
$R^1 = R^2 = C_6H_5$
$R^1 = c\text{-}C_6H_{11}$; $R^2 = CH_3$

**[–SO–N=CR$^2$–O–C(=NR$^1$)–NR$^1$–]**

The title compounds III in the reaction scheme (p. 106) along with 1,2,4-thiadiazetidine 1-oxides IV are thought to be intermediates in the 1:1 reaction of carbodiimides I with N-sulfinylcarboxamides II in ether at room temperature. The reactions afford oily products which could neither be crystallized after prolonged standing at −20°C nor distilled without decomposition. Pyrolysis of the oily residues at 110°C under reduced pressure gives volatile

Table 42

Derivatives of 5,6-Dihydro-1,2,3,5-oxathiadiazin-4(3H)-one 2-Oxide, [–SO–N(COR)–CO–NR$^1$–CHR$^2$–O–] [1, 2, 3].
(References on p. 104)

| No. | $R^1$ | $R^2$ | R | yield in % | m.p. in °C | IR data (ν in cm$^{-1}$) C=O | S=O | $^1$H NMR spectrum (standard TMS; δ in ppm) |
|---|---|---|---|---|---|---|---|---|
| 1 | $C_6H_5$ | 4-$(CH_3)_2NC_6H_4$ | $CCl_3$ | 65 | 150 to 155 (dec.) | 1680 | 1210 | [a] |
| 2 | $C_6H_5$ | 4-$(CH_3)_2NC_6H_4$ | $C_6H_5$ | 46 | 158 to 160 (dec.) | 1680 | 1205 | in $(CD_3)_2CO$: 3.03 (s, $(CH_3)_2N$), 6.77 (s, OCH), 7.55 to 8.2 (m, $C_6H_5$) |
| 3 | $C_6H_5$ | 4-$(C_2H_5)_2NC_6H_4$ | $C_6H_5$ | 51 | 85 to 90 (dec.) | 1670 | 1210 | in $CH_2Cl_2$: 1.20 (t) and 3.40 (q, $(C_2H_5)_2N$), 6.90 (s, OCH), 7.50 to 8.12 (m, $C_6H_5$) |
| 4 | $C_6H_5$ | 4-$NO_2C_6H_4$ | $CCl_3$ | 60 | 105 to 107 | 1680 | 1190 | in $CH_2Cl_2$: 6.75 (s, OCH), 7.65 (m, arom. H) |
| 5 | 4-$CH_3OC_6H_4$ | $C_6H_5$ | $CCl_3$ | 56 | 126 to 128 | 1675 | 1190 | in $CH_2Cl_2$: 3.18 (s, $CH_3O$), 6.70 (s, OCH), 7.70 (m, arom. H) |
| 6 | 4-$CH_3OC_6H_4$ | 4-$(C_2H_5)_2NC_6H_4$ | $CCl_3$ | — | — | — | — | [a] |

[a] No $^1$H NMR spectrum; slightly soluble with decomposition in most solvents.

liquids consisting of the compounds V, VI, and VII and a residual solid which was separated into the compounds VIII, IX, and X (see the following reaction scheme):

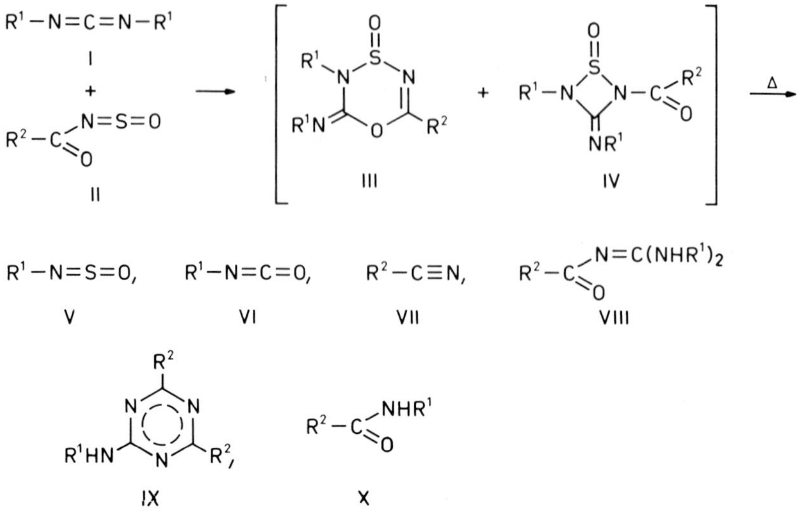

$R^1-N=C=N-R^1$ (I) + $R^2-C(=O)-N=S=O$ (II) → [III + IV] →(Δ)

$R^1-N=S=O$, V; $R^1-N=C=O$, VI; $R^2-C\equiv N$, VII; $R^2-C(=O)-N=C(NHR^1)_2$ VIII

IX; $R^2-C(=O)-NHR^1$ X

| starting materials | oily products | thermolysis products |
|---|---|---|
| I ($R^1$ = c-$C_6H_{11}$) + II ($R^2$ = $C_6H_5$) | III and IV | V, VI, VII, VIII, IX, X |
| I ($R^1$ = $C_6H_5$) + II ($R^2$ = $C_6H_5$) | III and IV | V, VI, VII, X |
| I ($R^1$ = c-$C_6H_{11}$) + II ($R^2$ = $CH_3$) | III and IV | V, VI, VII |

**Reference:**

Minami, T., Fukuda, M., Abe, M., Agawa, T. (Bull. Chem. Soc. Japan **46** [1973] 2156/9).

## 9.8 $SN_2CP_2$ Ring

### 9.8.1 Derivative of 3,3,5,5-Tetrahydro-1H-1,2,6,3,5-thiadiazadiphosphorine

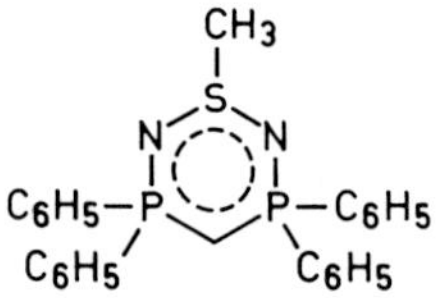

**[$\dot{-}$S($CH_3$)$\dot{-}$N$\dot{-}$P($C_6H_5$)$_2$$\dot{-}$CH$\dot{-}$P($C_6H_5$)$_2$$\dot{-}$N$\dot{-}$]**

The compound is obtained by treating the corresponding hydrobromide (II, see p. 108) with liquid ammonia. It is a crystalline solid, m.p. 140 to 142°C, and is readily soluble in $CHCl_3$ or benzene [1].

The crystals are orthorhombic, space group Pbnm–$D_{2h}^{16}$ (No. 62), with a = 9.473(4), b = 11.551(5), and c = 21.453(9) Å; Z = 4. $D_x$ = 1.30 g/cm³. R = 0.063 for 2450 independent reflections. The positional parameters and temperature factors were given [2]. The molecular structure is shown in **Fig. 19**.

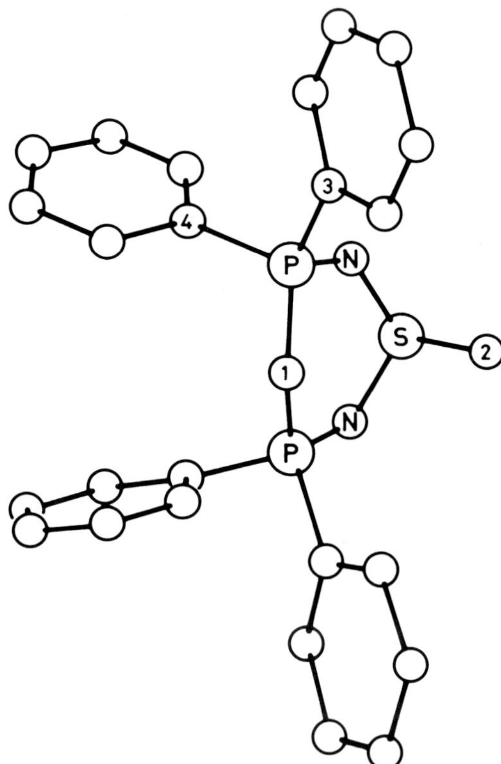


Fig. 19. Molecular structure of $[\overset{..}{-}S(CH_3)\overset{..}{-}N\overset{..}{-}P(C_6H_5)_2\overset{..}{-}CH\overset{..}{-}P(C_6H_5)_2\overset{..}{-}N\overset{..}{-}]$ without H atoms.

The six-membered $SN_2CP_2$ ring is puckered and adopts a boat conformation. The S and C atoms are located out of the plane running through the N and P atoms by 0.550 and 0.456 Å, respectively. The coordination at the phosphorus atom is distorted tetrahedral. All bond lengths within the $SN_2CP_2$ ring are between those of a single and double bond [2]. The principal bond lengths and bond angles are given in Table 43.

Table 43

Selected Bond Lengths and Bond Angles in $[\overset{..}{-}S(CH_3)\overset{..}{-}N\overset{..}{-}P(C_6H_5)_2\overset{..}{-}CH\overset{..}{-}P(C_6H_5)_2\overset{..}{-}N\overset{..}{-}]$

| bond length in Å | | bond angle in ° | |
|---|---|---|---|
| S–N | 1.612(3) | N–S–N | 113.2(2) |
| P–N | 1.628(3) | N–S–C(2) | 105.7(2) |
| P–C(1) | 1.715(3) | S–N–P | 119.3(2) |
| S–C(2) | 1.827(7) | P–C(1)–P | 116.5(3) |
| P–C(3) | 1.804(3) | C(1)–P–N | 113.2(2) |
| P–C(4) | 1.815(4) | C(1)–P–C(4) | 114.1(2) |
| | | C(3)–P–C(4) | 103.0(2) |
| | | N–P–C(3) | 111.4(2) |
| | | N–P–C(4) | 105.9(2) |
| | | C(1)–P–C(3) | 108.8(2) |

**References:**

[1] Appel, R., Hänssgen, D. (Angew. Chem. **79** [1967] 577; Angew. Chem. Intern. Ed. Engl. **6** [1967] 560).
[2] Weiss, J. (Acta Cryst. B **30** [1974] 2888/91).

## 9.8.2 Derivative of 3,4,5,5-Tetrahydro-1H-1,2,6,3,5-thiadiazadiphosphorinium Bromide

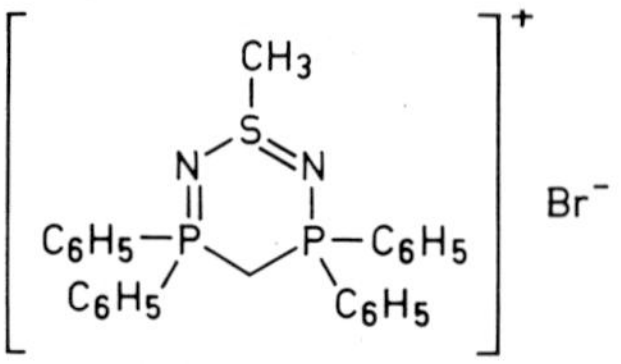

$$\left[\begin{array}{c} CH_3 \\ | \\ S \\ N{=}\ \ \ N \\ C_6H_5{-}P\ \ \ P{-}C_6H_5 \\ C_6H_5\ \ \ C_6H_5 \end{array}\right]^+ Br^-$$

**$[{-}S(CH_3){=}N{-}P(C_6H_5)_2{-}CH_2{-}P(C_6H_5)_2{=}N{-}]^+Br^-$**

Compound II forms by splitting off $CH_3Br$ from dibromide I at 110 to 120°C in high vacuum.

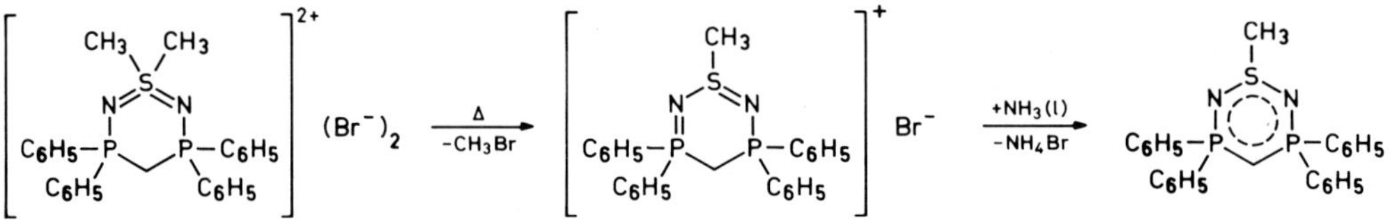

$$[\text{I}]^{2+}(Br^-)_2 \xrightarrow[-CH_3Br]{\Delta} [\text{II}]^+ Br^- \xrightarrow[-NH_4Br]{+NH_3(l)} \text{III}$$

I II III

The cation has been characterized by elemental analysis, ionic weight, and $^1H$ NMR.

The solid, m.p. 255 to 257°C (dec.), is soluble in polar organic solvents (e.g., alcohols or DMSO) and is insoluble in ether and hydrocarbons. It reacts with liquid ammonia to form the neutral compound III (see p. 106).

**Reference:**

Appel, R., Hänssgen, D. (Angew. Chem. **79** [1967] 577; Angew. Chem. Intern. Ed. Engl. **6** [1967] 560).

# 10 Seven Atom S–N–C and S–N–C–O Ring Systems

## 10.1 $S_3N_3C$ Ring

### 10.1.1 Derivatives of 1,3,5,2,4,6-Trithia(3-$S^{IV}$)triazepine

R = H, $COOCH_3$, and $CONHNH_2$

**[=S=N–S–N=CR–S–N=]** (Reduced formula in the text $S_3N_3CR$)

**$S_3N_3CH$**

The parent compound $S_3N_3CH$ of this heterocyclic system has not been synthesized until recently. Preliminary RHF MNDO and ab initio SCF MO calculations on the unsubstituted ring predicted a structure being very similar to planar [10]annulenes [1].

**$S_3N_3C(COOCH_3)$**

The compound was obtained in minor yields (5 to 6%) from the reaction of $S_4N_4$ with acetylenecarboxylates, $R'C{\equiv}CCOOCH_3$, R′ = H, $C_6H_5$, $COOCH_3$ (1:2 mole ratio) in refluxing toluene for 6 to 8 h. Colorless needles, m.p. 82 to 83°C, were isolated by column chromatography on silica gel using n-hexane and benzene as eluents [2], see also [3, 4]. An ambiguous partial structure was first assigned to the compound [2]. An X-ray diffraction study [1] led to completion and refinement of this structure.

Crystals are monoclinic, space group Pn (standard setting Pm) -$C_s^1$ (No. 6), with a = 3.816(2), b = 14.682(9), c = 6.725(6) Å, and β = 93.43(6)°; Z = 2. V = 376 Å$^3$, $D_x$ = 1.86 g/cm$^3$. R = 0.067 for 588 observed reflections [1].

The molecular structure is shown in **Fig. 20**.

The seven-membered $S_3N_3C$ ring is planar with a maximum deviation from the least-squares plane of 0.009 Å (N(6)). The S–N bond lengths (1.57 to 1.60 Å, see Fig. 20) are nearer to double (1.55 Å) than to single (1.67 Å) bonds. The planarity of this system results in greatly enlarged angles at the nitrogen atoms. The angles are 141°, 135°, and 140° at N(2), N(4), and N(6), respectively.

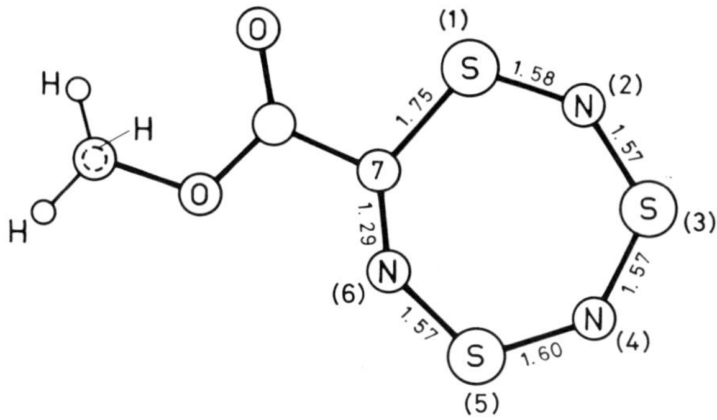


Fig. 20. Molecular structure of $S_3N_3C(COOCH_3)$.

The ring can, by a count of the delocalizable electrons, be considered as a 10π aromatic system. This view is strongly supported by the measured bond lengths, the UV spectra, the thermal and chemical stability of the ring, and MO calculations.

The greater stability (see below) of the trithiatriazepine, compared to the trithiadiazepines (see p. 117), is understandable since dipolar structures will undoubtedly make important contributions to these structures and in the title compound another nitrogen site is available for negative charge [1].

In the IR spectrum (KBr) the absorption band at 1700 $cm^{-1}$ has been assigned to the C=O vibration. UV-visible spectra (in $C_2H_5OH$): $\lambda_{max}$ (in nm (log ε)) = 276(4.54), and 335(3.91) [2], 266(4.09), and 332(3.40), indicating a highly delocalized structure [1]. $^1H$ NMR spectrum ($CDCl_3$/TMS): δ = 4.00 ppm (s, $CH_3$). $^{13}C$ NMR spectrum ($CDCl_3$/TMS): δ (in ppm) = 53.7, 151.0, and 161.8 (unassigned). For the mass spectrum (70 eV) the base peak has been observed at m/e = 78, 12 further unassigned peaks were given [2].

$S_3N_3C(COOCH_3)$ is very stable. It shows no decomposition on boiling in xylene (138°C) for 24 h. In boiling Decalin (190°C) some of the trithiatriazepine still remained after 33 h. The compound also survived on boiling in toluene with triphenylphosphine. The compound is strikingly inert to 3-chloroperbenzoic acid, and did not react significantly with the peracid in boiling $CH_2Cl_2$ [1]. On treatment with excess hydrazine hydrate in refluxing $C_2H_5OH$, $S_3N_3C(CONHNH_2)$ is obtained [2].

**$S_3N_3C(CONHNH_2)$**

The compound has been obtained from $S_3N_3C(COOCH_3)$ (see above) on treatment with excess hydrazine hydrate in refluxing ethanol. Colorless needles, m.p. 214 to 216°C (from $CHCl_3$) [2].

**References:**

[1] Daley, S. T. A. K., Rees, C. W., Williams, D. J. (J. Chem. Soc. Chem. Commun. **1984** 55/7).
[2] Mataka, S., Takahashi, K., Yamada, Y., Tashiro, M. (J. Heterocycl. Chem. **16** [1979] 1009/15).
[3] Tashiro, M., Mataka, S., Takahashi, K. (Heterocycles **6** [1977] 933/9).
[4] Tashiro, M., Mataka, S., Takahashi, K., Yamada, Y. (Heterocycles **9** [1978] 126).

## 10.2 $S_3N_2C_2$ Ring

### 10.2.1 6,7-Dihydro-1,3,5,2,4-trithia(3-$S^{IV}$)diazepine

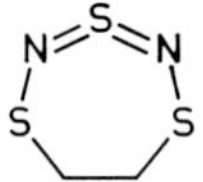


**[=S=N–S–$CH_2$–$CH_2$–S–N=]** (Reduced formula in the text $S_3N_2C_2H_4$)

The compound was prepared in 20% yield when dilute solutions of $ClSCH_2CH_2SCl$ and $(CH_3)_3SiN{=}S{=}NSi(CH_3)_3$ in $CH_2Cl_2$ were added slowly to a large volume of vigorously stirred $CH_2Cl_2$ under nitrogen [1].

The crystals of $S_3N_2C_2H_4$ are monoclinic, space group $C2/m-C^3_{2h}$ (No. 12), with a = 9.947(2), b = 8.807(2), c = 7.614(2) Å, and β = 114.97(2)°; Z = 4. V = 605 $Å^3$, $D_x$ = 1.68 g/$cm^3$. R = 0.033, $R_W$ = 0.040 for 376 reflections; the hydrogen atom positions were refined isotropically. The molecular structure (see **Fig. 21**) retains a planar S–N–S–N–S segment with a maximum deviation from the best plane of these atoms of 0.028 Å (for N(2)). The compound has a crystallographic two-fold axis passing through the apical sulfur and bisecting the C-C bond. Compared to the 1,3,5,2,4-trithia(3-$S^{IV}$)diazepine (see p. 116) there is a considerable lengthening and loss of double bond character in the N(1)-S(2) bond of the dihydro compound. The relatively large angles at nitrogen of 137° in the ring result from the $sp^3$ hybridization of the carbon atoms. These two carbon atoms lie symmetrically 0.47 Å above and below the S–N–S–N–S plane such that the methylene hydrogen atoms adopt an approximately cis and trans periplanar geometry, resulting in an H(e)CCH(e′) torsion angle of 31°. For selected bond lengths and bond angles, see Fig. 21 [2].

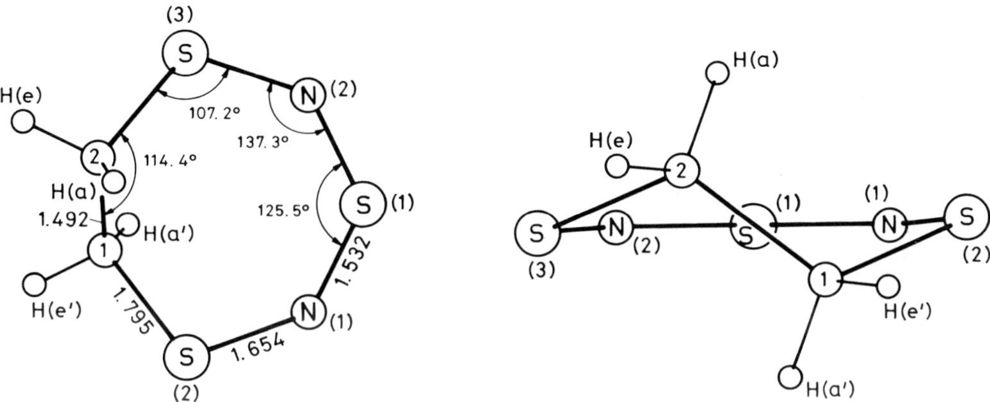


Fig. 21. Molecular structure of $S_3N_2C_2H_4$.
Bond lengths in Å, bond angles in °.

The orange crystals [1], highly volatile in air [2], melt at 30°C and have a boiling point of 45°C/0.3 Torr. UV spectrum ($C_2H_5OH$): $\lambda_{max}$ (in nm) = 208 (log $\varepsilon$ = 3.26), 260(3.15), 305(2.76), 408(3.26); IR spectrum (film): $\nu_{max}$ = 1390, 1138 $cm^{-1}$. $^1H$ NMR spectrum ($CDCl_3$/TMS, 90 MHz): $\delta$ = 3.71 ppm (s); $^{13}C$ NMR spectrum ($CDCl_3$/TMS): $\delta$ = 55.9 ppm [1]. In the 250 MHz $^1H$ NMR spectrum at room temperature is a sharp singlet, which shows characteristics of a partially exchange-broadened AA'BB' spin system only on cooling below 190 K, indicating a relatively low ($\leqq$40 kJ/mol) energy barrier for axial-equatorial hydrogen conversion. The MNDO calculated barrier is 55.6 kJ/mol, whereas a 4-31 G/MNDO calculation predicts a barrier of only 13.0 kJ/mol [2]. Mass spectrum: m/e = 152 $M^+$, 124 $M^+$-$C_2H_4$, 106 $M^+$-SN.

$S_3N_2C_2H_4$ is readily dehydrogenated with dichlorodicyanobenzoquinone in boiling dioxane to give 1,3,5,2,4-trithia(3-$S^{IV}$)diazepine (see p. 112) [1].

**References:**

[1] Morris, J. L., Rees, C. W., Rigg, D. J. (J. Chem. Soc. Chem. Commun. **1985** 396/7).
[2] Jones, R., Morris, J. L., Potts, A. W., Rees, C. W., Rigg, D. J., Rzepa, H. S., Williams, D. J. (J. Chem. Soc. Chem. Commun. **1985** 398/400).

## 10.2.2 1,3,5,2,4-Trithia(3-S$^{IV}$)diazepine and Derivatives

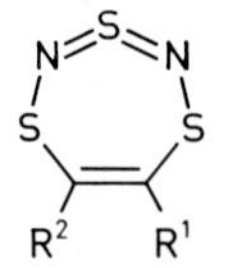

Parent compound: $R^1 = R^2 = H$
Derivatives: $R^1$, $R^2$ are compiled in Table 44.

**[=S=N–S–CR$^1$=CR$^2$–S–N=]** (Reduced formula in the text $S_3N_2C_2R^1R^2$)

### Formation. Properties

The parent compound ($R^1 = R^2 = H$) of this heterocyclic system is formed in 70% yield from dehydrogenation of 6,7-dihydro-1,3,5,2,4-trithia(3-S$^{IV}$)diazepine (see pp. 110/1) with the

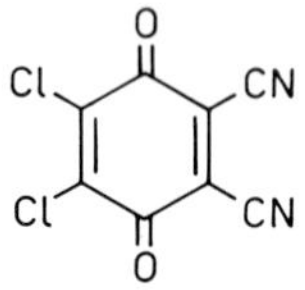

I

benzoquinone derivative (I) in boiling dioxane. An alternative route to the compound is by spontaneous loss of HCl when treating $ClSCHClCH_2SCl$, initially formed by chlorination of $HSCH_2CH_2SH$ with $Cl_2$ in $CCl_4$ at 0°C, with $(CH_3)_3SiN{=}S{=}NSi(CH_3)_3$; overall yield 30% [1].

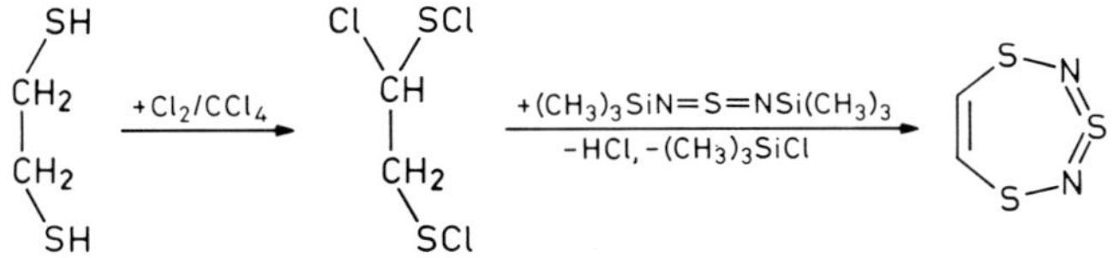

$$HSCH_2CH_2SH \xrightarrow{+Cl_2/CCl_4} ClS(Cl)CH{-}CH_2SCl \xrightarrow[-HCl,\,-(CH_3)_3SiCl]{+(CH_3)_3SiN=S=NSi(CH_3)_3} S_3N_2C_2H_2$$

$S_3N_2C_2H_2$ is a colorless, crystalline, volatile solid; m.p. 57°C [1].

Derivatives, 6-mono- and 6,7-disubstituted 1,3,5,2,4-trithia(3-S$^{IV}$)diazepines have been synthesized from $S_3N_2C_2H_2$ by electrophilic substitution reactions (method I), or by substituent exchange from the $Tl(OOCCF_3)_3$ derivative (method II) [1]. The methyl carboxylates with $R^1 = H$, $COOCH_3$, $C_6H_5$, and $R^2 = COOCH_3$ have been obtained in low yield from the reaction of $S_4N_4$ with the corresponding acetylenecarboxylates, $R^1C{\equiv}CCOOCH_3$ (method III) [3].

Reaction conditions, yields, and melting points of the trithiadiazepine derivatives produced are compiled in Table 44. Spectra of the parent compound $S_3N_2C_2H_2$ [1] and of some methyl carboxylate derivatives [3,4] are summarized in Table 45, p. 114.

### Structure

The structures of the parent compound $S_3N_2C_2H_2$ and of one of its derivatives, the 1,3,5,2,4-trithia(3-S$^{IV}$)diazepine-6,7-dicarboxylate ($R^1 = R^2 = COOCH_3$ (Table 44, No. 9)) have been determined by X-ray diffraction, the crystal data are given in Table 46, p. 115.

$S_3N_2C_2H_2$ has a striking planarity with maximum deviation from the least-squares plane of these atoms of 0.04 Å (for S(3)). The molecular structure is essentially symmetrical. Surprisingly, despite their planar geometry, the molecules do not adopt a parallel stacking arrangement in the crystal. The S–N bond lengths range from 1.54 to 1.60 Å, demonstrating a marked degree of partial double bond character. Ab initio and MNDO calculations also show a distinct pattern of $\pi$-energy levels characteristic of an aromatic system (S–N $\pi$-bond order 0.307). As a consequence of the molecular planarity, the angles at nitrogen are abnormally large, 137° and 138° [2].

Table 44

Reaction Conditions, Yields, and Melting Points of Derivatives of 1,3,5,2,4-Trithia(3-$S^{IV}$)diazepine, [=S=N–S–$CR^1$=$CR^2$–S–N=].

| No. | $R^1$ | $R^2$ | method | reactants | conditions | yield in % | m.p. in °C (recryst. from) | remarks | Ref. |
|---|---|---|---|---|---|---|---|---|---|
| 1 | H | Br | I | $S_3N_2C_2H_2$ + N-bromosuccinimide (1:1 mole ratio) | $CH_3CN$, room temperature | 88 | 31 | | [1] |
| 2 | H | I | II | $S_3N_2C_2H_2$ + aqueous KI | $CH_3CN$ | 80 | 93 | pale yellow crystals | [1] |
| 3 | H | $NO_2$ | I | $S_3N_2C_2H_2$ + $Cu(NO_3)_2 \cdot 3H_2O$ | acetic anhydride, 0°C | 90 | 84 | yellow crystals | [1] |
| 4 | H | $COOCH_3$ | II | $S_3N_2C_2H_2$ + CO/$CH_3OH$ | presence of $PdCl_2$, LiCl, and MgO | 70 | 111 | | [1] |
| | | | III | $S_4N_4$ + CH≡$CCOOCH_3$ | boiling toluene, 6 to 8 h | 5 | 109 to 111 (n-hexane) | colorless needles, isolated by column chromatography | [3] |
| 5 | H | CN | II | $S_3N_2C_2H_2$ + CuCN | boiling $CH_3CN$ | 85 | 71 | | [1] |
| 6 | H | $Tl(OOCCF_3)_2$ | I | $S_3N_2C_2H_2$ + $Tl(OOCCF_3)_3$ | refluxing in $CH_3CN$, 3 h | — | — | not isolated | [1] |
| 7 | Br | Br | I | $S_3N_2C_2H_2$ + N-bromosuccinimide (excess) | $CH_3CN$, room temperature | 50 | 89 | | [1] |
| 8 | $NO_2$ | $NO_2$ | I | $S_3N_2C_2H_2$ + $NO_2BF_4$(excess) | $CH_3CN$ | 54 | 60 | | [1] |
| 9 | $COOCH_3$ | $COOCH_3$ | III | $S_4N_4$ + $CH_3OOCC$≡C-$COOCH_3$ | boiling toluene, 6 to 8 h | 2 | 73.5 to 74 (n-hexane) | colorless needles, isolated by column chromatography | [3] |
| 10 | $C_6H_5$ | $COOCH_3$ | III | $S_4N_4$ + $C_6H_5C$≡C-$COOCH_3$ | boiling toluene, 6 to 8 h; heating at 90°C/1 Torr for 12 h | 2 | 84 to 85 (n-hexane) | pale yellow prisms | [3] |
| 11 | 1:1 complex of 10 with $CH_3OOCC$≡$CC_6H_5$ | | III | $S_4N_4$ + $C_6H_5C$≡C-$COOCH_3$ | boiling toluene, 6 to 8 h | — | 50 to 51 (n-hexane) | pale yellow needles, isolated by column chromatography | [3] |

Table 45

Spectra of Derivatives of 1,3,5,2,4-Trithia(3-$S^{IV}$)diazepine, [=S=N–S–$CR^1$=C($COOCH_3$)–S–N=] (No. 4 and 9 to 11 in Table 44, p. 113).

| No. | $R^1$ | IR (KBr) ν(CO) in $cm^{-1}$ | UV-visible ($C_2H_5OH$) $\lambda_{max}$ in nm (log ε) | $^1$H NMR ($CDCl_3$/TMS) δ in ppm | $^{13}$C NMR ($CDCl_3$/TMS) δ in ppm | MS (70 eV) |
|---|---|---|---|---|---|---|
| 4 | H | 1697 | 266 (4.09)<br>332 (3.40) | 3.98 (s, $CH_3$)<br>8.90 (s, CH) | — | 10 peaks (unassigned) given; base at m/e = 78 |
| 9 | $COOCH_3$ | 1725, 1705 | 274 (4.26)<br>334 (3.56) | 3.92 (*) or 4.01?), (s, $CH_3$) | 54.09, 140.7, 162.6 } unassigned | — |
| 10 | $C_6H_5$ | 1725 | — | 3.53 (*) or 3.51?), (s, $CH_3$)<br>7.43 (s, $C_6H_5$) | — | 10 peaks (unassigned) given; base at m/e = 124 |
| 11 | 1:1 complex of 10 with $CH_3OOCC{\equiv}CC_6H_5$ | 2205, 1730, 1710 | — | 3.53 (s, $CH_3$)<br>3.81 (s, $CH_3$)<br>7.2 to 7.7 (m, $C_6H_5$) | — | — |
| Ref. | | [3] | [4] | [3] | [3] | [3] |

*) The δ-value in parentheses is given in the text, the first δ-value is cited in the experimental part of the paper.

The molecular structure of the dicarboxylate is quite similar (first structural assumptions [3] were erroneous). It is planar with a maximum deviation of 0.019 Å (for C(1)) and symmetrical with delocalized ring bonds. The S–N bond lengths are, within statistical significance, the same as those observed for the parent compound. The C–S bond lengths increase from 1.684 to 1.694 Å in the parent compound to 1.709 to 1.718 Å in the dicarboxylate derivative. As a consequence of the planarity the angles at nitrogen are also abnormally large, 140° at N(1) and 138° at N(2) [4].

Principal bond lengths and bond angles of the parent $S_3N_2C_2H_2$ and its dicarboxylate derivative are presented in Table 47, p. 116; the molecular structure of $S_3N_2C_2H_2$ is shown in **Fig. 22**.

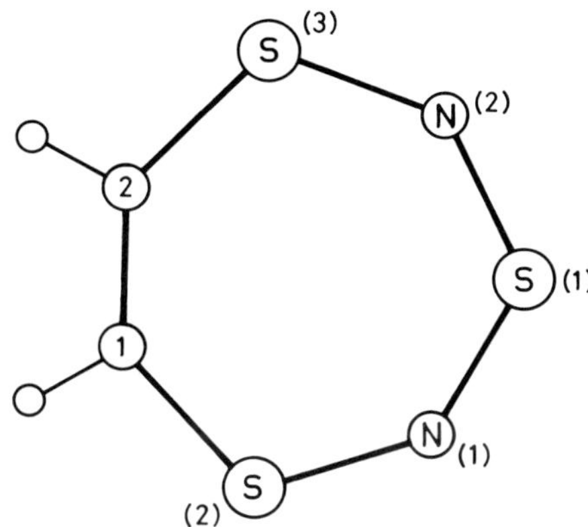


Fig. 22. Molecular structure of $S_3N_2C_2H_2$.

Table 46

Crystal Data of 1,3,5,2,4-Trithia(3-$S^{IV}$)diazepine, [=S=N–S–CH=CH–S–N=], and Some Derivatives.

| compound | (parent $S_3N_2C_2H_2$) | ($CH_3OOC$)₂ derivative | *) benzo derivative |
|---|---|---|---|
| symmetry | monoclinic | monoclinic | monoclinic |
| space group | $P2_1/n-C^5_{2h}$ (No. 14) | $C2/c-C^6_{2h}$ (No. 15) | $C2/c-C^6_{2h}$ (No. 15) |
| a in Å | 6.883(1) | 19.676(3) | 12.167(3) |
| b in Å | 7.693(1) | 5.798(1) | 9.698(3) |
| c in Å | 10.521(2) | 18.646(5) | 7.099(2) |
| β in ° | 96.50(1) | 94.17(2) | 108.51(2) |
| Z | 4 | 8 | 4 |
| V in Å³ | 554 | 2122 | 794 |
| $D_x$ in g/cm³ | 1.81 | 1.67 | 1.68 |
| R | 0.045 | 0.067 | 0.032 |
| $R_w$ | 0.049 | — | 0.036 |
| number of reflections | 623 | 1224 | 497 |
| Ref. | [2] | [4] | [2] |

*) The compound is described on pp. 118/9.

Table 47

Selected Bond Lengths (in Å) and Bond Angles (in °) in 1,3,5,2,4-Trithia(3-$S^{IV}$diazepine, [=S=N–S–CH=CH–S–N=], and Some Derivatives. Numeration of atoms as in Fig. 22, p. 115.

| compound | $S_3N_2C_2H_2$ (parent ring) | $(CH_3OOC)_2C_2S_3N_2$ | *) benzo derivative |
|---|---|---|---|
| bond length | | | |
| S(1)–N(1) | 1.542(5) | 1.539(6) | 1.538(3) |
| N(1)–S(2) | 1.599(5) | 1.595(6) | 1.609(3) |
| S(2)–C(1) | 1.684(6) | 1.718(7) | 1.731(3) |
| C(1)–C(2) | 1.346(9) | 1.344(10) | 1.387(5) |
| C(2)–S(3) | 1.694(7) | 1.709(8) | |
| S(3)–N(2) | 1.599(5) | 1.584(6) | |
| N(2)–S(1) | 1.559(5) | 1.554(6) | |
| bond angle | | | |
| N(2)–S(1)–N(1) | 122.8(3) | | 122.4(2) |
| S(1)–N(1)–S(2) | 138.3(3) | 140 | |
| N(1)–S(2)–C(1) | 115.0(3) | | |
| S(2)–C(1)–C(2) | 135.6(5) | | |
| C(1)–C(2)–S(3) | 135.6(5) | | 133.5(1) |
| C(2)–S(3)–N(2) | 115.7(3) | | 116.9(1) |
| S(3)–N(2)–S(1) | 136.8(3) | 138 | 138.3(2) |
| Ref. | [2] | [4] | [2] |

*) The compound is described on pp. 118/9.

**Electronic Structure**

The vertical ionization potentials of the parent $S_3N_2C_2H_2$ from its He(I)$\alpha$ photoelectron spectrum are given in the following table together with their orbital assignments on the basis of MNDO calculations [2].

| band | vertical ionization energies in eV | symmetry assignment |
|---|---|---|
| 1 | 8.58 | $3a_2(\pi)$ |
| 2 | 10.00 | $5b_2(\pi)$ |
| 3 | 11.42, 11.67, 12.11 | $13b_1(\sigma)$, $4b_2(\pi)$, $17a_1(\sigma)$ |
| 4 | 13.06 | $2a_2(\pi)$ |
| 5 | 13.99, 14.32 | $16a_1(\sigma)$, $12b_1(\sigma)$ |
| 6 | 16.06 | $3b_2(\pi)$, $11b_1(\sigma)$, $10b_1(\sigma)$ |

From the calculations, the HOMO consists predominantly of an out-of-phase $\pi$ overlap of sulfur and nitrogen lone pair orbitals. The form of the first photoelectron band, which shows a weak vibrational progression (1490 $cm^{-1}$) and a stable cation, is in accordance with ionization from this largely nonbonding $\pi$-type orbital. The low first ionization potential of $S_3N_2C_2H_2$ (8.58 eV) indicates that the radical cation should be readily generated. Valence shell orbital energies calculated by the MNDO, STO-4-31G, and STO-3G* methods are listed in the paper [2].

#### Chemical Reactions

The parent compound $S_3N_2C_2H_2$ is thermally stable, being unchanged in boiling 1,2-dichlorobenzene (180°C) for over 2 d, but is decomposed on irradiation (300 nm) in light petroleum. It is inert toward protic acids, e.g., $CH_3COOH$, aqueous HCl, and Lewis acids such as $BF_3$, $AlCl_3$, and toward $N(C_2H_5)_3$, but it is instantly destroyed by aqueous NaOH. The compound is only very slowly consumed by triphenylphosphine in boiling toluene or by 3-chloroperbenzoic acid in boiling $CH_2Cl_2$. It undergoes some standard electrophilic aromatic substitution reactions. Thus it forms the 6-bromo and the 6,7-dibromo derivatives (see Table 44, p. 113) when treated with corresponding amounts of N-bromosuccinimide. Analogously, the 6-nitro and the 6,7-dinitro derivatives are formed from reactions with $Cu(NO_3)_2 \cdot 3H_2O$ in acetic anhydride or with an excess of $NO_2BF_4$ in $CH_3CN$, respectively. With $Tl(OOCCF_3)_3$ in $CH_3CN$, a bis(trifluoroacetato)thallium derivative is formed. Attempts to acylate and formylate the trithiadiazepine failed. $S_3N_2C_2H_2$ shows no tendency to undergo cycloaddition reactions with a range of electron-rich and electron-poor $2\pi$ and $4\pi$ components nor to form charge-transfer complexes with, e.g., picric acid, tetracyanoethylene, or 2,4,7-trinitro-9-fluorenylidenemalononitrile. It was found to inhibit totally the catalytic hydrogenation of 4-nitroacetophenone, a reaction which otherwise goes to completion [1].

The nitro derivative $S_3N_2C_2H(NO_2)$ (Table 44, No. 3) is resistant toward catalytic hydrogenation (Pd-C, Pd-C-$H_2SO_4$, $PtO_2$) to form the amine [1]. A solution of the thallium derivative (Table 44, No. 6) in $CH_3CN$ reacts with aqueous KI, CuCN in boiling $CH_3CN$, or CO in $CH_3OH$ in the presence of $PdCl_2$, LiCl, and MgO to give the 6-iodo-, 6-cyano-, and 6-methoxycarbonyl-substituted 1,3,5,2,4-trithia(3-$S^{IV}$)diazepines, respectively [1]. $S_3N_2C_2(COOCH_3)_2$ (Table 44, No. 9) showed no decomposition on boiling in xylene (138°C) for 24 h. In boiling Decalin (190°C) the compound was decomposed within 5.5 h. It was strikingly inert to 3-chloroperbenzoic acid (even in boiling $CH_2Cl_2$) and to triphenylphosphine. The compound was slowly decomposed by triphenylphosphine in boiling toluene [4]. The complex compound $S_3N_2C_2$-$(C_6H_5)(COOCH_3) \cdot CH_3OOCC{\equiv}CC_6H_5$ (Table 44, No. 11) yields $S_3N_2C_2(C_6H_5)(COOCH_3)$ (Table 44, No. 10) when heated at 90°C/1 Torr for 12 h [3].

#### References:


[1] Morris, J. L., Rees, C. W., Rigg, D. J. (J. Chem. Soc. Chem. Commun. **1985** 396/7).

[2] Jones, R., Morris, J. L., Potts, A. W., Rees, C. W., Rigg, D. J., Rzepa, H. S., Williams, D. J. (J. Chem. Soc. Chem. Commun. **1985** 398/400).

[3] Mataka, S., Takahashi, K., Yamada, Y., Tashiro, M. (J. Heterocycl. Chem. **16** [1979] 1009/15).

[4] Daley, S. T. A. K., Rees, C. W., Williams, D. J. (J. Chem. Soc. Chem. Commun. **1984** 55/7).

### 10.2.3 1,3,5,2,4-Benzotrithiadiazepine-3-$S^{IV}$

**[=S=N–S–{1,2-$C_6H_4$}–S–N=]** (Reduced formula in the text $S_3N_2C_6H_4$)

The compound was prepared in 50 [1] and in 41% [2] yield when dilute equimolar solutions of 1,2-$(ClS)_2C_6H_4$ and $(CH_3)_3SiN{=}S{=}NSi(CH_3)_3$ in $CH_2Cl_2$ were added slowly to a large volume of vigorously stirred $CH_2Cl_2$ in an inert gas atmosphere [1, 2]. The solvent was removed in vacuum to yield an orange solid. Heating this residue at 60°C/0.01 Torr yielded a yellow sublimate, which was further purified by column chromatography and recrystallization. The compound has also been obtained from its norbornadiene adduct on dissolving in $CH_2Cl_2$ or $CHCl_3$ at 25°C [2].

The crystal structure was determined by X-ray diffraction. Data are given in Table 46, p. 115; for selected bond lengths and bond angles, see Table 47, p. 116. The molecule has the same striking planarity as the parent compound $S_3N_2C_2H_2$ (see p. 112) with a maximum deviation from the best plane of 0.032 Å for S(3). It has a crystallographic two-fold axis passing through the apical sulfur and bisecting the C–C bond. The molecules pack with parallel overlap, the seven-membered ring of one directly overlying the six-membered ring of another and vice versa. The interplanar separation is small at 3.54 Å. The angles at nitrogen are abnormally large, 138° [3]. The S–N bond lengths demonstrate the marked degree of partial double bond character. (This delocalization is also supported by ab initio and MNDO calculations.) From the bond lengths it was concluded that the compound possesses compact N–S–N units which are suggestive of the diimide formulations I and II.

I II III

An MNDO molecular orbital calculation on the compound involving full geometry optimization within $C_{2v}$ symmetry reveals the $\pi$ bond orders C–S 0.257 (0.260 in [3]), N–S(C) 0.262, and N–S–N (av) 0.606 that provide a satisfying correlation with the experimental bond distances [2]. A pronounced bond alternation is also induced in the carbocyclic ring; for MNDO $\pi$ bond orders, see [2, 3]. There should also be some involvement of the quinoid structure III. An orbital mixing of the SNSNS and $C_6$ components or a somewhat intermediate situation between quinoid and diimide formulation is in agreement with MNDO calculations. The E(HOMO) value of $-8.501$ eV is consistent with the relatively low oxidation potential of the compound, the LUMO energy of $-1.736$ eV is consistent with a remarkable susceptibility to reduction, in spite of its being an "electron-rich" (14-electron-) $\pi$ system [2].

The first ionization potential of the compound has been derived from its He(I)$\alpha$ photoelectron spectrum to be 7.90 eV, indicating that the radical cation should be readily generated [3].

In comparison with the parent compound $S_3N_2C_2H_2$, the increase in the C–S bond length and the accompanying decrease in the MNDO $\pi$ bond order from 0.307 to 0.260 cause a destabilization of the aromatic trithiadiazepine system [3]. Thus $S_3N_2C_6H_4$ is much more reactive than $S_3N_2C_2H_2$ toward reagents such as $P(C_6H_5)_3$ and 3-chloroperbenzoic acid [3].

The bright yellow crystals [1] and yellow needles of $S_3N_2C_6H_4$ [2] melt at 78 and 78 to 80°C, respectively.

IR spectrum ($CHCl_3$): $\nu_{max}$=1455, 1150 $cm^{-1}$ [1]; (Nujol mull, 1600 to 250 $cm^{-1}$ region): 14 absorption bands are listed in the paper [2]. UV spectrum ($\lambda_{max}$ in nm, (log $\varepsilon$)) in $C_2H_5OH$: 252(4.25), 292(4.02), 365(3.63) [1]; in $CH_2Cl_2$: 256(4.3), 294(4.0), 347(sh, 3.5), 384(3.7); the lowest energy absorption band ($\lambda_{max}$=384 nm) has been tentatively assigned to a $3a_2$ (HOMO)-$5b_1$ (LUMO) excitation [2]. The absorption spectrum in $CH_3CN$ (given in the paper) shows three bands at 252, 291, and 379 nm and two shoulders at 233 and 347 nm. The Magnetic Circular Dichroism (MCD) spectrum in $CH_3CN$ (given in the paper) shows five bands located at these energies and reveals a sixth band at about 208 nm. The linear dichroism spectrum in a stretched (~75%) polyethylene sheet at room temperature shows that the transitions at 233 and 379 nm are polarized perpendicular and the transitions at 252, 291, and 347 nm are polarized parallel to the orientation axis. The UV and MCD spectra have been discussed in relation to the perimeter model and $\pi$-electron calculations [4]. $^1$H NMR spectrum ($CDCl_3$/TMS): AA′BB′ multiplet (18 lines detected), $\delta_A$=7.78, $\delta_B$=7.29 ppm, $J_{AA'}$=0.54, $J_{AB'}$=1.17, $J_{BB'}$=6.80, $J_{AB}$=8.70 Hz [1]. Another $^1$H spectrum is reported [2], but solvent and standard are not given. The (relatively high) ratio of the J values indicate a greater bond localization into the quinoid structure [2]. $^{13}$C NMR spectrum ($CDCl_3$/TMS): $\delta$=121.8, 123.0, 147.2 ppm [1], 121.7, 122.8, 147.0 ppm [2].

Electron impact mass spectrum (70 eV): m/e = 200 $M^+$ (79%), 154 $M^+$–NS (100), 108 $M^+$–$S_2N_2$ (28) [2], see also [1].

Electrochemical oxidation at a rotating platinum electrode and electrochemical reduction at a dropping mercury electrode carried out in $CH_3CN$ with 0.1M $(C_2H_5)_4NClO_4$ as supporting electrolyte give the half wave potentials $E_{1/2}$(ox) = + 1.36 V and $E_{1/2}$(red) = − 0.83 V referenced to the saturated calomel electrode.

The compound is extremely stable with respect to both thermolysis and hydrolysis; it is also air-stable [2]. The compound is decomposed at room temperature by $P(C_6H_5)_3$ in toluene or by 3-chloroperbenzoic acid in $CH_2Cl_2$ [1]. With an excess of norbornadiene in ether, a 1:1 addition compound is formed [2].

**References:**


[1] Morris, J. L., Rees, C. W., Rigg, D. J. (J. Chem. Soc. Chem. Commun. **1985** 396/7).

[2] Cordes, A. W., Hojo, M., Koenig, H., Noble, M. C., Oakley, R. T., Pennington, W. T. (Inorg. Chem. **25** [1986] 1137/45).

[3] Jones, R., Morris, J. L., Potts, A. W., Rees, C. W., Rigg, D. J., Rzepa, H. S., Williams, D. J. (J. Chem. Soc. Chem. Commun. **1985** 398/400).

[4] Klein, H.-P., Oakley, R. T., Michl, J. (Inorg. Chem. **25** [1986] 3194/201).


### 10.2.4 [1,2,3]Dithiazolo[5,4-g]-2,4,1,3,5-benzodithiatriazepine-2-$S^{IV}$

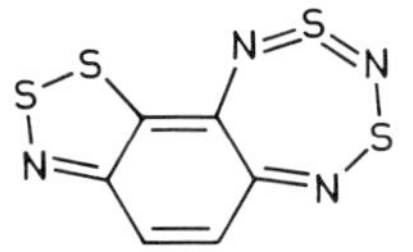

**[=S=N–S–N={$C_6H_2NS_2$}–N=]**

The heterocycle was formed when a solution of the bifunctional 4-phenylene bis(sulfur diimide), $(CH_3)_3SiN{=}S{=}N(C_6H_4)N{=}S{=}NSi(CH_3)_3$, in $CH_2Cl_2$ was added over 5 h to a stirred solution of an equimolar amount of $SCl_2$ in $CH_2Cl_2$. After 24 h the mixture was worked up to give metallic golden needles in 1% yield; m.p. 254 to 257°C (recrystallized from hot toluene).

An X-ray diffraction study shows that crystals of the compound are orthorhombic, space group Pbca-$D_{2h}^{15}$ (No. 61), with a = 7.1334(8), b = 14.761(2), and c = 16.677(2) Å; Z = 8. V = 1756.0(3) $Å^3$. $D_x$ = 1.95 g/$cm^3$. R = 0.044, $R_w$ = 0.027 for 1203 reflections [1].

The molecular structure of the compound is shown in **Fig. 23**.

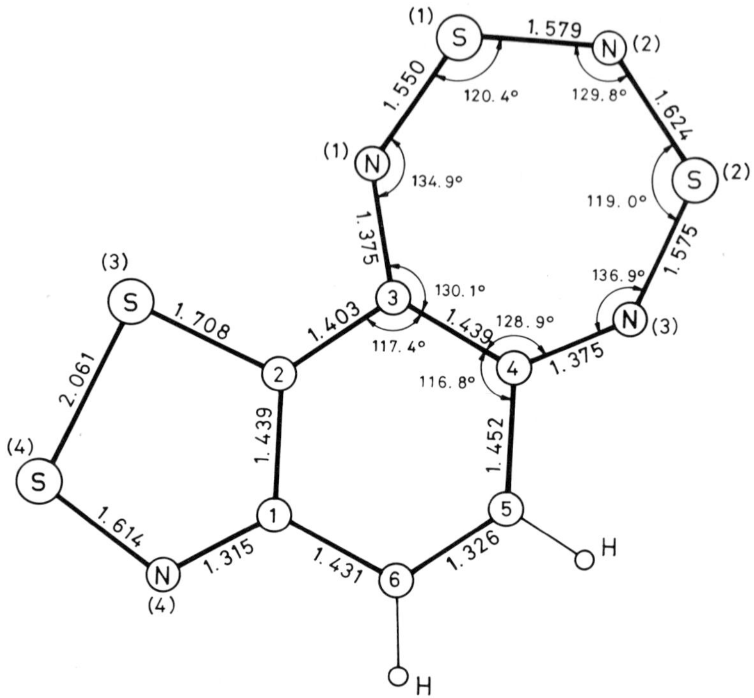


Fig. 23. Molecular structure of [=S=N–S–N={$C_6H_2NS_2$}–N=] with selected bond lengths in Å and bond angles in °.

The molecule is essentially planar (σ = 0.001 Å), the two largest deviations from the mean molecular plane occuring at S(3) (0.030 Å) and N(1) (0.027 Å). The slight buckling of the perimeter at these two points probably reflects the rather short nonbonded contact S(3)···N(1) (2.850(4) Å), which is considerably shorter than the sum of van der Waals' radii for sulfur and nitrogen (3.35 Å). The structural parameters within the rings indicate a delocalized π system whose ground state cannot be adequately represented by a single valence bond structure such as (A). Other likely resonance contributors are (B) and (C).

A B C D

The dipolar formulation (D) provides perhaps the most useful basis for describing the principal features of the molecular structure. According to (D) the molecule consists of a cyclic 5-atom 6 π-electron dithiazolium cation fused to a cyclic 7-atom 10 π-electron $S_2N_3C_2^-$ anion by means of the C(2)-C(3) bond. The π system of this anion can be viewed as isoelectronic with the thiotriazyl cation $S_4N_3^+$. The mean S–N distances in the seven-membered ring closely resemble those found in $S_4N_3^+$ salts. Principal bond lengths and bond angles of the molecule are given in Fig. 23; for the atomic coordinates, see the paper. The arrangement of the molecules in the unit cell provides further evidence in favor of the dipolar representations (B), (C), and (D). The molecules form stacks in the a direction, the packing being such that the five-membered $S_2NC_2$ ring of one molecule lies approximately over the seven-membered $S_2N_3C_2$ ring of the molecule below it. This pattern repeats itself throughout the stack, producing an alternating sequence of molecules, i.e., the positively charged end of one molecule is sandwiched between the negative ends of its two nearest neighbors, and vice versa. Thus, the structure represents an example of a self-complexing molecule. The planes of consecutive molecules in the stack are parallel and make an angle of 18° with the a axis. The interplanar separation is uniform at 3.39 Å, i.e., there is no pairing of the molecules [1].

Extended Hückel (EH) molecular orbital calculations on a planar projection of the molecule refine the valence bond picture and substantiate the charge distribution suggested by (D). Thus, a summation of the EH $\pi$ charge densities over the nine occupied $\pi$ orbitals (see **Fig. 24**) reveals a total $\pi$ charge density on the $S_2NC_2$ and $S_2N_3C_2$ rings of 6.4 and 9.7 $\pi$ electrons, respectively, i.e., there is a net $\pi$ charge transfer of about 0.6 electrons from the small to the large ring [1].

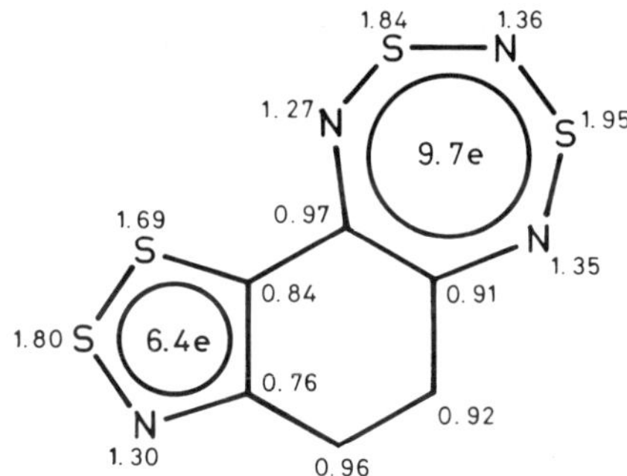


Fig. 24. Extended Hückel $\pi$ charge densities. Circled numbers represent total $\pi$ charge densities in the different rings.

UV-visible spectrum ($CH_2Cl_2$): $\lambda_{max}$ (in nm (log $\varepsilon$)) = 342(4.0), 366(4.0), 582(4.0), 625(4.0) [1]. A reinvestigation gave bands at 352 and 585 nm and shoulders at 301, 318, 385, and 714 nm [2]. The MCD spectrum clearly reveals the presence of eight bands. The first six bands are centered at the above energies, and the last two bands are located near 242 and 265 nm, respectively. The UV [1, 2] and MCD [2] spectra (given in the papers) have been discussed in relation to the perimeter model [1, 2].

Whereas the structural evidence suggests the possibility of intermolecular charge-transfer interactions in the solid state, there are no indications of such phenomena in dilute solution. The deep blue color of the compound in solution differs markedly from that in the solid state (metallic bronze), and the visible spectrum shows no deviations from the Beer-Lambert law [1].

**References:**

[1] Codding, P. W., Koenig, H., Oakley, R. T. (Can. J. Chem. **61** [1983] 1562/6).
[2] Klein, H.-P., Oakley, R. T., Michl, J. (Inorg. Chem. **25** [1986] 3194/201).

## 10.3 $S_2N_3C_2$ Ring

### 10.3.1 1,3,2-Dithiazolo[4,5-e][1,3,2,4,7]dithiatriazepine

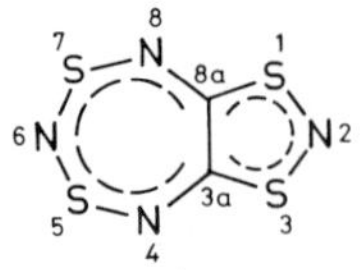

[-S-N-S-N-{C-S-N-S-C}-N-] (Reduced formula in the text $S_4N_4C_2$)

The compound was formed in 24% yield by heating $S_4N_4$ with $C_6H_5S(O)CH{=}CH_2$ (2:1 mole ratio) in toluene for 6 h, along with the main products sulfur und diphenyl disulfide. Replacing $C_6H_5S(O)CH{=}CH_2$ with $C_6H_5SO_2CH{=}CH_2$ gave the product in 4% yield. A possible mechanism of formation is discussed in the paper.

An X-ray diffraction study showed the crystals to be monoclinic, space group $P2_1/a$ (standard setting $P2_1/c$)-$C^5_{2h}$ (No. 14), with a = 7.144(1), b = 16.826(2), c = 11.271(2) Å, and β = 105.67(1)°; Z = 8. V = 1304 $Å^3$, $D_x$ = 2.13 g/cm³. R = 0.033, $R_w$ = 0.038. The molecular structure of the compound is shown in **Fig. 25,** the stacking of the molecules in the crystal in **Fig. 26**. The compound contains two crystallographically independent molecules, A and B. Both molecules are essentially planar with approximate $C_{2v}$ symmetry, and all the equivalent bond lengths are the same within statistical significance. The S–N bonds are all of similar length, though those of the five-membered ring are slightly longer than those of the seven-membered ring. Overall this indicates a completely delocalized 14 π structure, in accord with MNDO and ab initio MO calculations which show a distinct pattern of π energy levels charactistic of an aromatic species.

The molecules form continuous, parallel, overlapping stacks (Fig. 26), each stack being comprised of only one type of molecule (either all A or all B). Furthermore the mode of stacking for type A molecules is different from that for type B. Adjacent molecules in type A stacks are rotated by ~80° in plane with respect to each other, whereas those of type B are rotated by 180°. The minimum interplanar atomic separations in the A stacks are 3.26 Å between C(8a) and N(4) and 3.48 Å between C(8a) and N(6); in the B stacks they are 3.46 Å between N(4′) and S(1′) and 3.50 Å between C(3a′) and S(1′). The atoms in type A have a noticeably larger maximum deviation from their least-squares plane (0.027 Å for N(8)) than those in type B (0.007 Å for S(3′)). Bond lengths are given in Fig. 25, bond angles in Table 48.

Table 48
Bond Angles of $S_4N_4C_2$.

| bond angle in ° | A | B | bond angle in ° | A | B |
|---|---|---|---|---|---|
| N(2)–S(1)–C(8a) | 101.8(2) | 101.6(2) | N(4)–S(5)–N(6) | 118.7(2) | 118.4(2) |
| S(1)–N(2)–S(3) | 113.8(2) | 114.2(2) | S(5)–N(6)–S(7) | 133.6(2) | 133.2(2) |
| N(2)–S(3)–C(3a) | 101.6(2) | 101.7(2) | N(6)–S(7)–N(8) | 118.8(2) | 119.1(2) |
| S(3)–C(3a)–N(4) | 116.7(2) | 117.0(3) | S(7)–N(8)–C(8a) | 131.9(3) | 132.8(3) |
| S(3)–C(3a)–C(8a) | 111.6(3) | 111.3(2) | S(1)–C(8a)–C(3a) | 111.2(2) | 111.2(3) |
| N(4)–C(3a)–C(8a) | 131.8(3) | 131.7(3) | S(1)–C(8a)–N(8) | 116.0(3) | 117.1(2) |
| C(3a)–N(4)–S(5) | 132.3(2) | 133.1(3) | C(3a)–C(8a)–N(8) | 132.8(3) | 131.7(3) |

The compound, a green-black crystalline solid with a metallic luster, melts at 142 to 143°C. The solution IR spectrum is very simple with only four absorptions above 600 cm⁻¹ at 1488,

895, 885, and 625 $cm^{-1}$. The UV spectrum, with long wavelength maxima at $\lambda_{max}=322$ nm (log $\varepsilon=3.35$) and 492 nm (3.31), suggests a highly symmetrical and delocalized structure. $^{13}C$ NMR spectrum: $\delta=161.8$ ppm(s).

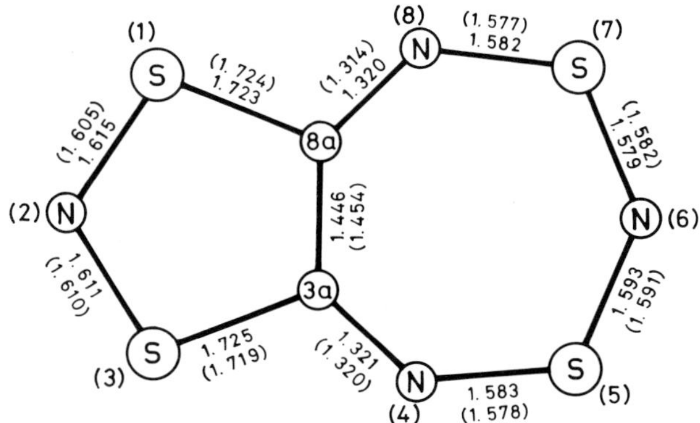


Fig. 25. Molecular structure of $S_4N_4C_2$. Bond lenghts in Å; the values for the second molecule are given in parentheses.

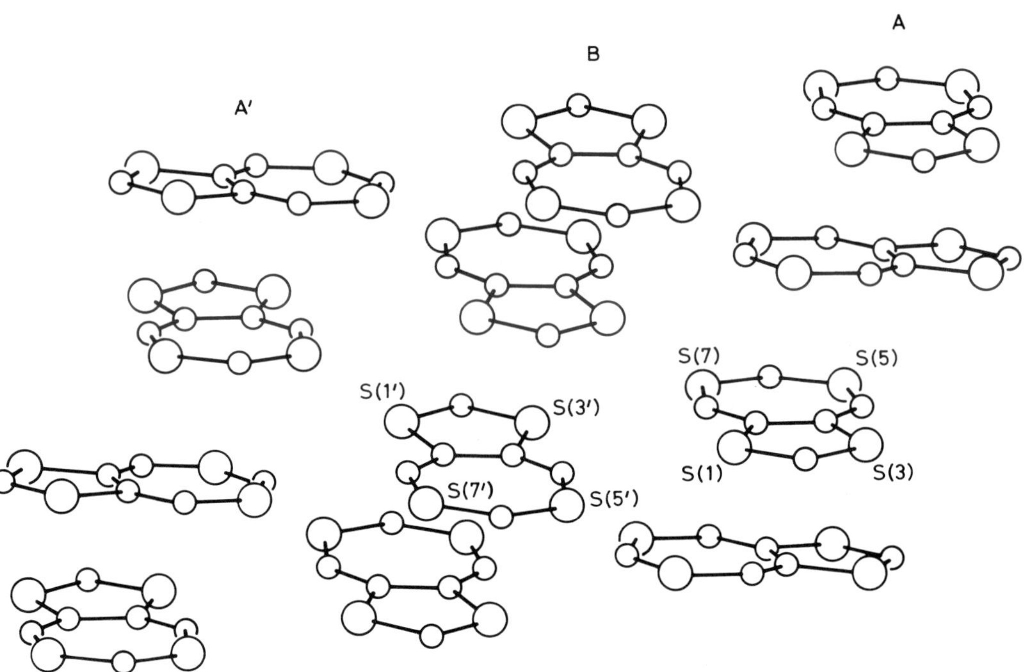


Fig. 26. The stacking of the molecules A and B in the crystal.

The compound, once isolated, is stable in boiling benzene, but it is extensively decomposed under these conditions in the presence of either $S_4N_4$ or $C_6H_5S(O)CH{=}CH_2$. The compound does not form charge-transfer complexes with electron acceptors like picric acid or 2,4,7-trinitro-9-fluorenylidenemalononitrile.

**Reference:**

Jones, R., Morris, J. L., Rees, C. W., Williams, D. J. (J. Chem. Soc. Chem. Commun. **1985** 1654/6).

## 10.4 $SN_2C_4$ Ring

### 10.4.1 3,4,5,6-Tetrahydro-1,2,7-thia(1-$S^{IV}$)diazepine

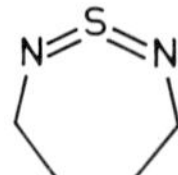


**[=S=N–$(CH_2)_4$–N=]** (Reduced formula in the text $SN_2(CH_2)_4$)

Monocyclic $SN_2C_n$ rings (n = 4, 5, 6) are reported to form in the reaction of $(4\text{-}CH_3C_6H_4SO_2\text{–}N\text{=})_2S$ with diamines $H_2N(CH_2)_nNH_2$ (1:1 mole ratio) in dry $CCl_4$ under an $N_2$ atmosphere. Thus $SN_2(CH_2)_4$ forms in 34% yield in the reaction of $(4\text{-}CH_3C_6H_4SO_2\text{–}N\text{=})_2S$ with $H_2N(CH_2)_4NH_2$.

$SN_2(CH_2)_4$ as well as $SN_2(CH_2)_5$ and $SN_2(CH_2)_6$ cannot be purified by distillation [1, 2]. The products polymerize, even under vaccum. They were purified only by shaking the $CCl_4$ solution with water; evaporation of the water yields yellow oils. Elemental analysis was not possible because the substances decompose slowly at room temperature and spontaneously at higher temperatures.

The yellow oils are not sensitive toward water. With dilute acids or bases the corresponding diamines form. $^1H$ NMR (broad signals) and IR spectra are listed in the paper [2].

**References:**

[1] Kresze, G., Grill, H. (Tetrahedron Letters **1969** 4117/20, 4118).
[2] Grill, H. (Diss. T.H. München 1970, pp. 1/91, 29/37, 74/7).

## 10.5 $SN_2C_3O$ Ring

### 10.5.1 5-Benzoyl-2,6-diphenyl-6,7-dihydro-5H-1,4,3,5-oxathiadiazepine 4-Oxide

**[–SO–N($COC_6H_5$)–CH($C_6H_5$)–$CH_2$–O–C($C_6H_5$)=N–]**

The title compound is obtained in 41% yield from the reaction (1 h) of $C_6H_5CONSO$ with $C_6H_5$–$\underline{CHCH_2O}$ (1:1 mole ratio) in refluxing benzene in the presence of $(C_2H_5)_4NBr$. The colorless prisms melt at 171 to 172°C with decomposition.

The $^1H$ NMR spectrum ($CDCl_3$/TMS) exhibits three doublets of doublets at δ (in ppm) = 4.75 ($H^3$), 5.64 ($H^2$), and 6.25 ($H^1$), with coupling constants $J(H^2H^3) = 8$, $J(H^1H^2) = 2$, and $J(H^1H^3) = 7$ Hz, besides a signal from the aromatic protons (15 H).

IR spectrum (KBr): ν = 1660 (CO), 1630 $cm^{-1}$ (C=N).

Mass spectrum (70 eV): m/e = 390 $M^+$, 389 $M^+$–H, 342 $M^+$–SO, 238 $M^+$–SO–$C_6H_5CH{=}CH_2$, 223 $M^+$–$C_6H_5CONSO$ (base peak), 193 $M^+$–$C_6H_5CONSO$–$CH_2O$, and 105 $C_6H_5CO^+$.

Hydrolysis of the compound with HCl in methanol and the reaction with $N_2H_4 \cdot H_2O$ in ethanol yield $C_6H_5CONH_2$ and $C_6H_5CONHC(C_6H_5){=}CHCH_2OH$.

**Reference:**

Tsuge, O., Mataka, S. (Bull. Chem. Soc. Japan **44** [1971] 2836/9).

# 11 Eight Atom S–N–C Ring Systems

## 11.1 $S_3N_4C$ Ring

### 11.1.1 Salt of 7-Trifluoromethyl-3H-1,3,5,2,4,6,8-trithia(5-$S^{IV}$)tetrazocine Ion(1+)

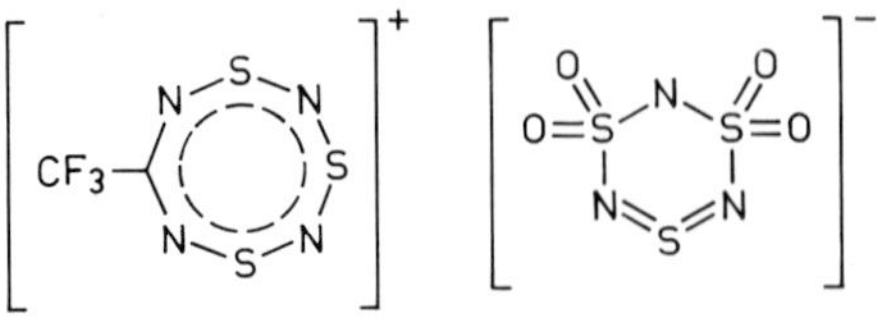

**[$\overset{..}{-}$(S$\overset{..}{-}$N)$_3$$\overset{..}{-}$C($CF_3$)$\overset{..}{-}$N$\overset{..}{-}$]$^+$$S_3N_3O_4^-$** (Reduced formula in the text $S_3N_4C(CF_3)^+S_3N_3O_4^-$)

The salt is formed from the complex reaction of $S_3N_3Cl_3$ with $CF_3CN$ and $SO_2$ (~1:3:8 mole ratio) at 55°C in a steel autoclave. After 40 h the volatile products were sublimed at 120°C ($10^{-5}$ mbar), and the residue was washed several times with $SO_2$–$SO_2ClF$. Recrystallization from $CH_2Cl_2$ yields the salt in a yield of 4% (related to $S_3N_3Cl_3$).

Crystals of the salt are monoclinic, space group $P2_1/c$–$C_{2h}^5$ (No. 14), with a = 14.590(1), b = 6.8093(6), c = 13.842(2) Å, and β = 96.491(5)°; Z = 4. V = 1366 Å$^3$, $D_x$ = 2.117 g/cm$^3$. R = 0.0379, $R_w$ = 0.0469 for 2297 independent reflections. The molecular structure of the cation is shown in **Fig. 27**.

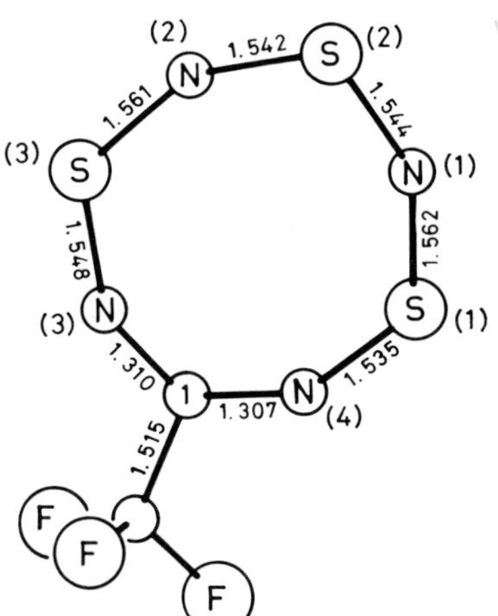


Fig. 27. Molecular structure of the $S_3N_4C(CF_3)^+$ cation. Bond lengths in Å.

The $S_3N_4C(CF_3)^+$ cation consists of a nearly planar eight-membered ring with C(1), S(2) and S(1), S(3) lying only a little above and below the plane through the four N atoms (by 0.062, 0.072, −0.077, −0.090 Å, respectively). This slight folding at S(1)/S(3), however, does not lead to a transannular S–S interaction (S···S = 4.008 Å).

The light ocherous solid melts at 111°C with decomposition. For the IR spectrum (Nujol), 29 unassigned absorption bands have been recorded. $^{19}F$ NMR spectrum (liquid $SO_2$/internal $CFCl_3$): $\delta(CF_3) = -67.03$ ppm.

**Reference:**

Höfs, H.-U., Hartmann, G., Mews, R., Sheldrick, G. M. (Angew. Chem. **96** [1984] 1001/2; Angew. Chem. Intern. Ed. Engl. **23** [1984] 988/9).

### 11.1.2 3,7-Disubstituted 3H-1,3,5,2,4,6,8-Trithiatetrazocine

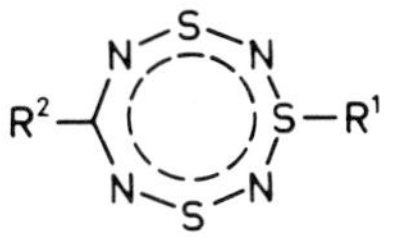

$R^1 = Cl$; $R^2 = N(CH_3)_2$, t-$C_4H_9$
$R^1 = N{=}P(C_6H_5)_3$, $N{=}As(C_6H_5)_3$; $R^2 = C_6H_5$

**[∸S∸N∸SR¹∸N∸S∸N∸CR²∸N∸]** (Reduced formula in the text $S_3N_4CR^1R^2$)

**Survey**

The formation reactions of the eight-membered ring compound are shown in the following scheme; details are given for specific compounds.

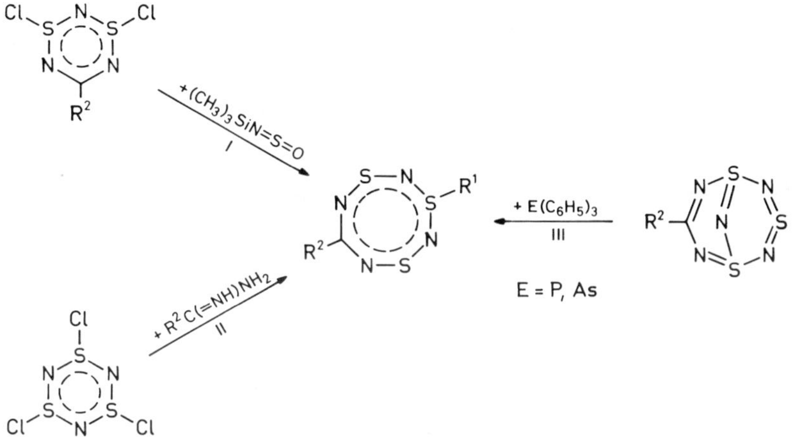

+ $(CH_3)_3SiN{=}S{=}O$ I

+ $R^2C({=}NH)NH_2$ II

+ $E(C_6H_5)_3$ III

E = P, As

In the neutral $S_3N_4C$ ring compounds (but not in the ion, see p. 126) the distance between the two opposite S atoms is small (2.378 to 2.432 Å); the ring is folded along this S···S bridge. The conformation of the $S_3N_2$ part of the molecule depends on $R^1$. It is nearly planar with $R^1 = Cl$; it is envelope-shaped with $R^1 = N{=}P(C_6H_5)_3$ or $N{=}As(C_6H_5)_3$. The $S_2N_2C$ part of the molecule is almost planar. Details are given in the description of specific compounds.

**$S_3N_4CCl(N(CH_3)_2)$**

The eight-membered ring compound was prepared in 32% yield [1] from the reaction of the six-membered ring compound $S_2N_3CCl_2(N(CH_3)_2)$ (see p. 9) with $(CH_3)_3SiN{=}S{=}O$ (1:2 mole ratio) in $CCl_4$ at 23°C. After 24 to 27 days, the solvent was removed in vacuum to leave a solid residue which was extracted with $CH_3CN$ and cooled at −30°C for 3 days to give yellow crystals of the title compound [1, 2]. The mechanism of formation was discussed [1].

X-ray diffraction studies show the crystals to be monoclinic, space group $P2_1/n$ (standard setting $P2_1/c$)–$C^5_{2h}$ (No. 14), with a = 8.938(2), b = 11.947(1), c = 8.485(1) Å, and β = 91.387(8)°; Z = 4. V = 905.7(2) $Å^3$, $D_x$ = 1.79 $g/cm^3$. R = 0.045, $R_w$ = 0.041 for 834 reflections. The final atomic coordinates of the non-hydrogen atoms are given in the paper [1]. The molecular structure of the compound with selected bond lengths and bond angles is shown in **Fig. 28**, p. 128.

The molecule has a folded structure with a cross-ring S–S distance of 2.432(3) Å and the chlorine atom in an endo position. The planar geometry at C(1) and N(5) and the short C(1)–N(5) distance (1.31(1) Å) indicate strong π bonding between the exocyclic nitrogen and

C(1). The $S_3N_4C$ ring system can be considered as a bicyclic system. The five-membered $S_2N_2C$ ring is almost planar, the largest deviation being 0.09 Å for the carbon atom. Larger deviations from planarity are observed for the $S_3N_2$ ring with S(2) being 0.17 Å above and N(1) and N(2) 0.13 Å below the plane. The variation in S–N bond distances is consistent with some localization of positive charge on S(2) [1], see also [2].

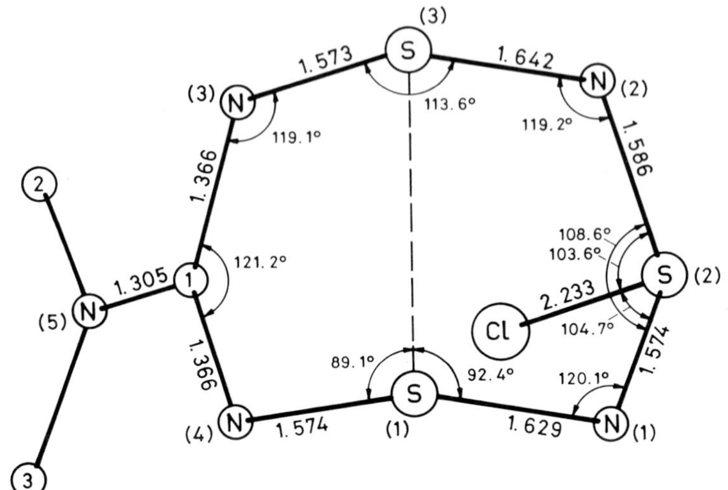


Fig. 28. Molecular structure of $S_3N_4CCl(N(CH_3)_2)$. Bond lengths in Å, bond angles in °.

For the IR spectrum (Nujol), 17 absorption bands between 1590 and 320 $cm^{-1}$ are listed in the paper [1]. $^1H$ NMR spectrum ($CDCl_3$/TMS): δ = 3.20 [1], 3.17 ppm [2].

### $S_3N_4CCl(C_4H_9\text{-}t)$

The compound is formed from the reaction of 2,2-dimethylpropanamidine, t-$C_4H_9C(=NH)NH_2$, with $S_3N_3Cl_3$. Yellow needles result by recrystallization from $CH_3CN$.

The compound crystallizes in the tetragonal space group $P4_2/n\text{-}C_{4h}^4$ (No. 86), with a = 17.856(2) and c = 6.646(1) Å; Z = 8. V = 2118.9(8) $Å^3$, $D_x$ = 1.61 $g/cm^3$. R = 0.049 for 1201 reflections at T = 294 K.

Atomic coordinates and equivalent isotropic temperature factors are given in the paper; for selected bond lengths and bond angles, see Table 49. The molecular structure of the compound is shown in **Fig. 29**.

The molecule has a folded butterfly structure, with almost planar SNCNS and SNSNS fragments, and the Cl atom is endo. The transannular S–S contact distance (2.378(2) Å) is the shortest such distance in any thiazyl compound [3].

### $S_3N_4C(N{=}P(C_6H_5)_3)(C_6H_5)$

The reaction of triphenylphosphine with an equimolar amount of $S_3N_5C(C_6H_5)$ (see p. 149) in toluene afforded an orange solution which slowly faded during 24 h to pale yellow and produced a yellow crystalline precipitate of the compound in an overall yield of 91%. It melts at 194 to 196°C. $^{31}P$ NMR spectrum: δ = 21.4 ppm. The compound crystallizes in the monoclinic space group $P2_1/c\text{-}C_{2h}^5$ (No. 14), with a = 13.957(3), b = 9.242(3), c = 19.473(4) Å, and β =

102.80(2)°; Z=4. V=2449.4 Å$^3$ and $D_x$=1.40 g/cm$^3$. R=0.038 for 2430 observed reflections. The $S_3N_4C$ ring is folded by 114.5°, with an almost planar SNCNS fragment and an endo-S-envelope SNSNS fragment. The $(C_6H_5)_3P{=}N$ substituent at S is endo and the transannular S···S contact is 2.415(1) Å, similar to the structure of the endo-$(C_6H_5)_3As{=}N$ substituted compound, see Fig. 30a, p. 131. Selected bond lengths and bond angles are given in Table 50, p. 130 [5].

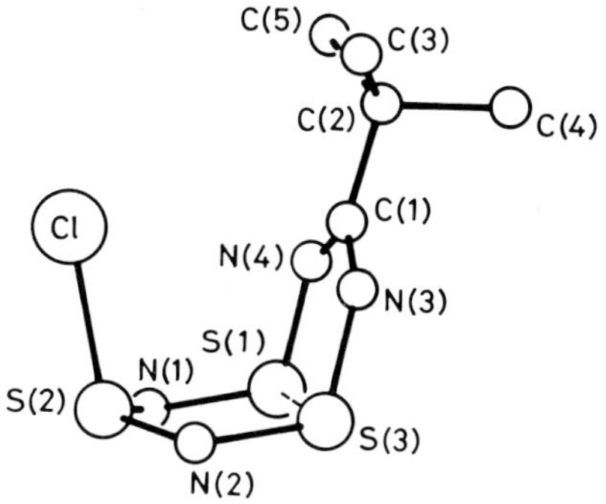

Fig. 29. Molecular structure of $S_3N_4CCl(C_4H_9\text{-t})$.

Table 49

Selected Bond Lengths and Bond Angles of $S_3N_4CCl(C_4H_9\text{-t})$. Estimated standard deviations in parentheses.

| bond length in Å | | bond angle in ° | |
|---|---|---|---|
| S(1)–N(1) | 1.636(4) | N(1)–S(2)–N(2) | 109.0(2) |
| S(1)–N(4) | 1.605(3) | N(1)–S(1)–N(4) | 112.7(2) |
| S(2)–N(1) | 1.579(4) | N(2)–S(3)–N(3) | 112.5(2) |
| S(2)–N(2) | 1.575(4) | S(1)–N(1)–S(2) | 118.9(2) |
| S(3)–N(2) | 1.633(3) | S(1)–N(4)–C(1) | 119.3(3) |
| S(3)–N(3) | 1.596(3) | S(2)–N(2)–S(3) | 119.6(2) |
| N(3)–C(1) | 1.337(4) | S(3)–N(3)–C(1) | 118.3(3) |
| N(4)–C(1) | 1.323(4) | N(3)–C(1)–N(4) | 123.6(3) |
| S(1)···S(3) | 2.378(2) | N(3)–C(1)–C(2) | 117.1(3) |
| S(2)–Cl | 2.218(2) | N(4)–C(1)–C(2) | 119.1(3) |
| C(1)–C(2) | 1.506(5) | Cl–S(2)–N(1) | 103.7(1) |
| | | Cl–S(2)–N(2) | 105.2(2) |

**$S_3N_4C(N{=}As(C_6H_5)_3)(C_6H_5)$**

The compound is formed when $S_3N_5C(C_6H_5)$ (see p. 149) is warmed with an equimolar amount of triphenylarsine in acetonitrile. The yellow crystalline precipitate of $S_3N_4C(N{=}As(C_6H_5)_3)(C_6H_5)$ is formed immediately. Yield 87%. The same material is produced when the reaction is performed in toluene at room temperature; however, a second product, an orange-red crystalline material with the same elemental composition, is also obtained in low (<5%) yield. The two products (yellow and orange-red) can be separated manually.

From a single crystal X-ray analysis of each product it was established that the orange-red and yellow solids are, respectively, the exo and endo isomers of $S_3N_4C(N{=}As(C_6H_5)_3)(C_6H_5)$ [4]. The crystal data of the two isomers are given in Table 51, p. 130.

Table 50

Selected Bond Lengths and Bond Angles of $S_3N_4C(N{=}P(C_6H_5)_3)(C_6H_5)$ [5].

| bond length in Å | | bond angle in ° | |
|---|---|---|---|
| S(1)–N(1) | 1.589(3) | S(1)–N(1)–S(2) | 118.9(2) |
| S(1)–N(4) | 1.629(3) | S(1)–N(4)–C(1) | 118.2(3) |
| S(2)–N(1) | 1.638(3) | S(2)–N(2)–S(3) | 119.0(2) |
| S(2)–N(2) | 1.647(3) | S(3)–N(3)–C(1) | 117.5(2) |
| S(3)–N(2) | 1.591(3) | N(1)–S(1)–N(4) | 114.6(1) |
| S(3)–N(3) | 1.630(3) | N(1)–S(2)–N(2) | 101.6(2) |
| N(3)–C(1) | 1.334(4) | N(2)–S(3)–N(3) | 115.0(2) |
| N(4)–C(1) | 1.323(4) | N(3)–C(1)–N(4) | 125.9(3) |
| S(1)···S(3) | 2.415(1) | N(1)–S(2)–N(5) | 106.6(1) |
| S(2)–N(5) | 1.612(2) | N(2)–S(2)–N(5) | 107.1(1) |
| P–N(5) | 1.601(2) | P–N(5)–S(2) | 121.0(2) |
| C(1)–C(phenyl) | 1.481(4) | N–C(1)–C(phenyl)(avg) | 117(3) |

Table 51

Crystal Data of endo- and exo-$S_3N_4C(N{=}As(C_6H_5)_3)(C_6H_5)$ [4].

| | endo-$S_3N_4C(N{=}As$-$(C_6H_5)_3)(C_6H_5)$ | exo-$S_3N_4C(N{=}As$-$(C_6H_5)_3)(C_6H_5)$ |
|---|---|---|
| symmetry | monoclinic | triclinic |
| space group | $P2_1/c$-$C^5_{2h}$ (No. 14) | $P\bar{1}$-$C^1_i$ (No. 2) |
| a in Å | 13.841(2) | 9.133(1) |
| b in Å | 9.499(2) | 11.700(3) |
| c in Å | 19.655(2) | 14.439(3) |
| α in ° | — | 89.87(2) |
| β in ° | 104.54(2) | 84.93(1) |
| γ in ° | — | 82.49(2) |
| Z | 4 | 2 |
| V in Å$^3$ | 2501 | 1524 |
| $D_x$ in g/cm$^3$ | 1.49 | 1.42 |
| R | 0.033 | 0.073 |
| $R_w$ | 0.040 | 0.096 |
| number of reflections used | 2646 | 3088 |

The molecular structures of the two isomers are shown in **Fig. 30**. The small differences observed between the corresponding bonds in the two isomers mirror almost exactly the differences between the two halves of exo- and endo-$S_4N_4(N{=}P(C_6H_5)_3)_2$ (see "Sulfur-Nitrogen Compounds" 2, 1985, p. 240). The intramolecular dimensions are generally similar to those reported for $S_3N_4CCl(N(CH_3)_2)$ (see p. 128) and $S_3N_4CCl(C_4H_9$-t) (see p. 129). The exocyclic S–N and N–As distances are comparable to those found in $S_3N_3$–N=As$(C_6H_5)_3$ (see "Sulfur-Nitrogen Compounds" 2, 1985, p. 85). Principal bond lengths and bond angles of the two isomers are given in Table 52.

Table 52

Selected Bond Lengths and Bond Angles of exo- and endo-$S_3N_4C(N{=}As(C_6H_5)_3)(C_6H_5)$ [4].

| | bond length in Å | | | bond angle in ° | |
|---|---|---|---|---|---|
| | endo | exo | | endo | exo |
| S(1)–N1() | 1.589(4) | 1.574(6) | S(1)–N(1)–S(2) | 118.4(2) | 116.9(3) |
| S(1)–N(4) | 1.630(4) | 1.655(6) | S(1)–N(4)–C(1) | 118.5(3) | 118.1(5) |
| S(2)–N(1) | 1.647(4) | 1.683(6) | S(2)–N(2)–S(3) | 118.6(2) | 115.4(3) |
| S(2)–N(2) | 1.651(4) | 1.703(6) | S(3)–N(3)–C(1) | 117.4(3) | 117.0(5) |
| S(3)–N(2) | 1.519(4) | 1.574(6) | N(1)–S(1)–N(4) | 114.6(2) | 114.0(3) |
| S(3)–N(3) | 1.631(4) | 1.649(6) | N(1)–S(2)–N(2) | 100.8(2) | 99.2(3) |
| N(3)–C(1) | 1.342(5) | 1.340(9) | N(1)–S(2)–N(5) | 106.7(2) | 105.7(3) |
| N(4)–C(1) | 1.321(5) | 1.319(8) | N(2)–S(3)–N(3) | 115.6(2) | 114.0(3) |
| S(1)···S(3) | 2.419(2) | 2.420(3) | N(3)–C(1)–N(4) | 125.6(4) | 126.5(7) |
| S(2)–N(5) | 1.612(3) | 1.576(6) | N(2)–S(2)–N(5) | 108.1(2) | 104.9(3) |
| As–N(5) | 1.755(3) | 1.769(5) | As–N(5)–S(2) | 115.9(1) | 120.7(3) |
| As–C(avg) | 1.911(9) | 1.91(1) | N–As–C(avg) | 110(5) | 111(8) |
| | | | C–As–C(avg) | 109(2) | 108(1) |

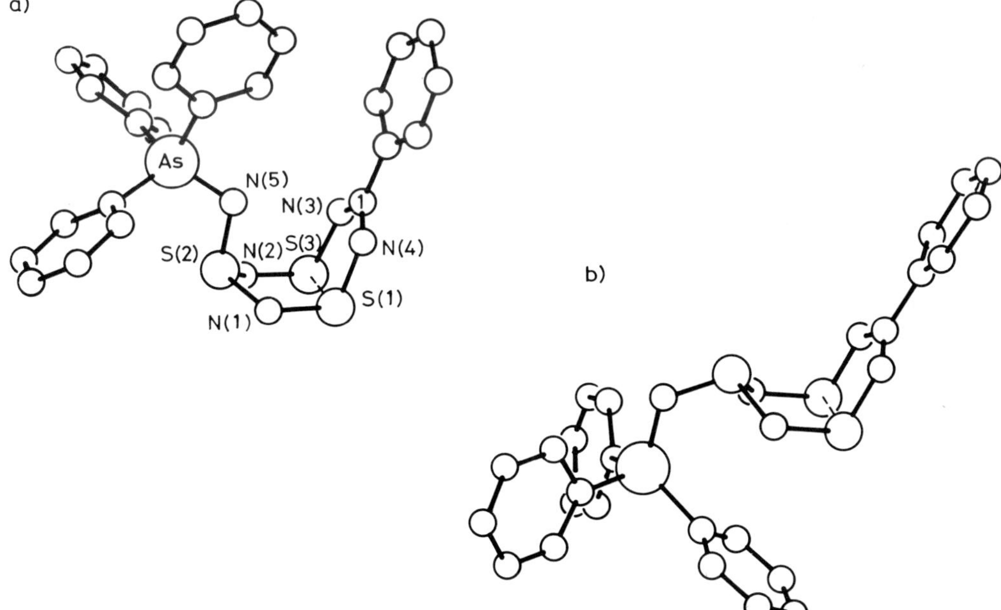


Fig. 30. Molecular structure of $S_3N_4C(N{=}As(C_6H_5)_3)(C_6H_5)$; a) endo isomer, b) exo isomer.

The orange-red crystals of the exo isomer melt at 158 to 159°C with decomposition. UV spectrum ($CH_2Cl_2$): $\lambda_{max}$ = 422 nm ($\varepsilon = 1 \times 10^3$ L·mol$^{-1}$·cm$^{-1}$). The yellow crystals of the endo isomer melt at 163 to 165°C with decomposition. When the exo isomer is dissolved in $CH_3CN$ at room temperature, the red color of the solution slowly fades, and a yellow precipitate of the endo isomer (identified by IR analysis) is formed. This irreversible conversion establishes that the endo isomer is more thermodynamically stable. The occurrence of the isomerization at room temperature suggests a low activation energy; however, the nature of the transition state is by no means obvious [4].

**Refences:**


[1] Chivers, T., Richardson, J. F., Smith, N. R. M. (Inorg. Chem. **25** [1986] 272/5).

[2] Chivers, T., Richardson, J. F., Smith, N. R. M. (Mol. Cryst. Liquid Cryst. **125** [1985] 319/27).

[3] Boeré, R. T., Oakley, R. T., Cordes, A. W. (Acta Cryst. C **41** [1985] 1686/7).

[4] Boeré, R. T., Cordes, A. W., Craig, S. L., Graham, J. B., Oakley, R. T., Privett, J. A. J. (J. Chem. Soc. Chem. Commun. **1986** 807/8).

[5] Boeré, R. T., Ferguson, G., Oakley, R. T. (Acta Cryst. C **42** [1986] 900/2).


## 11.2 $S_4N_2C_2$ Ring

### 11.2.1 6a,7,10,10a-Tetrahydro-7,10-methano-1,2,4,6,3,5-benzotetrathiadiazocine-4-$S^{IV}$

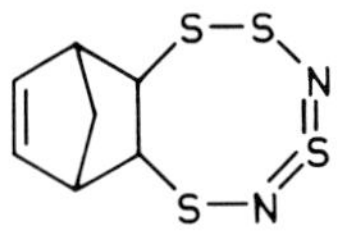

**[=S=N–S–{2,3-$C_7H_8$}–S–S–N=]**

The title compound forms as an oil by the reaction of equimolar amounts of $S_4N_2$ and norbornadiene in $CH_2Cl_2$ for 10 h, followed by evaporation to dryness. The oil was purified by chromatography and recrystallization leaving orange crystals in 15% yield.

An X-ray crystal structure determination revealed that the olefin addition cleaved the six-membered ring structure of $S_4N_2$ producing the eight-membered heterocycle which is formed via the exo addition of an S–S $\sigma$ bond of $S_4N_2$ across a $\pi$ bond of norbornadiene.

Crystals are monoclinic, space group $P2_1/c$-$C^5_{2h}$ (No. 14), with a = 6.127(1), b = 17.369(1), c = 9.580(1) Å, and $\beta$ = 106.74(1)°; Z = 4. V = 1003.8(5) Å$^3$, $D_x$ = 1.64 g/cm$^3$. R = 0.039, $R_w$ = 0.055 for 1588 reflections. Atomic coordinates for all atoms are given in the paper. The molecular structure of the compound is shown in **Fig. 31**.

The $S_4N_2C_2$ ring exhibits no unusually short intermolecular contacts. The ring possesses an approximately planar (to within 0.15 Å) S(1)–S(4)–N(5)–S(3)–N(6)–S(2) sequence with the S(1)–C(9)–C(7)–S(2) unit (which is itself planar to within 0.04 Å) folded out of this plane to produce a dihedral angle of 74.5°. Within the $S_4N_2C_2$ ring there are variations in the N–S distances similar to those observed in $S_4N_2$ (see "Sulfur-Nitrogen Compounds" 2, 1985, p. 64). Thus, the bonds to S(3) (mean 1.536 Å) are significantly shorter than those to S(4) and S(2) (mean 1.662 Å). The S(1)–S(4) bond (2.017 Å) is slightly shorter than in $S_4N_2$ (2.061 Å). The S(1)–S(2) separation (3.156 Å) is shorter than expected for a normal nonbonded contact. The endocyclic angles at sulfur and nitrogen are all somewhat larger than in $S_4N_2$ itself, as would be expected following the expansion of the ring.

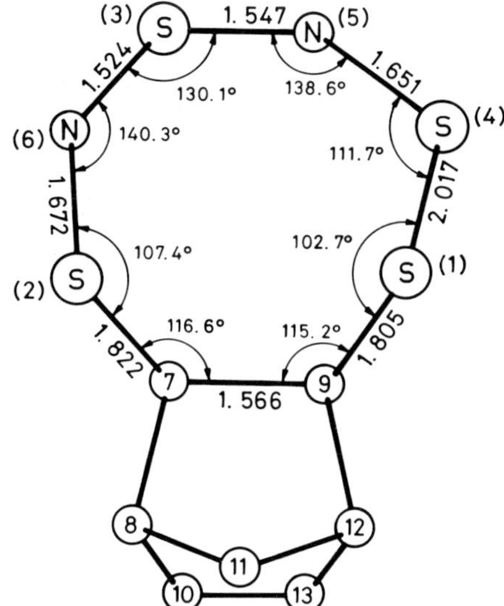


Fig. 31. Molecular structure of $[=S=N-S-\{2,3-C_7H_8\}-S-S-N=]$. Bond lengths in Å, bond angles in °.

The orange micaceous plates melt at 92 to 94°C. UV-visible spectrum ($CH_2Cl_2$): $\lambda_{max}$ (in nm) = 266 ($\varepsilon = 4 \times 10^3$ L · $mol^{-1}$ · $cm^{-1}$), 362 ($3 \times 10^3$), 420 ($2 \times 10^3$). $^1H$ NMR spectrum ($CDCl_3$/TMS): δ (in ppm) = 6.45, 6.33 (m, H 7, H 9), 3.25, 3.21 (s, H 8, H 12), 2.81, 2.53 (d of d, H 10, H 13, J = 1.8, 6.6 Hz), 1.82, 1.68 (AB, H 11, H 11′, $J_{11-11'}$ = 9.5 Hz).

The compound is indefinitely stable in air as a solid and in hexane solution.

**Reference:**

Koenig, H., Oakley, R. T., Cordes, A. W., Noble, M. C. (Can. J. Chem. **61** [1983] 1185/8).

## 11.3 $S_2N_4C_2$ Ring [5,7C]

### 11.3.1 1,3-Dichloro-5,7-bis(dimethylamino)-1H,3H-1,3,2,4,6,8-dithiatetrazocine

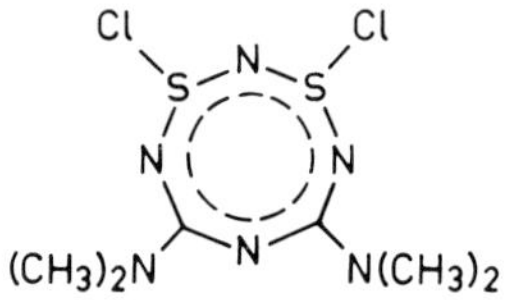

**$[\overset{..}{-}(SCl\overset{..}{-}N)_2\overset{..}{-}(C(N(CH_3)_2)\overset{..}{-}N)_2\overset{..}{-}]$** (Reduced formula in the text $S_2N_4C_2Cl_2(N(CH_3)_2)_2$ [5, 7C])

The compound forms by quickly adding a solution of $(CH_3)_2NCN$ in $CCl_4$ to a solution of $S_3N_3Cl_3$ (~12:1 mole ratio) in $CCl_4$ at ca. 65°C. An immiscible red oil is formed. The reaction mixture is heated at reflux for ca. 3 h, and the solvent is then removed under vacuum. The sticky yellow-brown residue is extracted with $CCl_4$ at 23°C leaving a yellow-brown solid, which is recrystallized from $CH_3NO_2$–$(C_2H_5)_2O$ (1:2) at −20°C to give $S_2N_4C_2Cl_2(N(CH_3)_2)_2$ [5, 7C]. A further crop with an identical IR spectrum is produced from the red-brown $CCl_4$ extract after 7 d at −30°C. The overall yield is ca. 85% based on $S_3N_3Cl_3$ [1], see also [2].

For the IR spectrum (Nujol), 26 absorption bands between 1591 and 422 $cm^{-1}$ are listed in the paper. NMR spectra ($CDCl_3$/TMS), $^1H$: δ (in ppm) = 2.88 (s, 6H), 3.26 (s, 6H); $^{13}C$: δ (in ppm) = 36.65 (s), 40.46 (s). Mass spectrum (70 eV): m/e (relative intensity) = 232 $M^+ - 2Cl$(2), and 7 further fragments (unassigned) [1].

The compound forms moisture-sensitive yellow crystals and is thermally unstable and undergoes a quantitative ring contraction in $CDCl_3$ or $CH_3CN$ at 23°C to give $S_2N_3CCl_2$-$(N(CH_3)_2)$[5,7C] (see p. 7) and $(CH_3)_2NCN$ as determined by $^1H$ NMR spectroscopy. This behavior is in contrast to that of the related phosphazene-thiazyl ring system $S_2N_4P_2(C_6H_5)_4$-$Cl_2$[5,7P] (see "Sulfur-Nitrogen Compounds" 3, 1987, p. 34) [1, 2].

**References:**

[1] Chivers, T., Richardson, J. F., Smith, N. R. M. (Inorg. Chem. **25** [1986] 47/51).
[2] Chivers, T., Richardson, J. F., Smith, N. R. M. (Mol. Cryst. Liquid Cryst. **125** [1985] 319/27).

## 11.4 $S_2N_4C_2$ Ring [3,7C]

### 11.4.1 1,5,2,4,6,8-Dithiatetrazocine

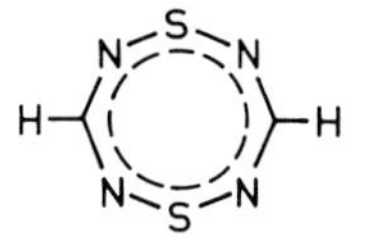

**[(∸S∸N∸CH∸N∸)₂]** (Reduced formula in the text $S_2N_4C_2H_2$ [3, 7C])

Attempts to prepare $S_2N_4C_2H_2$ [3,7C] analogously to the derivatives failed [1]. The molecule was treated theoretically by model calculations for the planar ($D_{2h}$) structure (a) and two bent ($C_{2v}$) structures, i.e., ring folding along the S···S (b) or along the C···C axis (c). Geometry optimization was carried out using restricted Hartree-Fock theory and an STO-3G basis set (C–H bond length assumed to be 1.08 Å). Total energies (for basis sets STO-3G, STO-3G + d, i.e., d functions on S, and 4-31 G) obtained at the STO-3G geometries and at the experimental geometries adopted from the planar $C_6H_5$ and the puckered $N(CH_3)_2$ derivatives (see pp. 138/9) exhibit the relative stabilities of the various conformations. All calculations except those with the STO-3G basis predict the planar conformation (a) to be more stable than the puckered ones (and (b) more stable than (c)). The lowest total energies were obtained at the STO-3G geometries and for the 4-31G basis: $E_T = -1088.09202$ (a), $-1088.04741$ (b), $-1088.02269$ (c) a.u., i.e., conformation (a) is more stable by 28 and 43.5 kcal/mol than (b) and (c), respectively [2]. Similar calculations for structures (a) and (b), assuming the experimental geometries (cf. above) slightly idealized to give $D_{2h}$ and $C_{2v}$ symmetry, also show the planar form to be favored over the bent one by 1.8 (STO-3G), 12.7 (STO-3G + d), 45.8 (4-31 G), and 38.0 (4-31 G + d) kcal/mol (lowest energy $E_T = -1088.24941$ a.u.) [3]. The stability of (a) relative to (b) is also reflected by the results of MNDO calculations [2, 4].

Information on bonding in (a) and (b) was provided by population analyses and by correlating the energies and distributions of the highest occupied valence orbitals and lowest virtual orbitals of (a) and (b). The planar $S_2N_4C_2$ ring was shown to be a 10π system, and the orbitals are those expected from the interaction of two N–S–N allyl-type fragments perturbed by the interaction with the C 2$p_z$ orbital, whereas the puckered ring can be considered as two 4π systems delocalized over the N–C–N fragments. On passing from (a) to (b) ($D_{2h} \rightarrow C_{2v}$) the highest occupied MO, $2b_{1u}$ ($\pi$), which is S–N antibonding, transforms into the more stable S–S $\sigma$ bonding orbital $7a_1$, and the S–S antibonding $\pi$ orbital $1b_{2g}$ (S···S = x axis) destabilizes by approximately the same amount into the $5b_1$ orbital, which essentially arises from S–S $\sigma$ antibonding and N $\pi$ lone pair combinations [2, 3]. Significant also is the stabilization of the S–N and C–N antibonding $\sigma$ orbital $5a_g$ into the S–S $\sigma$ bonding orbital $3a_1$ and the diminution of the S–N or C–N antibonding character in those $\sigma$ orbitals which arise from N lone pair combinations [3].

The balance between the counteracting forces in the $S_2N_4C_2$ ring is easily perturbed, e.g., by the influence of substituents, leading to a transannular S–S interaction and thus ring puckering in the case of $\pi$-electron donating substituents [2] (see pp. 138/40).

**References:**


[1] Wenkert, D. (Diss. Harvard Univ. 1979, pp. 1/222; Diss. Abstr. Intern. B **40** [1979] 1728).
[2] Gleiter, R., Bartetzko, R., Cremer, D. (J. Am. Chem. Soc. **106** [1984] 3437/42).
[3] Millefiori, S., Millefiori, A., Granozzi, G. (Inorg. Chim. Acta **90** [1984] L55/L58).
[4] Oakley, R. T. (Can. J. Chem. **62** [1984] 2763/8).


### 11.4.2 3,7-Diamino-1,5,2,4,6,8-dithiatetrazocine

$H_2N$–$C(N{=}S{=}N)_2C$–$NH_2$ (ring structure)

**[(⩦S⩦N⩦C(NH$_2$)⩦N⩦)$_2$]** (Reduced formula in the text $S_2N_4C_2(NH_2)_2$ [3,7C])

Synthesis of $S_2N_4C_2(NH_2)_2$ [3,7C] has not been reported so far but the molecule has been treated theoretically as a model substance adopting the experimental geometries of the planar $C_6H_5$ and the puckered $N(CH_3)_2$ derivatives (see p. 138) for the $S_2N_4C_2$ ring (slightly idealized to give $D_{2h}$ and $C_{2v}$ symmetries), the relative stabilities of the two conformations have been studied by ab initio SCF MO calculations (cf. p. 134). For the STO-3G and STO-3G + d basis sets the nonplanar conformation was found to be more stable by 23.1 and 16.2 kcal/mol, respectively, whereas for the 4-31G and 4-31G + d basis the planar form was favored by 16.3 and 7.0 kcal/mol, respectively. (Analogous to the unsubstituted $S_2N_4C_2H_2$ [3,7C] molecule, population analysis and a correlation diagram for the highest valence and lowest virtual orbitals were also included in the discussion on the stability.) Thus, for the vapor phase of $S_2N_4C_2(NH_2)_2$ [3,7C] a very fine equilibrium between the planar and nonplanar forms may be expected, which is easily removed in condensed media by solid state effects or solute-solvent interactions [8]. An MNDO study (partially optimized geometry) favored the planar conformation, however, by only 4.6 kcal/mol [9].

For the interpretation of the photoelectron spectrum and the electronic absorption spectrum of $S_2N_4C_2(N(CH_3)_2)_2$ [3,7C] (see pp. 141/2, 140) theoretical results for $S_2N_4C_2(NH_2)_2$ [3,7C] were used, i.e., MO energies (ab initio SCF MO, 4-31G basis) [3] and transition energies and oscillator strengths (CNDO calculation including configuration interaction) [8], respectively.

### 11.4.3 Other 3,7-Disubstituted 1,5,2,4,6,8-Dithiatetrazocines

R = $N(CH_3)_2$, $N(C_6H_5)_2$, t-$C_4H_9$, $C_6H_5$, $C_6H_4OCH_3$-4, $C_6H_4COOC_2H_5$-4

**[(∸S∸N∸CR∸N∸)$_2$]** (Reduced formula in the text $S_2N_4C_2R_2$ [3,7C])

#### Preparation

The derivatives with R = $N(CH_3)_2$, $N(C_6H_5)_2$, t-$C_4H_9$, $C_6H_5$, $C_6H_4OCH_3$-4, and $C_6H_4COOC_2H_5$-4 could be prepared. Attempts to prepare derivatives with R = $C_6H_4NO_2$-4 and $C_6H_4N(CH_3)_2$-4 failed. Derivatives with two different substituents are unknown. The derivatives are formed in a complex reaction in minor yields from the reaction of appropriate amidines (or their hydrochlorides) with sulfur dichloride in the presence of excess 1,5-diazabicyclo[5.4.0]undec-5-ene (DBU) [1, 2, 3] according to:

$$2\ R{-}C(=NH)NH_2 + 3\ SCl_2 \xrightarrow[-6\ DBU\cdot HCl]{+6\ DBU} S_2N_4C_2R_2 + S$$

In a typical experiment (for R = $C_6H_5$) a solution of freshly distilled sulfur dichloride in absolute $CH_2Cl_2$ was added dropwise under $N_2$ in an ice-water bath to a solution of benzamidine and DBU in absolute $CH_2Cl_2$ ($C_6H_5C(=NH)NH_2:SCl_2:DBU = 2:3:6$ mole ratio). After stirring for 1 h at room temperature the resulting red-brown reaction mixture was washed with water and saturated brine, dried over $Na_2SO_4$, evaporated under vacuum, and further purified chromatographically. Repeated crystallization of the eluate from ethyl acetate and ethyl acetate-benzene afforded a crystalline product [1, 2]. The other derivatives are prepared analogously. Yields, melting points, and NMR and mass spectra are given in Table 53.

Table 53

Yields, Melting Points, and NMR and Mass Spectra of 3,7-Disubstituted 1,5,2,4,6,8-Dithiatetrazocines, $S_2N_4C_2R_2$ [3,7C]. (For UV-visible spectra see p. 140, IR absorptions on p. 139.) (References on p. 143.)

| R | yield in % | appearance m.p. in °C (recrystallized from) | NMR spectra ($CDCl_3$/TMS, δ in ppm) mass spectra | Ref. |
|---|---|---|---|---|
| $N(CH_3)_2$ | 54 | yellow plates<br>212 with sublimation<br>(ethyl acetate) | $^1$H NMR (100 MHz):<br>δ = 3.15 (s, $CH_3$)<br>$^{13}$C NMR (25.2 MHz):<br>δ = 38.921, 179.951<br>MS (20°C): m/e = 232 $M^+$, another 6 peaks (unassigned) given | [1,2] |
| $N(C_6H_5)_2$ | 6 | yellow needles<br>230 | MS: m/e = 480 $M^+$, no other peaks given | [1] |
| t-$C_4H_9$ | 13.7 | yellow powder<br>108 to 109 | $^1$H NMR (60 MHz): δ = 1.55 (s)<br>$^{13}$C NMR (75.46 MHz):<br>δ = 1.7, 30.1, 144.8 | [3] |

Table 53 (continued)

| R | yield in % | appearance m.p. in °C (recrystallized from) | NMR spectra ($CDCl_3$/TMS, δ in ppm) mass spectra | Ref. |
|---|---|---|---|---|
| $C_6H_5$ | 7.4 [2] | bright yellow shiny plates 225 to 226 (ethyl acetate-benzene); reshaping and sublimation of crystals at ~220 | MS (90°C): m/e = 298 $M^+$, another 11 peaks (unassigned) given [2] | [1, 2] |
| $C_6H_4OCH_3$-4 | 3 | yellow crystals 268 ($CH_3CN$) | $^1$H NMR: δ = 3.90 (s, $OCH_3$); 7.05 (HCCCO), J = 9.0 Hz; 8.45 (HCCO) [1] | [1, 2] |
| $C_6H_4COOC_2H_5$-4 | 6 | yellow crystals 305 ($CH_3CN$) | $^1$H NMR: δ = 1.44 (t, $CO_2CH_2C\underline{H}_3$), 4.44 (q, $CO_2C\underline{H}_2CH_3$), $^3J$ = 7 Hz; 8.23 to 8.64 (arom. H) [1] | [1, 2] |

**Crystal Structure**

X-ray diffraction analyses of single crystals have been performed for three of these compounds (R = $N(CH_3)_2$, $C_6H_5$, and $C_6H_4OCH_3$-4). The crystallographic data are listed in Table 54.

Table 54
Crystal Data of 3,7-Disubstituted 1,5,2,4,6,8-Dithiatetrazocines, $S_2N_4C_2R_2$ [3,7C].
(References on p. 143.)

| R | $C_6H_5$ | $C_6H_4OCH_3$-4 | $N(CH_3)_2$ |
|---|---|---|---|
| symmetry | monoclinic | orthorhombic | triclinic |
| space group | $P2_1/c-C^5_{2h}$ (No. 14) | $Pbca-D^{15}_{2h}$ (No. 61) | $P1-C^1_1$ (No. 1) [2] |
| a in Å | 10.901(3) | 6.3967(4) | 8.469(3) |
| b in Å | 5.694(2) | 34.486(5) | 8.453(3) |
| c in Å | 12.200(3) | 7.3826(4) | 8.075(3) |
| α in ° | | | 105.64(3) |
| β in ° | 117.56(3) | | 113.00(3) |
| γ in ° | | | 82.94(3) |
| V in $Å^3$ | 671.3 [1] | | 512.3 [1] |
| Z | 2 | 4 | 2 |
| molecular symmetry | $C_i$ | $C_i$ | nearly $C_{2v}$ |
| $D_x$ in $g/cm^3$ | 1.476 | | 1.506 |
| $D_m$ in $g/cm^3$ | 1.473 | 1.46 [1] | 1.48 |
| R | 0.037 | 0.041 | 0.067 |
| unique reflections | 1441 | 951 | 1527 |
| Ref. | [1, 2] | [2]; somewhat different values in [1] | [1, 2] |

**Molecular Structure**

From X-ray diffraction data it was shown that the ring geometries of the derivatives were quite different, depending on the nature of the substituents. In the diphenyl derivative, the dithiatetrazocine ring is perfectly planar with only the phenyl groups slightly twisted out of the $S_2N_4C_2$ plane (at an angle of 9.7°). All S–N bond lengths in the heterocycle were found to be equal as were those of its C–N bonds. There are no significant differences in the geometry of the molecule with $R=C_6H_4OCH_3$-4; the 4-methoxyphenyl rings are distorted out of the plane of the eight-membered ring. In contrast, the bis(dimethylamino) derivative shows a remarkable difference in its molecular shape. It is folded along an axis passing through the sulfur atoms with an interplanar angle of 101°, and except for the methyl hydrogens all atoms lie in two planes intersecting this axis. The average S–N and C–N bond distances in $S_2N_4C_2(N(CH_3)_2)_2$ [3,7C] are longer than the corresponding values found for the diphenyl and bis(4-methoxyphenyl) derivatives. They lie between the typical values for the corresponding single and double bonds [1, 2].

For the planar molecules the remarkably short distance between the sulfur atoms suggests partial bonding. The bond lengths as well as the electronic spectra suggest a delocalized, aromatic 10π-electron system [1, 2].

The folded structure of $S_2N_4C_2(N(CH_3)_2)_2$ [3,7C], in contrast to the planar structures for $S_2N_4C_2R_2$ [3,7C], with $R=C_6H_5$ and $C_6H_4OCH_3$-4, has been explained in terms of the effect of the π donor properties of the $N(CH_3)_2$ group on the π electronic structure of the $S_2N_4C_2$ ring [2]. Evidence for a nonplanar ring conformation of $S_2N_4C_2(N(CH_3)_2)_2$ [3,7C] was also provided by comparing its experimental UV spectrum (in 96% ethanol) [2] with the theoretical UV spectra (CNDO-CI method) calculated for the planar and nonplanar conformations of $S_2N_4C_2(NH_2)_2$ [3,7C] as a model substance (however, solvent effects on the planar-nonplanar equilibrium (cf. p. 135) may be possible) [8]. A strong stabilization upon folding of the $S_2N_4C_2$ ring in the $N(CH_3)_2$ derivative, which cannot be explained solely by crystal packing effects, was indicated by the theoretical energy difference of 25.9 kcal/mol; ab initio SCF MO calculations (STO-3G basis) were carried out for the $N(CH_3)_2$ derivative at its experimental geometry (accommodated to $C_{2v}$ symmetry) and for a hypothetical planar form with the experimental ring geometry of the $C_6H_5$ derivative (accommodated to $D_{2h}$ symmetry) and the same exocyclic C–N distance as in the $N(CH_3)_2$ derivative [9]. General considerations are made for the parent compound $S_2N_4C_2H_2$ [3,7C] on p.134. The molecular structures of the derivatives with $R=C_6H_5$ and $R=N(CH_3)_2$ are shown in **Fig. 32** and **Fig. 33**, p. 140, for selected bond lengths and bond angles of the X-ray diffraction studied derivatives, see Table 55.

Table 55
Selected Bond Lengths (in Å) and Bond Angles (in °) of 3,7-Disubstituted 1,5,2,4,6,8-Dithiatetrazocines, $S_2N_4C_2R_2$ [3,7C] [2].

| R | $C_6H_5$ | $C_6H_4OCH_3$-4 | $N(CH_3)_2$*) |
|---|---|---|---|
| bond length | | | |
| S(1)–N(1)<br>N(4)–S(1) | 1.559 | 1.559(2) | 1.605<br>1.601 |
| N(1)–C(1)<br>C(2)–N(4) | 1.319 | 1.330(3) | 1.350<br>1.347 |
| C(1)–N(2)<br>N(3)–C(2) | 1.327 | 1.328(3) | 1.348<br>1.346 |

Table 55 (continued)

| R | $C_6H_5$ | $C_6H_4OCH_3$-4 | $N(CH_3)_2$*) |
|---|---|---|---|
| N(2)–S(2)<br>S(2)–N(3) | 1.569 | 1.564(2) | 1.609<br>1.604 |
| C(1)–C(5)[N(5)]<br>C(2)–C(6)[N(6)] | 1.495 | 1.497(3) | 1.343<br>1.354 |
| S(1)···S(2) | 3.79 | | 2.428 |
| bond angle | | | |
| N(4)–S(1)–N(1)<br>N(3)–S(2)–N(2) | 127.0(2) | 127.2(1) | 114.3(2)<br>113.3(2) |
| S(1)–N(1)–C(1)<br>C(2)–N(4)–S(1) | 140.9(2) | 142.2(2) | 117.7(3)<br>117.3(3) |
| N(1)–C(1)–N(2)<br>N(3)–C(2)–N(4) | 129.2(3) | 128.5(2) | 124.0(4)<br>124.8(4) |
| C(1)–N(2)–S(2)<br>S(2)–N(3)–C(2) | 142.9(2) | 142.2(2) | 117.7(3)<br>117.7(3) |
| N(1)–C(1)–C(5)[N(5)]<br>N(4)–C(2)–C(6)[N(6)] | 115.1(3) | 115.9(2) | 117.8(4)<br>117.7(4) |
| N(2)–C(1)–C(5)[N(5)]<br>N(3)–C(2)–C(6)[N(6)] | 115.7(3) | 115.7(2) | 117.9(4)<br>117.2(4) |

*) Standard deviations of the distances are 0.004 to 0.006 Å.

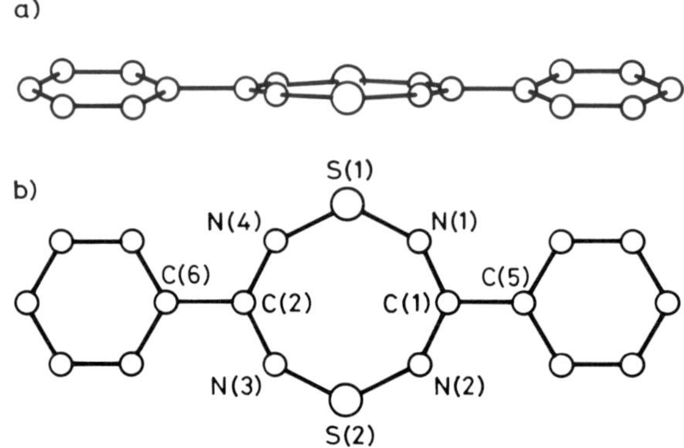


Fig. 32. Molecular structure of $S_2N_4C_2(C_6H_5)_2$ [3,7C].

**IR Absorption**

Unassigned IR bands in the ~2900 to ~690 $cm^{-1}$ range, which have been recorded for KBr pellets of $S_2N_4C_2R_2$ [3,7C], are listed and/or figured in the original paper for R = t-$C_4H_9$ [3], $C_6H_5$, $C_6H_4OCH_3$-4, $C_6H_4COOC_2H_5$-4, $N(CH_3)_2$ [1, 2], and $N(C_6H_5)_2$ [1]. For $S_2N_4C_2(C_6H_4COO$-$C_2H_5$-$4)_2$, ν(CO) equals 1709 $cm^{-1}$ in KBr [1].

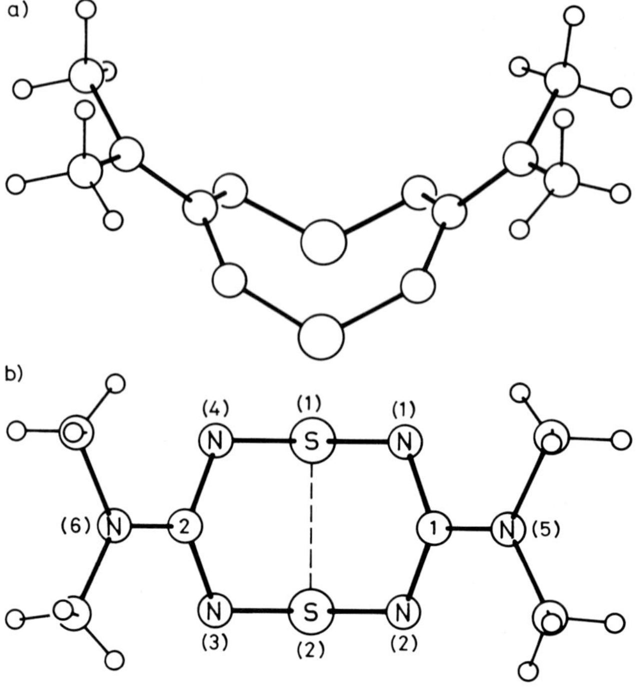


Fig. 33. Molecular structure of $S_2N_4C_2(N(CH_3)_2)_2$ [3,7C].

**UV-Visible Absorption and Magnetic Circular Dichroism (MCD)**

Information on the ambient temperature UV-visible spectra of deeply (yellow) colored $S_2N_4C_2R_2$ [3,7C] compounds is listed below:

| R | solvent | $\lambda_{max}$ in nm ($\varepsilon$ in $L \cdot mol^{-1} \cdot cm^{-1}$) | Ref. |
|---|---|---|---|
| $N(CH_3)_2$ | 96% $C_2H_5OH$ | 280 (inflection; 3850), 229(23500), 219 (inflection; 22000) | [2] |
| | | 276 | [1] |
| $N(C_6H_5)_2$ | $C_2H_5OH$ | 268, 242 | [1] |
| $t\text{-}C_4H_9$ | $CH_3OH$ | 363(9775), 277(2035), 243(16020), 220(6380) | [3] |
| | n-hexane | 362, 279, 250 (shoulder), 235 | [4] |
| $C_6H_5$ | 96% $C_2H_5OH$ | 409(7300), 318(29000), 306.5(37500), 294.5(29600), 281(26200), 271.5(26500) | [2] |
| | $CH_3CN$ | 415, 309, 271, 250*⁾, 211 | [3, 4] |
| | ethyl acetate-$C_2H_5OH$ 1:19 | 408(7580), 317(33630), 306(39180), 294(30820), 280(27400), 272(27730) | [1] |
| | cyclohexane | 411, 318, 307, 294, 281, 272 | [1] |
| $C_6H_4OCH_3$-4 | dioxane | 433(4700), 334(35300), 322(48700), 311(39200), 293(27800), 285(26200) | [1, 2] |
| $C_6H_4COOC_2H_5$-4 | dioxane | 402(7930), 338(38600), 326(48600), 276(22900) | [1, 2] |

*⁾ Revealed from linear dichroism.

Polarized absorption spectra of the t-$C_4H_9$ and $C_6H_5$ derivatives in stretched polyethylene sheets were recorded at room temperature with the incident light polarized along the long (y) and along the short (x) axes of the molecules. The following wavenumbers $\nu_{max}$ (in $10^3$ $cm^{-1}$), polarization directions p ($\perp$ or $\parallel$ with respect to the y axis) and log $\varepsilon$ values (in cyclohexane) were obtained:

| | band ........ | A | B | C | D | E | F |
|---|---|---|---|---|---|---|---|
| R = t-$C_4H_9$ | $\nu_{max}$ ......... | 27.4 | 35.7 | 36.6 | 42.7 | | |
| | p ........... | $\perp$ | $\parallel$ | $\perp$ | $\parallel$ | | |
| | log $\varepsilon$ ........ | 4.1 | 3.3 (B, C) | | 4.3 | | |
| R = $C_6H_5$ | $\nu_{max}$ ......... | 23.4*) | 31.1*) | 36.0 | 36.0 | 43.5 | 44.0 |
| | | 24.0*) | 32.5*) | | | | |
| | | 24.5*) | 33.9*) | | | | |
| | p ........... | $\perp$ | $\parallel$ | $\parallel$ | $\perp$ | $\perp$ | $\parallel$ |
| | log $\varepsilon$ ........ | 3.9 | 4.5 | 4.4 (C, D) | | 4.4 (E, F) | |

*) Vibrational fine structure.

Calculations of $\pi$-electron transitions by the PPP-CI method (Pariser-Parr-Pople method including configuration interaction) reproduced quite well the energies, relative intensities, and the polarization behavior of bands A, B, D of the t-$C_4H_9$ derivative (model calculations for the $CH_3$ derivative) and all bands observed for the $C_6H_5$ derivative; band C of the t-$C_4H_9$ derivative, not reproduced by the PPP-CI model, was tentatively assigned to a $\pi^* \leftarrow n$ transition [3].

In the region of the UV absorption bands, the MCD spectrum of the t-$C_4H_9$ derivative exhibits four bands with signs (in order of increasing energy) +, +, −, and−, that of the $C_6H_5$ derivative three bands with the signs +, −, and +; analyses in terms of the perimeter model [10] and $\pi$-electron PPP calculations were offered [4].

**Photoelectron Spectrum**

The X-ray photoelectron spectra (XPS) of $S_2N_4C_2R_2$ [3,7C] with R = $N(CH_3)_2$ and $C_6H_5$ (pellets on Au substrates) where recorded at room temperature and pressures of (2 to 4) $\times 10^{-9}$ Torr using the AlK$\alpha$ line at 1486.6 eV. For the core levels C1s, N1s (endocyclic carbon/ nitrogen), and S2p the following binding energies were obtained (in eV):

| R | C1s | N1s | S2p |
|---|---|---|---|
| $N(CH_3)_2$ | 287.3 | 397.7 | 165.5 |
| $C_6H_5$ | 286.6 | 398.7 | 165.4 |

Core level satellites observed for the $C_6H_5$ derivative were related to "shake-up" transitions originating from the highest occupied orbital (HOMO), $4b_{1u}(\pi)$, in the case of the C1s and S2p peaks, and from the second-highest occupied orbital, $2a_{1u}(\pi)$, in the case of the N1s peak.

In the region of the valence-level binding energies (0 to 30 eV) the following peaks were observed (in eV):

| R | A | B | C | D | E | F | G | H |
|---|---|---|---|---|---|---|---|---|
| $N(CH_3)_2$ | 22.6 | — | 16.9 | 14.5 | 10.1 | 6.5 to 5.5 | 3.5 | ? |
| $C_6H_5$ | 21.8 | 20.0 | 17.0 | 13.3 | 9.6 | 6.8 | 4.4 | 2.3 |

Theoretically simulated valence XPS (ab initio SCF MO calculation, STO-3G basis) suggested the following interpretation for the spectrum of the $C_6H_5$ derivative: peaks A and B result from MOs with strong N2s character, C and D are related to σ N2s-C2s and N2s-S3s interactions, and E to C2s-C2s interactions; peaks F and G result from σ N2p-S3p contributions, and the HOMO (π orbital) is related to peak H. For the $N(CH_3)_2$ derivative, the agreement between experimental and simulated spectra was considerably worse, which was attributed to the presence of additional valence levels originating from impurities [9].

The HeI photoelectron spectra (UVPS) were recorded up to 16 eV for three $S_2N_4C_2R_2$ [3,7C] derivatives with $R = N(CH_3)_2$, t-$C_4H_9$, and $C_6H_5$ with the samples heated to 108, 70, and 120°C, respectively. The spectrum of the t-$C_4H_9$ derivative exhibits four peaks between 8 and 11 eV clearly separated from strongly overlapping bands above 11 eV. The spectra of the other two molecules show a broad band in the low-energy region (8 to 10 eV) which results from the overlap of several peaks. The following vertical ionization potentials were measured (in eV):

| | | | | | | | |
|---|---|---|---|---|---|---|---|
| $R = N(CH_3)_2$ ............ | 8.15 | 8.5 | 8.8 | 10.6 | 10.95 | | |
| R = t-$C_4H_9$ ............. | 8.39 | 9.2 | 9.68 | 10.25 | | | |
| $R = C_6H_5$ ............... | 9.0 | 9.2 | 9.3 | 9.6 | 9.9 | 11.1 | 12.2 |

Interpretations of the UVPS were attempted by considering the orbital energies (Koopmans' theorem) obtained by various MO calculations. For the t-$C_4H_9$ derivative, the observed four ionization potentials could be correlated with five MOs, $b_{1u}(\pi)$, $a_u(\pi)$, $b_{1g}(n)/b_{3u}(n)$, $b_{2u}(n)$, predicted for the planar $S_2N_4C_2H_2$ [3,7C] molecule ($D_{2h}$ symmetry) by ab initio SCF MO (STO-3G and 4-31G basis sets) calculations. For the $C_6H_5$ derivative experimental results and MO energies calculated by the semiempirical MNDO and Pariser-Parr-Pople (PPP) methods were compared, and for the $N(CH_3)_2$ derivative comparison was drawn with the results of a model calculation for the $NH_2$ derivative (ab initio SCF MO, 4-31G basis, $C_{2v}$ symmetry). Clear assignments could not be given [3].

**Chemical Behavior**

$S_2N_4C_2(N(CH_3)_2)_2$ [3,7C] does not possess basic properties, as shown by titration with $HClO_4$ in $CH_3COOH$–$C_6H_5Cl$ [2]. The compound reacts with 2 mol of $Cl_2$ (as $Cl_2$ gas) to give the trichloride $S_2N_4C_2Cl(N(CH_3)_2)_2^+Cl_3^-$ [3,7C] (see p. 143) [5]. The reaction with $Br_2$ in $CCl_4$ or $CH_3CN$ yields $S_2N_4C_2Br(N(CH_3)_2)_2^+Br_3^-$ [3,7C] (see p. 144) in analogy to the trichloride [6, 7]. The compound reacts with $SO_2Cl_2$ (1:2 mole ratio) in $CCl_4$ at room temperature by oxidative addition of $Cl_2$ across the S–S bond to give the S,S′-dichlorinated compound $S_2N_4C_2Cl_2(N(CH_3)_2)_2$ [3,7C] (see p. 145). $R_f = 0.47$ (toluene-ethyl acetate 1:1).

$S_2N_4C_2(N(C_6H_5)_2)_2$ [3,7C] has no basic properties ($pK_B = 12$ to 13) [1].

$S_2N_4C_2(C_6H_5)_2$ [3,7C] also does not possess basic properties, as shown by titration with $HClO_4$ in $CH_3COOH$–$C_6H_5Cl$ [2] ($pK_B > 13$ [1]). It proved remarkably stable to heat, not changing even on prolonged refluxing in xylene at 220 to 240°C. With KOH (50°C/80 min) or HCl (50°C/2h) in aqueous dioxane, the compound was easily hydrolyzed to benzamidine; however, in nonpolar solvents (benzene, toluene, xylene), it proved unreactive toward substoichiometric amounts of various nucleophiles ($C_6H_5CH_2NH_2$, $C_6H_4SC(SLi)=N$, $((CH_3)_3Si)_2NH$, and $C_4H_9Li$). It was also resistant toward oxidation by 3-chloroperbenzoic acid in boiling $CH_2Cl_2$ [2]. $R_f = 0.69$ (toluene-ethyl acetate 1:1) [1, 2].

**References:**

[1] Wenkert, D. (Diss. Harvard Univ. 1979, pp. 1/222; Diss. Abstr. Intern. B **40** [1979] 1728).
[2] Ernest, I., Holick, W., Rihs, G., Schomburg, D., Shoham, G., Wenkert, D., Woodward, R. B. (J. Am. Chem. Soc. **103** [1981] 1540/4).
[3] Gleiter, R., Bartetzko, R., Cremer, D. (J. Am. Chem. Soc. **106** [1984] 3437/42).
[4] Klein, H.-P., Oakley, R. T., Michl, J. (Inorg. Chem. **25** [1986] 3194/201).
[5] Boeré, R. T., Cordes, A. W., Oakley, R. T., Reed, R. W. (J. Chem. Soc. Chem. Commun. **1985** 655/6).
[6] Chivers, T., Richardson, J. F., Smith, N. R. M. (Inorg. Chem. **25** [1986] 47/51).
[7] Chivers, T., Richardson, J. F., Smith, N. R. M. (Mol. Cryst. Liquid Cryst. **125** [1985] 319/27).
[8] Millefiori, S., Millefiori, A., Granozzi, G. (Inorg. Chim. Acta **90** [1984] L55/L58).
[9] Boutique, J. P., Riga, J., Verbist, J. J., Delhalle, J., Fripiat, J. G., Haddon, R. C., Kaplan, M. L. (J. Am. Chem. Soc. **106** [1984] 312/8).
[10] Michl, J. (J. Am. Chem. Soc. **100** [1978] 6801/11, 6812/8, 6819/24).

## 11.4.4 1-Halogeno-3,7-bis(dimethylamino)-1H,5H-1,5,2,4,6,8-dithiatetrazocine Ion (1+)

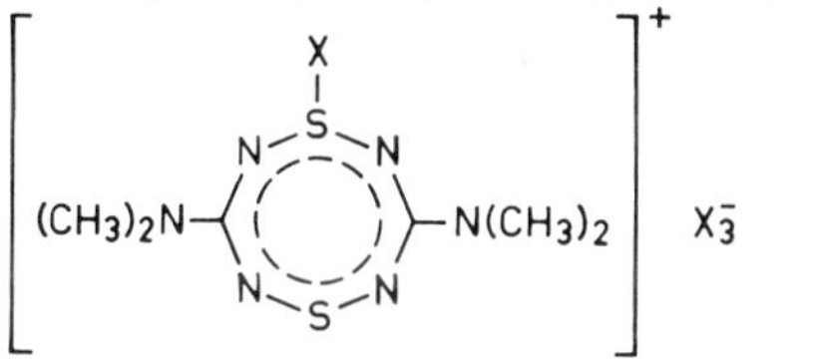

X = Cl, Br

**[∸SX∸N∸C(N(CH$_3$)$_2$)∸N∸S∸N∸C(N(CH$_3$)$_2$)∸N∸]$^+$X$_3^-$** (Reduced formula in the text $S_2N_4C_2X(N(CH_3)_2)_2^+X_3^-$ [3,7C])

**$S_2N_4C_2Cl(N(CH_3)_2)_2^+Cl_3^-$** [3,7C] is prepared by passing chlo-ine gas over a saturated solution of $S_2N_4C_2(N(CH_3)_2)_2$ (see p. 142) in $CH_3CN$ followed by cooling of the resulting solution to −30°C.

An X-ray diffraction study showed the crystals to be orthorhombic, space group Pnma-$D_{2h}^{16}$ (No. 62), with a = 23.644(3), b = 10.773(3), and c = 5.862(2) Å; Z = 4. V = 1493.1 Å$^3$, $D_x$ = 1.664 g/cm$^3$. R = 0.091. The molecular structure is shown in **Fig. 34**, p. 144. Selected bond lengths and bond angles are given in Table 56, p. 144.

The eight-membered $S_2N_4C_2$ ring is bisected by a crystallographic mirror plane containing both sulfur atoms with the trichloride anion also lying in this plane. The dihedral angle between the mean planes of the two $S_2N_2C$ units in the cation is 126°. Compared with the starting compound $S_2N_4C_2(N(CH_3)_2)_2$ (see p. 138) the S···S contact (3.491 Å) has opened up considerably and the S–N bonds are markedly shorter, whereas the N(1)–C(1) and N(2)–C(1) bonds remain virtually unchanged. The chlorine atom Cl(1) straddles the two sulfur atoms, but the S–Cl···S unit is noticeably asymmetric. This asymmetry in the solid state can be related to the effects of secondary bonding to the anion.

The orange, moisture-sensitive crystals decompose at 97 to 100°C. $^1$H NMR spectrum ($CDCl_3$/TMS, −90°C) : δ = 3.34 ppm (s, $CH_3$). The equivalence of the two methyl resonances at this temperature suggests either a rapid rotation of the $N(CH_3)_2$ groups about the respective C–N bonds or, more likely, a rapid exchange of the bridging chlorine between the two sulfur atoms, i.e., a symmetric structure on the NMR time scale [1].

Table 56
Selected Bond Lengths and Bond Angles of $S_2N_4C_2Cl(N(CH_3)_2)_2^+Cl_3^-$ [3,7C] [1].

| bond length in Å | | bond angle in ° | |
|---|---|---|---|
| S(1)–Cl(1) | 2.290(3) | N(1)–S(1)–N(1′) | 117.5(4) |
| S(1)–N(1) | 1.542(6) | S(1)–N(1)–C(1) | 131.0(5) |
| N(1)–C(1) | 1.363(7) | N(1)–C(1)–N(2) | 124.1(4) |
| N(2)–C(1) | 1.366(7) | C(1)–N(2)–S(2) | 135.0(4) |
| N(3)–C(1) | 1.277(7) | N(2)–S(2)–N(2′) | 123.3(4) |
| N(3)–C(2) | 1.501(7) | Cl(4)–Cl(3)–Cl(2) | 177.7(2) |
| N(3)–C(3) | 1.487(8) | | |
| S(2)–N(2) | 1.514(5) | | |
| S(2)···Cl(1) | 3.225(4) | | |
| S(2)···Cl(4) | 3.160(4) | | |

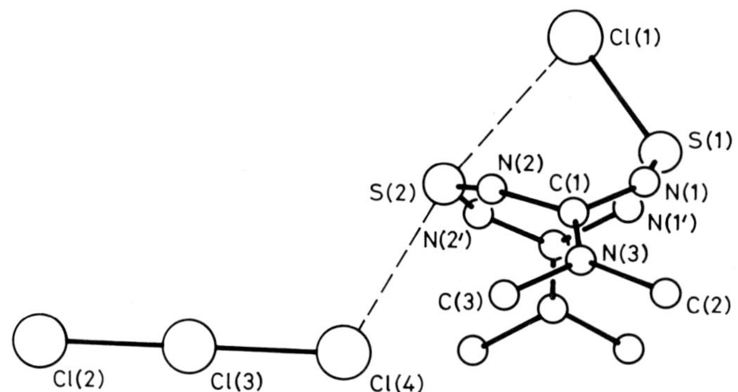


Fig. 34. Molecular structure of $S_2N_4C_2Cl(N(CH_3)_2)_2^+Cl_3^-$ [3,7C].

**$S_2N_4C_2Br(N(CH_3)_2)_2^+Br_3^-$** [3,7C] is formed in analogy to the chloro compound from the reaction of $S_2N_4C_2(N(CH_3)_2)_2$ (see p. 142) with $Br_2$ in $CCl_4$ or $CH_3CN$ as purple-black microcrystals [2] or dark red crystals [3]. Raman spectrum: $\nu_{sym}(Br_3^-)=162\ cm^{-1}$ [2].

**References:**

[1] Boeré, R. T., Cordes, A. W., Oakley, R. T., Reed, R. W. (J. Chem. Soc. Chem. Commun. **1985** 655/6).
[2] Chivers, T., Richardson, J. F., Smith, N. R. M. (Inorg. Chem. **25** [1986] 47/51).
[3] Chivers, T., Richardson, J. F., Smith, N. R. M. (Mol. Cryst. Liquid Cryst. **125** [1985] 319/27).

### 11.4.5 1,5-Dichloro-3,7-bis(dimethylamino)-1H,5H-1,5,2,4,6,8-dithiatetrazocine

**$[(\text{--}SCl\text{--}N\text{--}C(N(CH_3)_2)\text{--}N\text{--})_2]$** (Reduced formula in the text $S_2N_4C_2Cl_2(N(CH_3)_2)_2$ [3,7C])

The compound is formed in 18% yield when $SO_2Cl_2$ is added by syringe to a solution of $S_2N_4C_2(N(CH_3)_2)_2$ [3,7C] (see p. 142) (~2:1 mole ratio) in $CCl_4$ at 23°C. The solution changes color quickly from yellow to dark red-orange to give an orange precipitate after ca. 5 min [1], see also [2].

For the IR spectrum (Nujol), 12 unassigned absorption bands between 1571 and 542 $cm^{-1}$ are listed in the paper. NMR spectra ($CDCl_3$/TMS), $^1H$: $\delta=3.13$ ppm. $^{13}C$: $\delta=41.03$ ($CH_3$), 154.01 ppm ($NCN_2$). Electron impact mass spectrum (70 eV): m/e (relative intensity) = 232 $M^+$–2Cl (7%), and 8 further fragments (unassigned) [1].

The compound is thermally stable [1, 2]. It readily reacts with a slight excess of $SbCl_5$ in $CCl_4$ to form the dication $S_2N_4C_2(N(CH_3)_2)_2^{2+}$ [3,7C] [2].

**References:**

[1] Chivers, T., Richardson, J. F., Smith, N. R. M. (Inorg. Chem. **25** [1986] 47/51).
[2] Chivers, T., Richardson, J. F., Smith, N. R. M. (Mol. Cryst. Liquid Cryst. **125** [1985] 319/27).

## 11.5 $SN_2C_5$ Ring

### 11.5.1 4,5,6,7-Tetrahydro-3H-1,2,8-thia (1-$S^{IV}$)diazocine

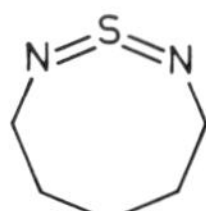


**$[=S=N\text{–}(CH_2)_5\text{–}N=]$** (Reduced formula in the text $SN_2(CH_2)_5$)

The unstable compound $SN_2(CH_2)_5$ is reported to form in the reaction of $H_2N(CH_2)_5NH_2$ with $(4\text{-}CH_3C_6H_4SO_2N=)_2S$ [1, 2]. Details are given on p. 124.

**References:**

[1] Kresze, G., Grill, H. (Tetrahedron Letters **1969** 4117/20, 4118).
[2] Grill, H. (Diss. T.H. München 1970, pp. 1/91, 29/37, 74/77).

# 12 Nine Atom S–N–C Ring Systems

## 12.1 $S_3N_5C$ Ring System

### 12.1.1 Derivatives of 1,3,5-Trithia(1,3,5-$S^{IV}$)-2,4,6,8,9-pentaazabicyclo[3.3.1]-nona-1(9),2,3,5,7-pentaene

R = $N(CH_3)_2$, $N(C_2H_5)_2$, $N(C_3H_7\text{-}i)_2$, $C_6H_5$

**[=\{1,3-$S_3N_3$\}–N=CR–N=]** (Reduced formula in the text $S_3N_5CR$)

#### 12.1.1.1 Survey

This bicyclic heterocyclic system has been prepared from the reaction of $S_2N_3CCl_2R$ with $(CH_3)_3SiN{=}S{=}NSi(CH_3)_3$ or $(CH_3)_3SiN{=}S{=}O$ and by treatment of $S_3N_3Cl_3$ with N,N,N'-tris(trimethylsilyl)benzamidine, see the following scheme.

+ $(CH_3)_3SiN{=}S{=}NSi(CH_3)_3$ ; + $(CH_3)_3SiN{=}S{=}O$ ; + $RC({=}NSi(CH_3)_3)N[Si(CH_3)_3]_2$

The structure of the ring system can be described in terms of a 1,3,2,4,6-dithiatriazine skeleton loosely bridged by an NSN fragment across the 1,3-sulfur atoms. The long S(2)–N(2) and S(1)–N(3) linkage as well as the small endocyclic angles at S(1), S(2), and N(1) can be attributed to a rehybridization occassioned by the presence of the bridge between the two sulfur atoms [4]. The structure of $S_3N_5C(C_6H_5)$ is shown in **Fig. 35**. In $S_3N_5C(N(C_3H_7\text{-}i)_2)$ the bonds connecting the –NSN– bridge to the six-membered ring are even longer than the corresponding distances in the phenyl derivative, suggesting a weaker interaction [7]. Principal bond lengths and bond angles of $S_3N_5CR$ with R = $N(C_3H_7\text{-}i)_2$ [7] and $C_6H_5$ [3] (see also [5]) are given in Table 57. IR, NMR, and mass spectra of the derivatives are summarized in Table 58, p. 148.

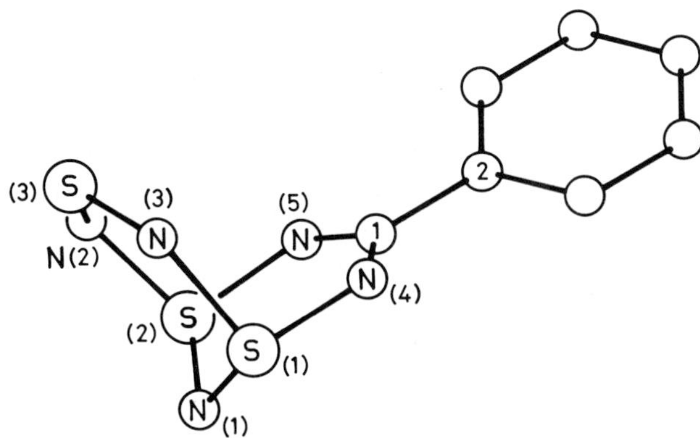


Fig. 35. Molecular structure of $S_3N_5C(C_6H_5)$.

Table 57

Selected Bond Lengths and Bond Angles of $S_3N_5CR$ (R = $N(C_3H_7\text{-}i)_2$ [7], $C_6H_5$ [3]).

bond length in Å

| | R = $N(C_3H_7\text{-}i)_2$ | R = $C_6H_5$ |
|---|---|---|
| S(1)–N(1) | 1.620(4) | 1.630(2) |
| S(1)–N(3) | 1.750(3) | 1.728(2) |
| S(1)–N(4) | 1.577(3) | 1.622(2) |
| S(2)–N(1) | 1.633(3) | 1.630(2) |
| S(2)–N(2) | 1.745(3) | 1.728(2) |
| S(2)–N(5) | 1.588(3) | 1.621(2) |
| S(3)–N(2) | 1.536(3) | 1.547(2) |
| S(3)–N(3) | 1.543(3) | 1.547(2) |
| N(4)–C(1) | 1.355(4) | 1.337(3) |
| N(5)–C(1) | 1.350(5) | 1.333(3) |
| C(1)–C(2) | — | 1.486(3) |
| C(1)–N*) | 1.341(4) | — |

bond angle in °

| | R = $N(C_3H_7\text{-}i)_2$ | R = $C_6H_5$ |
|---|---|---|
| N(1)–S(1)–N(3) | 104.4(2) | 106.1(1) |
| N(1)–S(1)–N(4) | 111.4(1) | 109.5(1) |
| N(3)–S(1)–N(4) | 104.3(1) | 102.5(1) |
| N(1)–S(2)–N(2) | 103.7(1) | 105.1(1) |
| N(1)–S(2)–N(5) | 112.1(2) | 110.2(1) |
| N(2)–S(2)–N(5) | 103.0(2) | 102.4(1) |
| N(2)–S(3)–N(3) | 118.8(2) | 119.0(1) |
| S(1)–N(1)–S(2) | 110.3(2) | 111.8(1) |
| S(2)–N(2)–S(3) | 119.3(2) | 119.7(1) |
| S(1)–N(3)–S(3) | 121.9(2) | 121.5(1) |
| S(1)–N(4)–C(1) | 119.4(3) | 118.8(2) |
| S(2)–N(5)–C(1) | 121.9(2) | 120.8(2) |
| N(4)–C(1)–N(5) | 127.0(3) | 129.9(2) |
| N(4)–C(1)–C(2) | — | 115.6(2) |
| N(5)–C(1)–C(2) | — | 114.5 |
| N(5)–C(1)–N*) | 116.7(3) | — |
| N(4)–C(1)–N*) | 116.2(3) | — |

*) Exocyclic N atom.

### 12.1.1.2 $S_3N_5C(N(CH_3)_2)$

The compound is formed from the reaction of $S_2N_3CCl_2(N(CH_3)_2)$ (see p. 7) with equimolar amounts of $(CH_3)_3SiN{=}S{=}NSi(CH_3)_3$ in $CCl_4$ by stirring at room temperature for 15 h. Removal of the solvent in vacuum gave an orange solid (78% yield) which was recrystallized from $CH_3CN$ at −25°C. The compound has also been observed (monitored by $^1H$ NMR spectroscopy) in the reaction of $S_2N_3CCl_2(N(CH_3)_2)$ with $(CH_3)_3SiNSO$ (1:2 mole ratio) in $CCl_4$ at 23°C for 27 d [1], see also [2]. IR, NMR, and mass spectra of the compound are listed in Table 58, p. 148. The compound reacts with equimolar amounts of $S_3N_3Cl_3$ in $CCl_4$ at 65°C to give $S_2N_3CCl_2(N(CH_3)_2)$ and $S_4N_4$ [1].

### 12.1.1.3 $S_3N_5C(N(C_2H_5)_2)$

The compound is formed analogously to the dimethylamino derivative from $S_2N_3CCl_2(N(C_2H_5)_2)$ and $(CH_3)_3SiN{=}S{=}NSi(CH_3)_3$ in $CCl_4$ at 23°C for 24 h. The solution was reduced in volume and after 16 h at −20°C orange crystals were isolated; yield 41%. IR, NMR, and mass spectra of the compound are listed in Table 58, p. 148 [1].

Table 58

IR, NMR, and Mass Spectra of Derivatives of 1,3,5-Trithia(1,3,5-$S^{IV}$)-2,4,6,8,9-pentaazabicyclo[3.3.1]nona-1(9),2,3,5,7-pentaene, $S_3N_5CR$.

| R | IR spectrum (Nujol, in $cm^{-1}$) $\nu(CN_2)$ | $\nu$(–N=S=N–) | NMR spectra ($CDCl_3$/TMS, $\delta$ in ppm) $^1H$ NMR | $^{13}C$ NMR | mass spectrum (EI, 70 eV) m/e (relative intensity in %) | Ref. |
|---|---|---|---|---|---|---|
| $N(CH_3)_2$ | 1537 | 1032, 1015, 547 | 3.03 (s, $CH_3$) | 35.87 ($CH_3$)<br>154.87 (ring) | 222 $M^+$ (0.5), 162 $M^+$–$SN_2$ (42), 148 (49), 116 (31), 102 (37), 92 (27), 78 (60), 70 (44), 69 (75), 46 (100) | [1] |
| $N(C_2H_5)_2$ | 1514 | 1030, 548 | 1.16 (t, $CH_3$),<br>3.47 (q, $CH_2$),<br>$^3J(CH_2,CH_3)=6.6$ Hz | 13.11 ($CH_3$)<br>41.13 ($CH_2$)<br>153.78 (ring) | 250 $M^+$ (1.5), 190 $M^+$–$SN_2$ (32), 176 (42), 98 (32), 78 (62), 72 (61), 64 (38), 46 (100) | [1] |
| $N(C_3H_7\text{-}i)_2$ | 1496 | 1017, 545 | 1.26 (d, $CH_3$),<br>4.25 (sept, CH),<br>$^3J(CH,CH_3)=7.2$ Hz | 20.50 ($CH_3$)<br>45.92 (CH)<br>153.81 (ring) | — | [1] |
| $C_6H_5$ | 28 unassigned bands in the 1600 to 250 $cm^{-1}$ region | | — | 165.8 (ring) | 195 $S_2N_3CC_6H_5^+$ (23),<br>181 $S_2N_2CC_6H_5^+$ (75),<br>149 $SN_2CC_6H_5^+$ (29),<br>135 $SNCC_6H_5^+$ (30),<br>103 $CNC_6H_5^+$ (100) | [4] |

### 12.1.1.4 $S_3N_5C(N(C_3H_7\text{-}i)_2)$

The compound is formed from the reaction of equimolar amounts of $S_2N_3CCl_2(N(C_3H_7\text{-}i)_2)$ (see p. 10) and $(CH_3)_3SiN{=}S{=}NSi(CH_3)_3$ in n-pentane at 23°C for 24 h. The solution was reduced in volume. After 72 h at −20°C orange crystals were isolated; yield ~42%. IR and NMR spectra of the compound are listed in Table 58 [1].

An X-ray diffraction study shows that crystals of the compound are triclinic, space group $P\bar{1}\text{-}C_i^1$ (No. 2) with a = 5.648(1), b = 9.162(2), c = 13.432(2) Å, α = 69.52(1)°, β = 79.58(2)°, and γ = 76.66(2)°; Z = 4. V = 629.7(5) $Å^3$, $D_x$ = 1.468 g/$cm^3$, R = 0.042, $R_w$ = 0.040. Atomic positions are listed in the paper [7]. The molecular structure of the compound is similar to that of the phenyl-substituted compound, see Fig. 35, p. 146. Selected bond lengths and bond angles are given in Table 57, p. 147.

### 12.1.1.5 $S_3N_5C(C_6H_5)$

The bicyclic compound has been prepared by slow addition of a solution of $C_6H_5C({=}NSi(CH_3)_3)N[Si(CH_3)_3]_2$, in $CH_2Cl_2$ or $CH_3CN$ to a stirred solution of equimolar amounts of $S_3N_3Cl_3$ in $CH_2Cl_2$ or $CH_3CN$ at 0°C. Subsequent removal of the solvent and recrystallization of the residue from hot $CH_3CN$ afforded golden plates of the title compound in a yield of 61%. The compound can also be reformed from its oxidation product $S_2N_3CCl_2(C_6H_5)$ (see p. 11) by treating with $(CH_3)_3SiN{=}S{=}NSi(CH_3)_3$ in $CH_2Cl_2$; yield 79% [3, 4].

An X-ray diffraction study of the compound showed crystals to be monoclinic, space group $P2_1/n$ (standard setting $P2_1/c$)-$C_{2h}^5$ (No. 14) with a = 5.958(1), b = 22.955(2), c = 7.427(1) Å, and β = 106.25°; Z = 4. V = 975.2 $Å^3$, $D_x$ = 1.740 g/$cm^3$. R = 0.030 for 1364 independent reflections [3], see also [5]. Fractional atomic coordinates and isotropic thermal parameters are given in the paper [5]. The molecular structure of $S_3N_5C(C_6H_5)$ is shown in Fig. 35, p. 146. Principal bond lengths and bond angles are listed in Table 57, p. 147. The mass spectrum is given in Table 58.

The yellow-orange plates (from hot $CH_3CN$) decompose at 136°C. The compound is air-stable. Oxidation with $Cl_2$ in $CCl_4$ yields $S_2N_3CCl_2(C_6H_5)$ along with small amounts of $SN_3C_2Cl(C_6H_5)_2$ (see p. 55) and 4-phenyl-1,2,3,5-dithiadiazolium chloride (I) [3, 4]. The reactions with triphenylphosphine and triphenylarsine produce the corresponding 3-imino-7-phenyl-1,3,5,2,4,6,8-trithiatetrazocines $S_3N_4C(N{=}E(C_6H_5)_3)(C_6H_5)$, E = P or As (see pp. 128, 129) [6].

$[C_6H_5CN_2S_2]^+ Cl^-$

I

**References:**


[1] Chivers, T., Richardson, J. F., Smith, N. R. M. (Inorg. Chem. **25** [1986] 272/5).
[2] Chivers, T., Richardson, J. F., Smith, N. R. M. (Mol. Cryst. Liquid Cryst. **125** [1985] 319/27).
[3] Boeré, R. T., Cordes, A. W., Oakley, R. T. (J. Chem. Soc. Chem. Commun. **1985** 929/30).
[4] Boeré, R. T., French, C. L., Oakley, R. T., Cordes, A. W., Privett, J. A. J., Craig, S. L., Graham, J. B. (J. Am. Chem. Soc. **107** [1985] 7710/7).
[5] Cordes, A. W., Oakley, R. T., Boeré, R. T. (Acta Cryst. C **41** [1985] 1833/4).
[6] Boeré, R. T., Cordes, A. W., Craig, S. L., Graham, J. B., Oakley, R. T., Privett, J. A. J. (J. Chem. Soc. Chem. Commun. **1986** 807/8).
[7] Chivers, T., Edelmann, F., Richardson, J. F., Smith, N. R. M., Treu, Jr., O., Trsic, M. (Inorg. Chem. **25** [1986] 2119/25).

## 12.1.2 Derivatives of 1,3,5-Trithia(1,3,5-$S^{IV}$)-2,4,6,8,9-pentaazabicyclo[3.3.1]nona-1,3,5(9)-trien-7-one Ion (1+) and Salts

$R^1 = R^2 = CH_3$; $R^1 = CH_3$, $R^2 = C_6H_5$

**[–{1,3-$S_3N_3$}–$NR^1$–CO–$NR^2$–]$^+$**

### 12.1.2.1 The Cation

The bicylic cation forms from the reaction of $S_3N_3Cl_3$ or $S_4N_4Cl_2$ with trimethylsilyl substituted urea derivatives [1, 2].

An X-ray study of [–{1,3-$S_3N_3$}–$NCH_3$–CO–$NCH_3$–]$^+AsF_6^-$ shows that the skeleton of the cation (see **Fig. 36**) consists of a six-membered ring of alternating S and N atoms bridged by an –N–CO–N– group. Alternatively, the cation may be viewed as consisting of a six-membered 1,3,2,4,6-dithiatriaza-cyclic ketone with the sulfur atoms bridged by a sulfur diimide group. The short S···S distances indicate strong bonding interactions. Three different planes can be drawn through the molecule: plane I: S(1)N(1)C(1)N(2)S(2); plane II: S(1)N(5)S(3)N(4)S(2); plane III: S(1)N(3)S(2). The angles between these planes differ considerably: ∢I and II = 76.2°, ∢I and III = 59.3°, and ∢II and III = 45.3°. Bond distances and bond angles are given in Table 59 [1, 2].

Table 59
Bond Lengths and Bond Angles of the [–{1,3-$S_3N_3$}–$NCH_3$–CO–$NCH_3$–]$^+$ Cation [1, 2].

| bond length in Å | | | | | |
|---|---|---|---|---|---|
| S(1)–S(2) | 2.703(2) | S(2)–S(3) | 2.830(2) | S(3)–N(5) | 1.523(5) |
| S(1)–S(3) | 2.802(2) | S(2)–N(2) | 1.625(5) | N(1)–C(1) | 1.370(6) |
| S(1)–N(1) | 1.653(4) | S(2)–N(3) | 1.599(5) | N(1)–C(2) | 1.487(7) |
| S(1)–N(3) | 1.602(5) | S(2)–N(4) | 1.703(5) | N(2)–C(1) | 1.400(6) |
| S(1)–N(5) | 1.708(5) | S(3)–N(4) | 1.540(5) | N(2)–C(3) | 1.493(8) |
| | | | | C(1)–O | 1.213(6) |

| bond angle in ° | | | | | |
|---|---|---|---|---|---|
| N(1)–S(1)–N(3) | 105.6(2) | N(4)–S(3)–N(5) | 119.3(3) | C(1)–N(2)–C(3) | 117.1(5) |
| N(1)–S(1)–N(5) | 102.0(2) | S(1)–N(1)–C(1) | 126.4(3) | S(1)–N(3)–S(2) | 115.2(3) |
| N(3)–S(1)–N(5) | 107.2(2) | S(1)–N(1)–C(2) | 116.1(4) | S(2)–N(4)–S(3) | 121.5(3) |
| N(2)–S(2)–N(3) | 104.2(2) | C(1)–N(1)–C(2) | 117.1(4) | S(1)–N(5)–S(3) | 120.1(3) |
| N(2)–S(2)–N(4) | 102.6(2) | S(2)–N(2)–C(1) | 125.0(3) | N(1)–C(1)–N(2) | 118.8(4) |
| N(3)–S(2)–N(4) | 108.7(2) | S(2)–N(2)–C(3) | 117.1(5) | N(1)–C(1)–O | 121.5(5) |
| | | | | N(2)–C(1)–O | 111.7(7) |

The $^1H$ NMR spectra of salts of the cation with $R^1 = R^2 = CH_3$ (see Table 60, p. 152) show only one signal for the $CH_3$ groups in the range 3.15 to 3.46 ppm (vs. TMS), indicating equivalence of the $CH_3$ groups at room temperature [1, 2].

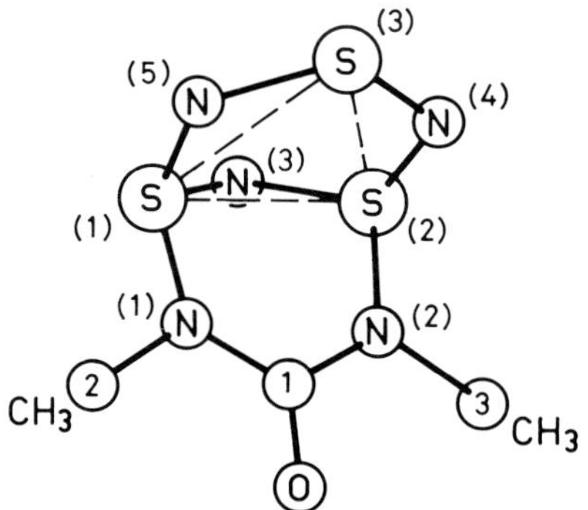


Fig. 36. Molecular structure of the $[-\{1,3\text{-}S_3N_3\}-NCH_3-CO-NCH_3-]^+$ cation.

The shift of the C=O stretching vibration from 1690 to 1695 $cm^{-1}$ in the chloro compound to 1705 to 1735 $cm^{-1}$ in the adducts with Lewis acids indicates that the electron density at the CO group increases and that the compounds have salt-like structures [1].

The occurrence of a weak molecular peak in the mass spectrum of $[-\{1,3\text{-}S_3N_3\}-NCH_3-CO-NC_6H_5-]^+Cl^-$ at m/e = 321 (1%), however, indicates that the ionic structure of this bicyclic system is only one resonance structure [1].

**References:**

[1] Roesky, H. W., Müller, T., Wehner, E., Rodek, E. (Chem. Ber. **113** [1980] 2802/7).
[2] Roesky, H. W., Müller, T., Rodek, E. (J. Chem. Soc. Chem. Commun. **1979** 439/40).

### 12.1.2.2 Salts of the $[-\{1,3\text{-}S_3N_3\}-NR^1-CO-NR^2-]^+$ Cation

**$[-\{1,3\text{-}S_3N_3\}-NR^1-CO-NR^2-]^+X^-$**

$R^1 = CH_3$, $R^2 = CH_3$, $C_6H_5$ (Anions are compiled in Table 60, p. 152)

The salts $[-\{1,3\text{-}S_3N_3\}-NCH_3-CO-NCH_3-]^+Cl^-$ and $[-\{1,3\text{-}S_3N_3\}-NCH_3-CO-NC_6H_5-]^+Cl^-$ form on stirring a suspension of $S_3N_3Cl_3$ or $S_4N_4Cl_2$ (see [2]) with $[(CH_3)_3SiNCH_3]_2CO$ [1, 2] or $(CH_3)_3SiN(CH_3)C(O)N(C_6H_5)Si(CH_3)_3$ [1] (1:1 mole ratio) in n-hexane for 12 or 6 h, respectively.

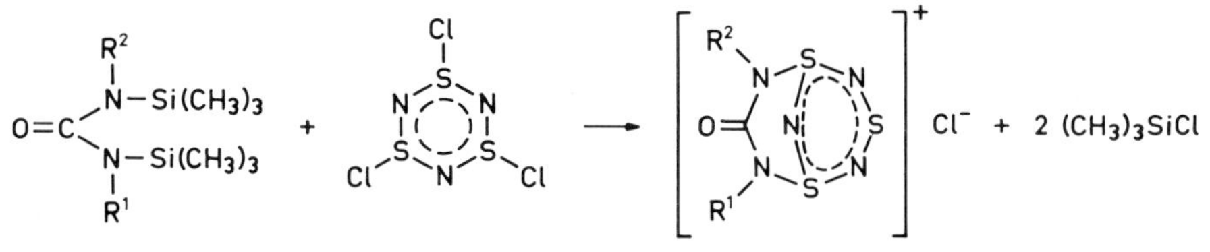

$$O{=}C[N(R^2){-}Si(CH_3)_3][N(R^1){-}Si(CH_3)_3] + S_3N_3Cl_3 \longrightarrow [-\{1,3\text{-}S_3N_3\}-NR^1-CO-NR^2-]^+Cl^- + 2\,(CH_3)_3SiCl$$

$R^1 = R^2 = CH_3$; $R^1 = CH_3$, $R^2 = C_6H_5$

Compounds with the anions $AsF_6^-$, $SbCl_6^-$, $SnCl_5^-$, or $TiCl_5^-$ form nearly quantitatively in the reaction of the chloride with excess $AsF_5$, $SbCl_5$, $SnCl_4$, or $TiCl_4$ in $CCl_4$. The compounds are very sensitive toward hydrolysis. They are compiled in Table 60, p. 152.

Table 60

Salts of the 1,3,5-Trithia(1,3,5-$S^{IV}$)-2,4,6,8,9-pentaazabicyclo[3.3.1]nona-1,3,5(9)-trien-7-one Ion (1+), $[-\{1,3\text{-}S_3N_3\}-NR^1-CO-NR^2-]^+ X^-$ [1,2]. Further information on compounds marked with an asterisk is given on p. 153.

| No. | X | $R^1$ | $R^2$ | yield in % | m.p. in °C | $^1H$ NMR spectrum (solvent(?)/ TMS, δ in ppm) | IR spectrum ν(C=O) in $cm^{-1}$ | mass spectrum (70 eV) m/e (relative intensity) | remarks |
|---|---|---|---|---|---|---|---|---|---|
| *1 | $Cl^-$ | $CH_3$ | $CH_3$ | 90 | 114 (dec.) | 3.15 (s, 6H) | 1690 | 244 $M^+ - Cl$ (54%), 28 $CO^+/N_2^+$ (100%) | yellow powder |
| *2 | $Cl^-$ | $CH_3$ | $C_6H_5$ | 88 | 100 (dec.) | 3.39 (s, 3H), 7.51 (m, 5H) | 1695 | 321 $M^+$ (1%), 286 $M^+ - Cl$ (8%), 119 $C_6H_5NCO^+$ (100%) | light brown precipitate |
| *3 | $AsF_6^-$ | $CH_3$ | $CH_3$ | 89 | 126 | 3.46 (s, 6H, in $CH_3CN$) | 1705 | unstable | lemon yellow powder |
| 4 | $SbCl_6^-$ | $CH_3$ | $CH_3$ | 98 | 128 (dec.) | 3.44 (s, 6H) | 1735 | 227 $SbCl_3^+$ (23%), 46 $NS^+$ (100%) | — |
| 5 | $SbCl_6^-$ | $CH_3$ | $C_6H_5$ | 99 | 107 (dec.) | 3.51 (s, 3H), 7.60 (m, 5H) | 1735 | 227 $SbCl_3^+$ (21%), 46 $NS^+$ (100%) | — |
| 6 | $SnCl_5^-$ | $CH_3$ | $CH_3$ | 89 | 143 (dec.) | 3.40 (s, 6H) | 1730 | 260 $SnCl_4^+$ (5%), 46 $NS^+$ (100%) | — |
| 7 | $TiCl_5^-$ | $CH_3$ | $CH_3$ | 95 | 132 (dec.) | 3.29 (s, 6H) | 1720 | 224 $M^+ - TiCl_5$ (57%), 46 $NS^+$ (100%) | — |

* Further information:

**[–{1,3-$S_3N_3$}–$NCH_3$–CO–$NCH_3$–]$^+Cl^-$** (Table 60, No. **1**). The salt reacts with $AsF_5$, $SbCl_5$, $SnCl_4$, and $TiCl_4$ to give the corresponding $AsF_6^-$, $SbCl_6^-$, $SnCl_5^-$, and $TiCl_5^-$ salts, respectively [1, 2]. Reactions with trimethylsilyl-substituted amines $(CH_3)_3SiNR_2$ yield the substitution products [–{1,3-$S_3N_3$($NR_2$-3)}–$NCH_3$–CO–$NCH_3$–] (R = $CH_3$, $C_6H_5$) (see below) [1].

**[–{1,3-$S_3N_3$}–$NCH_3$–CO–$NC_6H_5$–]$^+Cl^-$** (Table 60, No. **2**). The salt reacts with $SbCl_5$ to give the corresponding $SbCl_6^-$ salt [1].

**[–{1,3-$S_3N_3$}–$NCH_3$–CO–$NCH_3$–]$^+AsF_6^-$** (Table 60, No. **3**). X-ray diffraction measurements show that the crystals are orthorhombic, space group $P2_12_12_1$-$D_2^4$ (No. 19), with a = 21.077(6), b = 8.574(2), and c = 6.949(2) Å; Z = 4, $D_x$ = 2.18 g/cm$^3$ and R = 0.053 [1, 2]. Atomic parameters and anisotropic temperature factors were given [1]. The $^{19}$F NMR spectrum (solvent was not given, $CFCl_3$ as internal standard) shows a peak at δ = −65.67 ppm, J(AsF) = 935 Hz [1].

**References:**

[1] Roesky, H. W., Müller, T., Wehner, E., Rodek, E. (Chem. Ber. **113** [1980] 2802/7).
[2] Roesky, H. W., Müller, T., Rodek, E. (J. Chem. Soc. Chem. Commun. **1979** 439/40).

### 12.1.3 Derivatives of 3-Dialkylamino-6,8-dimethyl-1,3,5-trithia(1,3,5-$S^{IV}$)-2,4,6,8,9-pentaaza-bicyclo[3.3.1]nona-1,3,5(9)-trien-7-one

R = $CH_3$, $C_2H_5$

**[–{1,3-$S_3N_3$($NR_2$-3)}–$NCH_3$–CO–$NCH_3$–]**

**Preparation. Properties.** The dimethylamino derivative (compound **1**, R = $CH_3$) is obtained by adding $(CH_3)_3SiN(CH_3)_2$ to a solution of [–{1,3-$S_3N_3$}–$NCH_3$–CO–$NCH_3$–]$^+Cl^-$ in $CCl_4$ (1:1 mole ratio) at 0°C. The reaction mixture is then stirred at room temperature for 18 h. Evaporation of the $(CH_3)_3SiCl$ in vacuum leaves an orange-colored oil which crystallizes after some hours. Yield 59%, m.p. = 98°C.

The diethylamino derivative (compound **2**, R = $C_2H_5$) forms on adding $(CH_3)_3SiN(C_2H_5)_2$ to a suspension of [–{1,3-$S_3N_3$}–$NCH_3$–CO–$NCH_3$–]$^+Cl^-$ in n-hexane, followed by stirring at room temperature for 4 h. After removing the solvent at −20°C, the product precipitates as a yellow powder in 44% yield. M.p. = 30°C.

The compounds are water-stable and are readily soluble in nonpolar solvents.

**Spectra.** $^1$H NMR spectra (solvent not given, TMS internal standard, δ in ppm): **1** (R = $CH_3$): 2.42(s), 3.41(s); **2** (R = $C_2H_5$): 3.34(s), 1.24(t), 3.38(q), $^3J(CH_3, CH_2)$ = 7.5 Hz.

IR spectra (in Nujol): Both spectra show an absorption band at 680 cm$^{-1}$ which was assigned to the exocyclic S–N vibration. The C=O stretching vibration was observed at 1675 (**1**) and 1650 cm$^{-1}$ (**2**), respectively. Another 15 absorption bands for **1** and 19 for **2** are reported in the paper.

Mass spectra: The molecular peaks were observed with low intensity at m/e = 266 $M^+$ (2%) for **1** and m/e = 296 $M^+$ (3%) for **2**. The base peak in both spectra is at m/e = 46 $NS^+$ (100%).

**Structure.** On the basis of the $^1H$ NMR data and the absence of a fragment $(R_2N–S–N–C–O)^+$ in the mass spectra, it was assumed that the $–NR_2$ group is bonded to the S atom with the coordination number two.

**Reference:**

Roesky, H. W., Müller, T., Wehner, E., Rodek, E. (Chem. Ber. **113** [1980] 2802/7).

## 12.2 $SN_2C_6$ Ring System

### 12.2.1 3,4,5,6,7,8-Hexahydro-1,2,9-thia(1-$S^{IV}$)-diazonine

**[=S=N–(CH$_2$)$_6$–N=]** (Reduced formula in the text $SN_2(CH_2)_6$)

The unstable compound $SN_2(CH_2)_6$ is reported to form in 8% yield in the reaction of $(4\text{-}CH_3C_6H_4SO_2N{=})_2S$ with $H_2N(CH_2)_6NH_2$ [1, 2]. More information is given on p. 124.

**References:**

[1] Kresze, G., Grill, H. (Tetrahedron Letters **1969** 4117/20, 4118).
[2] Grill, H. (Diss. T.H. München 1970, pp. 1/91, 29/37, 74/7).

### 12.2.2 1,4,8,11-Tetrahydro-4a,7a-methano-4aH,7aH,12H-bisbenz[2,3]azirino-[1,2-b:2′,1′-e][1,2,6]thiadiazine 6-Oxide (1), 5,9-Methano-1H (or 14H)-[1,2,6]thiadiazino[2,3-a:6,5-a′]bisazepine 7-Oxide (2), 9,5-Metheno-5H-[1,2,6]thiadiazino[2,3-a:6,5-a′]bisazepine 7-Oxide (3)

**1** $S(=O)N_2C_{14}H_{16}$

**2** $S(=O)N_2C_{14}H_{12}$

**3** $S(=O)N_2C_{14}H_{10}$

The compounds **1**, **2**, and **3** are intermediates in the preparation of the annulene derivative II, see the scheme on p. 155.

Compound **1** is prepared in 65% yield by reacting 1,4,5,8-tetrahydro-9H,10H-anthracene-4a,9a:8a,10a-diimine (I) with equimolar amounts of $SOCl_2$ in the presence of 2 equivalents of $N(C_2H_5)_3$ in ether at 0°C and then at room temperature for 12 h. M.p. = 204 to 205°C.

Compound **2** forms in 35% yield by bromination of **1** with bromine (1:2 mole ratio) in $CH_2Cl_2$ at −78°C, followed by dehydrobromination with 6 equivalents of 2,6-dichlorobenzonitrile (DBN) in THF at −10°C for 2 h and then at room temperature for 24 h.

The annulene derivative **3** is obtained in 75% yield by treatment of **2** with an excess of 2,3-dichloro-5,6-dicyanobenzoquinone (DDQ) in refluxing benzene for 2 h. Acid hydrolysis of compound **3** with $CH_3COOH - 1N\ HCl$ (1:1) at 80°C for 30 min yields 1,6:8,13-diimino[14]annulene (II).

I $\xrightarrow[(C_2H_5)_3N]{SOCl_2}$ 1 $\xrightarrow[b)\ DBN]{a)\ Br_2}$ 2 $\xrightarrow{DDQ}$ 3 $\xrightarrow{CH_3COOH/1N\ HCl}$ II

**Reference:**

Vogel, E., Kuebart, F., Marco, J. A., Andree, R., Günther, H., Aydin, R. (J. Am. Chem. Soc. **105** [1983] 6982/3).

# Index

The index includes two different parts:

the "Formula Index" – starting on p. 157 and

the "Ring Index" – beginning on p. 211.

In the Formula Index the cyclic sulfur-nitrogen compounds described in "Sulfur-Nitrogen Compounds" Part 3 and Part 4 are arranged according to their empirical formulas. The first criterion used for indexing the compounds is the number of sulfur atoms, this is followed by the number of nitrogen atoms, and then by the remaining elements in alphabetical order, based on the number of atoms present in each case.

The first column of the Formula Index contains the empirical formulas of the isolated compounds, while the structural details are shown in the second column; the third column gives the page references. Salts or addition compounds can also be found under the specific ion or compound involved.

For derivatives of 1,2,5-thiadiazole the reduced ring formula 1,2,5-$SN_2C_2$ is not disjoined.

In the Ring Index the compounds are arranged according to their ring parents, which are divided first into monocyclic, bicyclic, tricyclic, tetracyclic, and hexacyclic systems, always beginning with the smallest rings.

The first column gives the ring analysis data and the structural formulas of the ring parents. The ring formulas are arranged in the same way as the empirical formulas. The second column contains the names of the unsubstituted ring parents as main entries, followed by additional modifying entries, for example, "Adducts and complexes", "Derivatives", "Ions", or "Salts". There, the specific compounds can be found in abbreviated form, for example, the 3-fluoro-2,5,5-tris(trifluoromethyl) derivative of the $SN_2CO$ ring ("1,3,2,4-Oxathiadiazole") is listed under the main entry

3H-1,3,2,4-Oxathiadiazole, 2,5-dihydro-
(=1,3$\lambda^4$,2,4-Oxathiadiazole, 2,5-dihydro-)

and the additional entry

Derivative:

(S: 3) –F; (N: 2) –$CF_3$; (C: 5) –$CF_3$, –$CF_3$

(The ring atom and location of the bonding site are given in parentheses.) The third column gives the page references.

The numbering and nomenclature of the ring parents correspond to IUPAC rules.

**$S_1$**

| | | | |
|---|---|---|---|
| $SN_2C_6H_8O_2$ | $1,2,5\text{-}SN_2C_2H\text{-}3\text{-}CH_2COOC_2H_5$ | **3** | 124 |
| $SN_2C_6H_{10}O$ | $1,2,5\text{-}SN_2C_2\text{-}3\text{-}OH\text{-}4\text{-}C_4H_9$ | **3** | 150/1 |
| | $1,2,5\text{-}SN_2C_2\text{-}3\text{-}OH\text{-}4\text{-}C_4H_9\text{-}i$ | **3** | 150/1 |
| $SN_2C_6H_{10}O_2$ | $1,2,5\text{-}SN_2C_2\text{-}3\text{-}OC_2H_5\text{-}4\text{-}OC_2H_5$ | **3** | 159 |
| $SN_2C_6H_{10}O_3$ | $[-SO-N{=}C(OC_2H_5)-C(OC_2H_5){=}N-]$ | **3** | 250/1, 253, 255/6 |
| $SN_2C_6H_{11}O^+$ | $[1,2,5\text{-}SN_2C_2H\text{-}2\text{-}C_3H_7\text{-}3\text{-}OCH_3]^+$ | | |
| | Salt: | | |
| | $[1,2,5\text{-}SN_2C_2H\text{-}2\text{-}C_3H_7\text{-}3\text{-}OCH_3]^+FSO_3^-$ | **3** | 243 |
| $SN_2C_6H_{12}$ | $[{=}S{=}N-(CH_2)_6-N{=}] \ = SN_2(CH_2)_6$ | **4** | 154 |
| $SN_2C_6H_{14}O$ | $[-SO-NC_2H_5-CH_2-CH_2-NC_2H_5-]$ | **3** | 264/5 |
| $SN_2C_6H_{18}OSi_2$ | $[-SO-NCH_3-Si(CH_3)_2-Si(CH_3)_2-NCH_3-]$ | **3** | 6/7 |
| $SN_2C_7ClH_7$ | $[\dot{-}S\dot{-}NCH_3\dot{-}\{1,2\text{-}C_6H_4\}\dot{-}N\dot{-}]^+Cl^-$ | **3** | 246, 248/9 |
| $SN_2C_7ClH_7O_4$ | $[\dot{-}S\dot{-}NCH_3\dot{-}\{1,2\text{-}C_6H_4\}\dot{-}N\dot{-}]^+ClO_4^-$ | **3** | 246, 248/9 |
| $SN_2C_7ClH_{11}O_2$ | $1,2,5\text{-}SN_2C_2\text{-}3\text{-}OCH_2CH(OH)CH_2Cl\text{-}4\text{-}C_2H_5$ | **3** | 157 |
| $SN_2C_7H_4O$ | $[{=}S{=}N-CO-\{1,2\text{-}C_6H_4\}-N{=}]$ | **4** | 64 |
| $SN_2C_7H_6O_2$ | $[{=}S{=}N-CO-\{\overset{\|}{C}-C(CH_3){=}C(CH_3)-O-\overset{\|}{C}\}-N{=}]$ | **4** | 70/1 |
| | $[-SO-NH-CO-\{1,2\text{-}C_6H_4\}-NH-]$ | **4** | 78/80 |
| $SN_2C_7H_7^+$ | $[\dot{-}S\dot{-}NCH_3\dot{-}\{1,2\text{-}C_6H_4\}\dot{-}N\dot{-}]^+$ | | |
| | Salts: | | |
| | $[\dot{-}S\dot{-}NCH_3\dot{-}\{1,2\text{-}C_6H_4\}\dot{-}N\dot{-}]^+Br^-$ | **3** | 246, 248/9 |
| | $[\dot{-}S\dot{-}NCH_3\dot{-}\{1,2\text{-}C_6H_4\}\dot{-}N\dot{-}]^+4\text{-}CH_3C_6H_4SO_3^-$ | **3** | 246, 248/9 |
| | $[\dot{-}S\dot{-}NCH_3\dot{-}\{1,2\text{-}C_6H_4\}\dot{-}N\dot{-}]^+CNS^-$ | **3** | 246, 248/9 |
| | $[\dot{-}S\dot{-}NCH_3\dot{-}\{1,2\text{-}C_6H_4\}\dot{-}N\dot{-}]^+Cl^-$ | **3** | 246, 248/9 |
| | $[\dot{-}S\dot{-}NCH_3\dot{-}\{1,2\text{-}C_6H_4\}\dot{-}N\dot{-}]^+ClO_4^-$ | **3** | 246, 248/9 |
| | $[\dot{-}S\dot{-}NCH_3\dot{-}\{1,2\text{-}C_6H_4\}\dot{-}N\dot{-}]^+HSO_4^-$ | **3** | 246, 248/9 |
| | $[\dot{-}S\dot{-}NCH_3\dot{-}\{1,2\text{-}C_6H_4\}\dot{-}N\dot{-}]^+I^-$ | **3** | 246, 248/9 |
| | $[\dot{-}S\dot{-}NCH_3\dot{-}\{1,2\text{-}C_6H_4\}\dot{-}N\dot{-}]^+2,4,6\text{-}(NO_2)_3C_6H_2O^-$ | **3** | 246, 248/9 |
| | $[\dot{-}S\dot{-}NCH_3\dot{-}\{1,2\text{-}C_6H_4\}\dot{-}N\dot{-}]^+2,4\text{-}(NO_2)_2C_6H_3SO_3^-$ | **3** | 246, 248/9 |

| Formula | Compound | Vol. | Page |
|---|---|---|---|
| $SN_2C_{10}H_{13}^+$ | $[∸S∸NC_4H_9∸\{1,2\text{-}C_6H_4\}∸N∸]^+$ | | |
| | Salts: | | |
| | $[∸S∸NC_4H_9∸\{1,2\text{-}C_6H_4\}∸N∸]^+I^-$ | **3** | 246, 249/50 |
| | $[∸S∸NC_4H_9∸\{1,2\text{-}C_6H_4\}∸N∸]^+2,4\text{-}(NO_2)_2C_6H_3SO_3^-$ | **3** | 246, 249/50 |
| $SN_2C_{10}H_{13}I$ | $[∸S∸NC_4H_9∸\{1,2\text{-}C_6H_4\}∸N∸]^+I^-$ | **3** | 246, 249/50 |
| $SN_2C_{10}H_{14}O$ | $[\text{-}SO\text{-}NCH_3\text{-}CH(CH_3)\text{-}CH_2\text{-}NC_6H_5\text{-}]$ | **3** | 266/7 |
| | $[\text{-}SO\text{-}NCH_3\text{-}CH_2\text{-}CH(CH_3)\text{-}NC_6H_5\text{-}]$ | **3** | 266/7 |
| $SN_2C_{10}H_{16}O_2$ | $[\text{-}SO\text{-}NCH_3\text{-}CH_2\text{-}CH_2\text{-}NCH_3\text{-}]\cdot C_6H_5OH$ | **3** | 266 |
| $SN_2C_{10}H_{18}O_3$ | $[\text{-}SO\text{-}N(C_4H_9\text{-}t)\text{-}CO\text{-}CO\text{-}N(C_4H_9\text{-}t)\text{-}]$ | **3** | 270/1 |
| $SN_2C_{10}H_{30}OSi_4$ | $[\text{-}SO\text{-}N(Si(CH_3)_3)\text{-}Si(CH_3)_2\text{-}Si(CH_3)_2\text{-}N(Si(CH_3)_3)\text{-}]$ | **3** | 6/7 |
| $SN_2C_{11}ClF_3H_{12}O$ | $[\text{-}SO\text{-}NCH_3\text{-}CH(CH_3)\text{-}CH_2\text{-}N(C_6H_3CF_3\text{-}5\text{-}Cl\text{-}2)\text{-}]$ | **3** | 266/7 |
| $SN_2C_{11}ClH_{11}O_2$ | $1,2,5\text{-}SN_2C_2\text{-}3\text{-}OCH_2CH(OH)CH_2Cl\text{-}4\text{-}C_6H_5$ | **3** | 157 |
| $SN_2C_{11}ClH_{13}O_2$ | $[\text{-}SO\text{-}NC_4H_9\text{-}C(C_6H_4Cl\text{-}4){=}N\text{-}O\text{-}]$ | **3** | 313/5 |
| | $[\text{-}SO\text{-}NC_2H_5\text{-}CO\text{-}\{1,2\text{-}C_6H_3Cl\text{-}4\}\text{-}NC_2H_5\text{-}]$ | **4** | 78/80 |
| $SN_2C_{11}F_3H_{13}O$ | $[\text{-}SO\text{-}NCH_3\text{-}CH(CH_3)\text{-}CH_2\text{-}N(C_6H_4CF_3\text{-}3)\text{-}]$ | **3** | 266/7 |
| $SN_2C_{11}H_{10}O_2$ | $1,2,5\text{-}SN_2C_2\text{-}3\text{-}OCH_2CHCH_2O\text{-}4\text{-}C_6H_5$ | **3** | 157 |
| $SN_2C_{11}H_{12}$ | $1,2,5\text{-}SN_2C_2\text{-}3\text{-}C_6H_5\text{-}4\text{-}C_3H_7$ | **3** | 129 |
| | $1,2,5\text{-}SN_2C_2\text{-}3\text{-}C_6H_5\text{-}4\text{-}C_3H_7\text{-}i$ | **3** | 129 |
| $SN_2C_{11}H_{12}O_2$ | $[\text{-}SO\text{-}NCH_3\text{-}CO\text{-}\{1,2\text{-}C_6H_3\text{-}2,3\text{-}\{\text{-}(CH_2)_3\text{-}N\}\}\text{-}]$ | **4** | 82 |
| $SN_2C_{11}H_{13}O^+$ | $[1,2,5\text{-}SN_2C_2\text{-}3\text{-}OC_2H_5\text{-}4\text{-}C_6H_5\text{-}5\text{-}CH_3]^+$ | | |
| | Salt: | | |
| | $[1,2,5\text{-}SN_2C_2\text{-}3\text{-}OC_2H_5\text{-}4\text{-}C_6H_5\text{-}5\text{-}CH_3]^+BF_4^-$ | **3** | 245 |
| $SN_2C_{11}H_{14}O_2$ | $[\text{-}SO\text{-}N(C_4H_9\text{-}t)\text{-}C(C_6H_5){=}N\text{-}O\text{-}]$ | **3** | 313/5 |
| $SN_2C_{11}H_{15}^+$ | $[∸S∸NC_5H_{11}∸\{1,2\text{-}C_6H_4\}∸N∸]^+$ | | |
| | Salts: | | |
| | $[∸S∸NC_5H_{11}∸\{1,2\text{-}C_6H_4\}∸N∸]^+I^-$ | **3** | 246, 249/50 |
| | $[∸S∸NC_5H_{11}∸\{1,2\text{-}C_6H_4\}∸N∸]^+2,4\text{-}(NO_2)_2C_6H_3SO_3^-$ | **3** | 246, 249/50 |
| | $[∸S∸N(C_5H_{11}\text{-}i)∸\{1,2\text{-}C_6H_4\}∸N∸]^+$ | | |
| | Salts: | | |
| | $[∸S∸N(C_5H_{11}\text{-}i)∸\{1,2\text{-}C_6H_4\}∸N∸]^+I^-$ | **3** | 246, 249/50 |
| | $[∸S∸N(C_5H_{11}\text{-}i)∸\{1,2\text{-}C_6H_4\}∸N∸]^+2,4\text{-}(NO_2)_2C_6H_3SO_3^-$ | **3** | 246, 249/50 |

| | | | |
|---|---|---|---|
| $SN_2C_{11}H_{15}I$ | $[\dot{-}S\dot{-}NC_5H_{11}\dot{-}\{1,2\text{-}C_6H_4\}\dot{-}N\dot{-}]^+I^-$ | **3** | 246, 249/50 |
| | $[\dot{-}S\dot{-}N(C_5H_{11}\text{-}i)\dot{-}\{1,2\text{-}C_6H_4\}\dot{-}N\dot{-}]^+I^-$ | **3** | 246, 249/50 |
| $SN_2C_{11}H_{16}O$ | $[-SO-NCH_3-CH(C_2H_5)-CH_2-NC_6H_5-]$ | **3** | 266/7 |
| | $[-SO-NC_2H_5-CH(CH_3)-CH_2-NC_6H_5-]$ | **3** | 266/7 |
| | $[-SO-NCH_3-CH(CH_3)-CH_2-N(C_6H_4CH_3\text{-}3)-]$ | **3** | 266/7 |
| $SN_2C_{11}H_{16}O_2$ | $[-SO-NCH_3-CH(CH_3)-CH_2-N(C_6H_4OCH_3\text{-}3)-]$ | **3** | 266/7 |
| $SN_2C_{11}H_{18}O_4$ | $[-SO-N(C_4H_9\text{-}t)-(CO)_3-N(C_4H_9\text{-}t)-]$ | **4** | 77 |
| $SN_2C_{11}H_{20}O_3$ | $[-SO-N(C_4H_9\text{-}t)-CO-CO-N(C(CH_3)_2C_2H_5)-]$ | **3** | 270/1 |
| $SN_2C_{12}ClH_{13}O_2$ | $1,2,5\text{-}SN_2C_2\text{-}3\text{-}OCH_2CH(OH)CH_2Cl\text{-}4\text{-}CH_2C_6H_5$ | **3** | 157 |
| $SN_2C_{12}ClH_{15}O_3$ | $[-SO-N(CH_2CH_2OC_2H_5)-CO-\{1,2\text{-}C_6H_3Cl\text{-}4\}-NCH_3-]$ | **4** | 78/80 |
| $SN_2C_{12}Cl_2H_{14}O_2$ | $[-SCl_2-N(C_4H_9\text{-}t)-CO-CO-NC_6H_5-]$ | **3** | 286 |
| $SN_2C_{12}H_8$ | $1,2,5\text{-}SN_2C_2H\text{-}3\text{-}(2'\text{-}C_{10}H_7)$ ($2\text{-}C_{10}H_7$ = 2-Naphthalenyl) | **3** | 129 |
| $SN_2C_{12}H_9IO_2$ | $[-SO-N(C_6H_4I\text{-}4)-CH(2\text{-}C_5H_4N)-O-]$ ($2\text{-}C_5H_4N$ = 2-Pyridinyl) | **3** | 63/4 |
| $SN_2C_{12}H_{10}O_2$ | $[-SO-NC_6H_5-CH(2\text{-}C_5H_4N)-O-]$ ($2\text{-}C_5H_4N$ = 2-Pyridinyl) | **3** | 63/4 |
| $SN_2C_{12}H_{10}O_5$ | $1,2,5\text{-}SN_2C_2\text{-}3\text{-}COOH\text{-}4\text{-}COC_6H_3(OCH_3)_2\text{-}2',5'$ | **3** | 181 |
| $SN_2C_{12}H_{12}O_2$ | $1,2,5\text{-}SN_2C_2\text{-}3\text{-}OCH_2\overline{CHCH_2O}\text{-}4\text{-}CH_2C_6H_5$ | **3** | 157 |
| $SN_2C_{12}H_{14}$ | $1,2,5\text{-}SN_2C_2\text{-}3\text{-}C_6H_5\text{-}4\text{-}C_4H_9$ | **3** | 130 |
| $SN_2C_{12}H_{14}O_3$ | $[-SO-N(C_4H_9\text{-}t)-CO-CO-NC_6H_5-]$ | **3** | 270, 272 |
| $SN_2C_{12}H_{16}O_2$ | $[-SO-N(C_4H_9\text{-}t)-CO-CH_2-NC_6H_5-]$ | **3** | 267/9 |
| $SN_2C_{12}H_{22}O_2$ | $[-SO-\{N-(CH_2)_{11}-C\}=N-O-]$ | **3** | 320/1 |
| $SN_2C_{12}H_{22}O_3$ | $[-SO-N(C(CH_3)_2C_2H_5)-CO-CO-N(C(CH_3)_2C_2H_5)-]$ | **3** | 270, 272 |
| $SN_2C_{13}ClH_{15}O_2$ | $[-SO-N(C_5H_9\text{-}c)-CO-\{1,2\text{-}C_6H_3Cl\text{-}4\}-NCH_3-]$ | **4** | 78/80 |
| $SN_2C_{13}Cl_2H_6O_5$ | $[-SO-N(C_6H_3Cl\text{-}2\text{-}NO_2\text{-}5)-CO-\{1,2\text{-}C_6H_3Cl\text{-}4\}-O-]$ | **4** | 95/6, 99 |
| $SN_2C_{13}H_{10}O_2$ | $[-SO-NC_6H_5-C(C_6H_5)=N-O-]$ | **3** | 313/4, 316/7 |
| $SN_2C_{13}H_{12}O_2$ | $[-SO-N(C_6H_4CH_3\text{-}4)-CH(2\text{-}C_5H_4N)-O-]$ ($2\text{-}C_5H_4N$ = 2-Pyridinyl) | **3** | 63/4 |
| $SN_2C_{13}H_{12}O_3$ | $[-SO-N(C_6H_4OCH_3\text{-}2)-CH(2\text{-}C_5H_4N)-O-]$ ($2\text{-}C_5H_4N$ = 2-Pyridinyl) | **3** | 63/4 |
| $SN_2C_{13}H_{12}O_5$ | $1,2,5\text{-}SN_2C_2\text{-}3\text{-}COOCH_3\text{-}4\text{-}COC_6H_3(OCH_3)_2\text{-}2',5'$ | **3** | 182 |

| | | | |
|---|---|---|---|
| $SN_3C_4H_8^+$ | $[1,2,5\text{-}SN_2C_2H\text{-}2\text{-}CH_3\text{-}3\text{-}NHCH_3]^+$ | | |
| | Salts: | | |
| | $[1,2,5\text{-}SN_2C_2H\text{-}2\text{-}CH_3\text{-}3\text{-}NHCH_3]^+Cl^-$ | **3** | 243/4 |
| | $[1,2,5\text{-}SN_2C_2H\text{-}2\text{-}CH_3\text{-}3\text{-}NHCH_3]^+FSO_3^-$ | **3** | 243/4 |
| $SN_3C_4H_{13}Si_2$ | $[{=}S{=}N{-}Si(CH_3)_2{-}NH{-}Si(CH_3)_2{-}N{=}]$ | **3** | 7 |
| $SN_3C_5ClH_2$ | [∸S∸N∸{C∸N∸CH∸CH∸CCl∸C}∸N∸] | **3** | 224/5 |
| | [∸S∸N∸{C∸CCl∸N∸CH∸CH∸C}∸N∸] | **3** | 225/6 |
| $SN_3C_5Cl_3F_3H_3O$ | [∸S($OCH_3$)∸N∸C($CCl_3$)∸N∸C($CF_3$)∸N∸] | **4** | 31/2 |
| $SN_3C_5Cl_6H_3O$ | [∸S($OCH_3$)∸N∸C($CCl_3$)∸N∸C($CCl_3$)∸N∸] | **4** | 31/2 |
| $SN_3C_5H_3$ | [∸S∸N∸{C∸N∸CH∸CH∸CH∸C}∸N∸] | **3** | 224/5 |
| | [∸S∸N∸{C∸CH∸N∸CH∸CH∸C}∸N∸] | **3** | 225 |
| $SN_3C_5H_3O$ | [∸S∸N∸{C∸C(OH)∸N∸CH∸CH∸C}∸N∸] | **3** | 225/6 |
| | [∸S∸N∸{C–CO–NH–CH=CH–C}∸N∸] | **3** | 225/6 |
| $SN_3C_5H_5O$ | [–SO–{N–CH=CH–CH=CH–C}=N–NH–] | **3** | 85/6 |
| | [–SO–{$N^+$–CH=CH–CH=CH–C}–NH–$N^-$–] | **3** | 85/6 |
| $SN_3C_5H_6O^+$ | [–SO–{N–CH=CH–CH=CH–C}=$NH^+$–NH–] | **3** | 86 |
| | [–SO–{$N^+$–CH=CH–CH=CH–C}–NH–NH–] | **3** | 86 |
| $SN_3C_5H_7O$ | $1,2,5\text{-}SN_2C_2\text{-}3\text{-}NH_2\text{-}4\text{-}OCH_2CH{=}CH_2$ | **3** | 146 |
| | $1,2,5\text{-}SN_2C_2H\text{-}3\text{-}CON(CH_3)_2$ | **3** | 174 |
| $SN_3C_5H_7O_2$ | $1,2,5\text{-}SN_2C_2H\text{-}3\text{-}NHCOOC_2H_5$ | **3** | 143 |
| | $1,2,5\text{-}SN_2C_2\text{-}3\text{-}CON(CH_3)_2\text{-}4\text{-}OH$ | **3** | 180/1 |
| $SN_3C_5H_9$ | $1,2,5\text{-}SN_2C_2H\text{-}3\text{-}NHC_3H_7$ | **3** | 142 |
| $SN_3C_5H_9O_2$ | $[{-}SO{-}N{=}C(OC_2H_5){-}C(NHCH_3){=}N{-}]$ | **3** | 250/1, 253, 255/6 |
| $SN_3C_5H_{15}Si_2$ | $[{=}S{=}N{-}Si(CH_3)_2{-}NCH_3{-}Si(CH_3)_2{-}N{=}]$ | **3** | 7 |
| $SN_3C_6ClH_8O$ | $1,2,5\text{-}SN_2C_2\text{-}3\text{-}Cl\text{-}4\text{-}NC_4H_8O$ ($NC_4H_8O$ = Morpholino) | **3** | 145/6 |
| $SN_3C_6ClH_{12}$ | $[1,2,5\text{-}SN_2C_2H\text{-}2\text{-}CH_3\text{-}3\text{-}NHC_3H_7]^+Cl^-$ | **3** | 243/4 |
| | $[1,2,5\text{-}SN_2C_2H\text{-}2\text{-}C_3H_7\text{-}3\text{-}NHCH_3]^+Cl^-$ | **3** | 243/4 |
| $SN_3C_6F_5H_{10}$ | $[{-}S({=}NC_2F_5){-}NCH_3{-}CH_2{-}CH_2{-}NCH_3{-}]$ | **3** | 276 |

| Formula | Compound | Vol. | Page |
|---|---|---|---|
| SN$_3$C$_9$H$_{20}$O$^+$ | [–SO–N(C$_3$H$_7$-i)∸C(∸N(CH$_3$)$_2$)∸N(C$_3$H$_7$-i)–]$^+$ | | |
| | Salt: | | |
| | [–SO–N(C$_3$H$_7$-i)∸C(∸N(CH$_3$)$_2$)∸N(C$_3$H$_7$-i)–]$^+$BF$_4^-$ | **3** | 62/3 |
| SN$_3$C$_{10}$ClH$_{10}$O$_4$ | [–SO–N(CH$_2$CH$_2$Cl)–CO–{1,2-C$_6$H$_3$NO$_2$-4}–NCH$_3$–] | **4** | 78/80 |
| SN$_3$C$_{10}$ClH$_{12}$O | [–SO–N(C$_6$H$_4$Cl-2)–C(=NH)–C(CH$_3$)$_2$–NH–] | **3** | 273 |
| | [–SO–N(C$_6$H$_4$Cl-3)–C(=NH)–C(CH$_3$)$_2$–NH–] | **3** | 273 |
| SN$_3$C$_{10}$ClH$_{14}$O | [–SO–NH–C(CH$_3$)$_2$–C(NHC$_6$H$_5$·HCl)=N–] | **3** | 260/1 |
| SN$_3$C$_{10}$ClH$_{20}$O$_2$ | [∸S(NH$_2$)∸N(C$_4$H$_9$-t)–CO–CO–N(C$_4$H$_9$-t)∸]$^+$Cl$^-$ | **3** | 282/3 |
| SN$_3$C$_{10}$ClH$_{22}$O | [–SO–NH–C(CH$_3$)$_2$–C(N(C$_2$H$_5$)$_3$)=N–]$^+$Cl$^-$ | **3** | 260 |
| SN$_3$C$_{10}$Cl$_2$H$_9$O$_3$ | [–SO–N(C$_6$H$_3$Cl$_2$-3,4)–N=C(COOCH$_3$)–NCH$_3$–] | **3** | 81/2 |
| SN$_3$C$_{10}$Cl$_2$H$_{13}$O | [–SO–NH–C(CH$_3$)$_2$–C(NHC$_6$H$_4$Cl-2·HCl)=N–] | **3** | 260/1 |
| SN$_3$C$_{10}$Cl$_3$H$_8$O | [∸S(OCH$_3$)∸N∸C(CCl$_3$)∸N∸C(C$_6$H$_5$)∸N∸] | **4** | 31/2 |
| SN$_3$C$_{10}$F$_3$H$_{10}$O$_2$ | [–SO–NCH$_3$–NCH$_3$–CO–N(C$_6$H$_4$CF$_3$-3)–] | **3** | 81, 83, 85 |
| SN$_3$C$_{10}$H$_5$O$_2$ | [∸S∸N∸{C–CO–NC$_6$H$_5$–CO–C}∸N∸] (C with :\| above) | **3** | 224 |
| SN$_3$C$_{10}$H$_9$O | 1,2,5-SN$_2$C$_2$H-3-CH$_2$CONHC$_6$H$_5$ | **3** | 124 |
| SN$_3$C$_{10}$H$_9$O$_6$ | [–SO–N(COOC$_2$H$_5$)–C(C$_6$H$_4$NO$_2$-3)=N–O–] | **3** | 313/4, 316/7 |
| | [–SO–N(COOC$_2$H$_5$)–C(C$_6$H$_4$NO$_2$-4)=N–O–] | **3** | 313/4, 316/7 |
| SN$_3$C$_{10}$H$_{12}^+$ | [1,2,5-SN$_2$C$_2$-3-NHCH$_3$-4-C$_6$H$_5$-5-CH$_3$]$^+$ | | |
| | Salt: | | |
| | [1,2,5-SN$_2$C$_2$-3-NHCH$_3$-4-C$_6$H$_5$-5-CH$_3$]$^+$BF$_4^-$ | **3** | 245 |
| SN$_3$C$_{10}$H$_{13}$O | [–SO–NC$_6$H$_5$–C(=NH)–C(CH$_3$)$_2$–NH–] | **3** | 273 |
| SN$_3$C$_{10}$H$_{13}$O$_3$ | [–SO–NCH$_3$–N(C$_6$H$_4$OCH$_3$-4)–CO–NCH$_3$–] | **3** | 81, 84/5 |
| SN$_3$C$_{10}$H$_{20}$O$_2^+$ | [∸S(NH$_2$)∸N(C$_4$H$_9$-t)–CO–CO–N(C$_4$H$_9$-t)∸]$^+$ | **3** | 282/3 |
| | Salts: | | |
| | [∸S(NH$_2$)∸N(C$_4$H$_9$-t)–CO–CO–N(C$_4$H$_9$-t)∸]$^+$2,4,6-(i-C$_3$H$_7$)$_3$C$_6$H$_2$SO$_3^-$ | **3** | 282/3 |
| | [∸S(NH$_2$)∸N(C$_4$H$_9$-t)–CO–CO–N(C$_4$H$_9$-t)∸]$^+$Cl$^-$ | **3** | 282/3 |
| SN$_3$C$_{10}$H$_{20}$O$_4$P | 1,2,5-SN$_2$C$_2$-3-OP(O)(OC$_2$H$_5$)$_2$-4-N(C$_2$H$_5$)$_2$ | **3** | 160 |

| Formula | Compound | Vol. | Page |
|---|---|---|---|
| $SN_3C_{10}H_{22}O^+$ | $[-SO-NH-C(CH_3)_2-C(N(C_2H_5)_3)=N-]^+$ | | |
| | Salt: | | |
| | $[-SO-NH-C(CH_3)_2-C(N(C_2H_5)_3)=N-]^+Cl^-$ | **3** | 260 |
| $SN_3C_{11}ClH_8O_2$ | $[-SO-\{\overset{\vert}{N}-CCl=CH-\{1,2\text{-}C_6H_4\}-\overset{\vert}{C}\}=N-N(COCH_3)-]$ | **3** | 88/9 |
| $SN_3C_{11}Cl_2H_{13}O_2$ | $[-SO-N(C_3H_7\text{-}i)-NCH_3-CO-N(C_6H_3Cl_2\text{-}3,4)-]$ | **3** | 81, 84/5 |
| $SN_3C_{11}Cl_3H_{10}O$ | $[\dot{-}S(OC_2H_5)\dot{-}N\dot{-}C(CCl_3)\dot{-}N\dot{-}C(C_6H_5)\dot{-}N\dot{-}]$ | **4** | 31/2 |
| $SN_3C_{11}H_9O_2$ | $[-SO-\{\overset{\vert}{N}-CH=CH-\{1,2\text{-}C_6H_4\}-\overset{\vert}{C}\}=N-N(COCH_3)-]$ | **3** | 88/9 |
| $SN_3C_{11}H_{13}O_2$ | $[-SO-NCH_3-N=C(CH_3)-CH_2-N(COC_6H_5)-]$ | **4** | 14, 16/7 |
| $SN_3C_{11}H_{13}O_4$ | $[-SO-NC_4H_9-C(C_6H_4NO_2\text{-}4)=N-O-]$ | **3** | 313/5 |
| $SN_3C_{11}H_{15}O_2$ | $[-SO-N(C_3H_7\text{-}i)-NC_6H_5-CO-NCH_3-]$ | **3** | 81, 84/5 |
| $SN_3C_{12}ClH_{14}O_3$ | $[-SO-N(C_6H_4Cl\text{-}3)-N=C(COOCH_3)-N(C_3H_7\text{-}i)-]$ | **3** | 81/2 |
| | $[-SO-N(C_6H_4Cl\text{-}4)-N=C(COOCH_3)-N(C_3H_7\text{-}i)-]$ | **3** | 81/2 |
| $SN_3C_{12}Cl_2H_{13}O_3$ | $[-SO-N(C_6H_3Cl_2\text{-}3,4)-N=C(COOCH_3)-N(C_3H_7\text{-}i)-]$ | **3** | 81/2 |
| $SN_3C_{12}Cl_3H_{10}P_2$ | $[\dot{-}SCl\dot{-}N\dot{-}(P(C_6H_5)Cl-N)_2\dot{-}]$ $= SN_3P_2(C_6H_5)_2Cl_3$ | **3** | 20/5 |
| $SN_3C_{12}H_9O_2$ | $[-SO-N(2\text{-}C_5H_4N)-N=C(C_6H_5)-O-]$ ($2\text{-}C_5H_4N$ = 2-Pyridinyl) | **3** | 309/10, 312 |
| $SN_3C_{12}H_9O_4$ | $[-SO-N(C_6H_4NO_2\text{-}4)-CH(2\text{-}C_5H_4N)-O-]$ ($2\text{-}C_5H_4N$ = 2-Pyridinyl) | **3** | 63/4 |
| $SN_3C_{12}H_{13}O_2$ | $[-SO-N=\{\overset{\vert}{C}-O-CH(C_6H_5)-CH(CH_3)-NCH_3-\overset{\vert}{C}\}=N-]$ | **3** | 258/9 |
| $SN_3C_{12}H_{24}O_2^+$ | $[\dot{-}S(NH_2)\dot{-}N(C(CH_3)_2C_2H_5)-CO-CO-N(C(CH_3)_2C_2H_5)\dot{-}]^+$ | **3** | 282/3 |
| | Salt: | | |
| | $[\dot{-}S(NH_2)\dot{-}N(C(CH_3)_2C_2H_5)-CO-CO-N(C(CH_3)_2C_2H_5)\dot{-}]^+2,4,6\text{-}(CH_3)_3C_6H_2SO_3^-$ | **3** | 282/3 |
| $SN_3C_{13}ClH_{12}O_2$ | $[-SO-NH-CO-\{\overset{\Vert}{C}-C(CH_3)=C(CH_3)-N(C_6H_4Cl\text{-}4)-\overset{\Vert}{C}\}-NH-]$ | **4** | 82/3 |
| $SN_3C_{13}ClH_{12}O_4$ | $[-SO-N(COC_6H_4Cl\text{-}4)-CO-C(-NC_4H_8O)=N-]$ ($NC_4H_8O$ = Morpholino) | **3** | 261/3 |
| $SN_3C_{13}ClH_{14}O_2$ | $[-SO-\{\overset{\vert}{N}-(CH_2)_5-\overset{\vert}{C}\}=N-N(COC_6H_4Cl\text{-}4)-]$ | **3** | 87 |
| $SN_3C_{13}F_3H_{14}O_3$ | $[-SO-N(C_6H_4CF_3\text{-}3)-N=C(COOCH_3)-N(C_3H_7\text{-}i)-]$ | **3** | 81/2 |
| $SN_3C_{13}H_9$ | $1,2,5\text{-}SN_2C_2\text{-}3\text{-}C_6H_5\text{-}4\text{-}(2'\text{-}C_5H_4N)$ ($2\text{-}C_5H_4N$ = 2-Pyridinyl) | **3** | 133/5 |
| $SN_3C_{13}H_9O_4$ | $[-SO-N(C_6H_4NO_2\text{-}2)-C(C_6H_5)=N-O-]$ | **3** | 313/4, 316/7 |

| Formula | Compound | Ref. |
|---|---|---|
| $SN_3C_{13}H_{13}O_4$ | [–SO–N($COC_6H_5$)–CO–C(–$NC_4H_8O$)=N–] ($NC_4H_8O$ = Morpholino) | **3** 261/3 |
| $SN_3C_{13}H_{15}O_2$ | [–SO–{N–$(CH_2)_5$–C}=N–N($COC_6H_5$)–] | **3** 87 |
| $SN_3C_{13}H_{15}O_3$ | [–SO–N($CH_2C_6H_5$)–CO–C(–$NC_4H_8O$)=N–] ($NC_4H_8O$ = Morpholino) | **3** 261/3 |
| $SN_3C_{13}H_{17}O$ | [–SO–$NCH_3$–N={1,2-$C_6H_9$}–$NC_6H_5$–] | **4** 14, 16/7 |
| $SN_3C_{13}H_{19}O_2$ | [–SO–NH–CO–{C–C($CH_3$)=C($CH_3$)–N($C_6H_{11}$-c)–C}–NH–] | **4** 82/3 |
| $SN_3C_{13}H_{21}O_2$ | [–SO–NH–CO–{C–C($CH_3$)=C($CH_3$)–$NC_6H_{13}$–C}–NH–] | **4** 82/3 |
| $SN_3C_{14}ClH_{10}$ | [∸SCl∸N∸C($C_6H_5$)∸N∸C($C_6H_5$)∸N∸] | **4** 53, 55/7 |
| $SN_3C_{14}ClH_{14}O_4$ | [–SO–N($CH_2COC_6H_4Cl$-4)–CO–C(–$NC_4H_8O$)=N–] ($NC_4H_8O$ = Morpholino) | **3** 261/3 |
| $SN_3C_{14}H_9O_2$ | 1,2,5-$SN_2C_2$-3-$C_6H_5$-4-$C_6H_4NO_2$-2′ | **3** 133/5 |
| | 1,2,5-$SN_2C_2$-3-$C_6H_5$-4-$C_6H_4NO_2$-4′ | **3** 133/5 |
| $SN_3C_{14}H_{10}^{\bullet}$ | [∸S∸N∸C($C_6H_5$)∸N∸C($C_6H_5$)∸N∸]$^{\bullet}$ = $SN_3C_2(C_6H_5)_2^{\bullet}$ | **4** 19 |
| $SN_3C_{14}H_{10}I$ | [∸SI∸N∸C($C_6H_5$)∸N∸C($C_6H_5$)∸N∸] | **4** 57/8 |
| $SN_3C_{14}H_{13}O_3$ | [–SO–N($CH_2C_6H_5$)–CO–C($NHCH_2$(2-$C_4H_3O$))=N–] (2-$C_4H_3O$ = 2-Furanyl) | **3** 261/3 |
| $SN_3C_{14}H_{17}O_2$ | [–SO–$NCH_3$–N={1,2-$C_6H_9$}–N($COC_6H_5$)–] | **4** 14, 16/7 |
| $SN_3C_{14}H_{17}O_3$ | [–SO–{N–$(CH_2)_5$–C}=N–N($COC_6H_4OCH_3$-4)–] | **3** 87 |
| $SN_3C_{14}H_{27}$ | 1,2,5-$SN_2C_2H$-3-$NHC_{12}H_{25}$ | **3** 142 |
| $SN_3C_{15}ClH_{12}$ | [∸S($CH_3$)∸N∸C($C_6H_5$)∸N∸C($C_6H_4Cl$-4)∸N∸] | **4** 58/9 |
| $SN_3C_{15}ClH_{12}O_3$ | [–SO–$NC_6H_5$–N=C($COOCH_3$)–N($C_6H_4Cl$-4)–] | **3** 81/2 |
| | [–SO–N($C_6H_4Cl$-4)–C(=$NCONHC_6H_5$)–$CH_2$–O–] | **3** 291/2 |
| $SN_3C_{15}Cl_2H_{11}$ | [∸S($CH_3$)∸N∸C($C_6H_4Cl$-4)∸N∸C($C_6H_4Cl$-4)∸N∸] | **4** 58/9 |
| $SN_3C_{15}H_{13}$ | [∸S($CH_3$)∸N∸C($C_6H_5$)∸N∸C($C_6H_5$)∸N∸] | **4** 58/9 |
| $SN_3C_{15}H_{19}O_3$ | [–SO–$NC_6H_5$–N={1,2-$C_6H_9$}–N($COOC_2H_5$)–] | **4** 14, 16/7 |
| $SN_3C_{15}H_{25}O_2$ | [–SO–N=C($CH_3$)–O–C(=N($C_6H_{11}$-c))–N($C_6H_{11}$-c)–] | **4** 104, 106 |
| $SN_3C_{15}H_{28}O^+$ | [–SO–N($C_6H_{11}$-c)∸C(∸N$(CH_3)_2$)∸N($C_6H_{11}$-c)–]$^+$ | |
| | Salt: | |
| | [–SO–N($C_6H_{11}$-c)∸C(∸N$(CH_3)_2$)∸N($C_6H_{11}$-c)–]$^+BF_4^-$ | **3** 62/3 |
| $SN_3C_{16}ClH_{14}O_3$ | [–SO–N($C_6H_4Cl$-4)–C(=$NCONHC_6H_5$)–CH($CH_3$)–O–] | **3** 291/2 |

| Formula | Structure | Vol. | Pages |
|---|---|---|---|
| $SN_3C_{22}F_6H_{15}$ | $[=S(C_6H_5)-NC_6H_5-C(C_6H_5)=N-C(CF_3)_2-N=]$ | **4** | 60/1 |
| $SN_3C_{22}H_{20}O_2^+$ | $[\dot{-}S(N(C_2H_5)C_6H_5)\dot{-}NC_6H_5-CO-CO-NC_6H_5\dot{-}]^+$ | **3** | 282/3, 285 |
| | Salt: | | |
| | $[\dot{-}S(N(C_2H_5)C_6H_5)\dot{-}NC_6H_5-CO-CO-NC_6H_5\dot{-}]^+BF_4^-$ | **3** | 282/3, 285 |
| $SN_3C_{23}H_{15}$ | $[\dot{-}S\dot{-}N\dot{-}\{\overset{:\vert}{C}\dot{-}C(C_6H_5)\dot{-}N\dot{-}C(C_6H_5)\dot{-}C(C_6H_5)\dot{-}\overset{:\vert}{C}\}\dot{-}N\dot{-}]$ | **3** | 225/6 |
| $SN_3C_{23}H_{15}O$ | $[=S=N-CO-\{\overset{\Vert}{C}-C(C_6H_5)=C(C_6H_5)-NC_6H_5-\overset{\Vert}{C}\}-N=]$ | **4** | 71/2 |
| $SN_3C_{23}H_{21}O$ | $[=S=N-CO-\{\overset{\Vert}{C}-C(C_6H_5)=C(C_6H_5)-N(C_6H_{11}\text{-c})-\overset{\Vert}{C}\}-N=]$ | **4** | 71/2 |
| $SN_3C_{23}H_{21}O_2$ | $[-S(=NC_6H_4CH_3\text{-}4)-N(C_6H_4CH_3\text{-}4)-CO-CO-N(C_6H_4CH_3\text{-}4)-]$ | **3** | 276, 279/80 |
| $SN_3C_{23}H_{21}O_4$ | $[-SO-N(COC_6H_5)-CO-NC_6H_5-CH(C_6H_4N(CH_3)_2\text{-}4)-O-]$ | **4** | 104/5 |
| $SN_3C_{23}H_{23}O$ | $[=S=N-CO-\{\overset{\Vert}{C}-C(C_6H_5)=C(C_6H_5)-NC_6H_{13}-\overset{\Vert}{C}\}-N=]$ | **4** | 71/2 |
| $SN_3C_{24}ClH_{20}P_2$ | $[\dot{-}SCl\dot{-}N\dot{-}(P(C_6H_5)_2\dot{-}N)_2\dot{-}]\ = SN_3P_2(C_6H_5)_4Cl$ | **3** | 20/6 |
| $SN_3C_{24}H_{17}O$ | $[=S=N-CO-\{\overset{\Vert}{C}-C(C_6H_5)=C(C_6H_5)-N(CH_2C_6H_5)-\overset{\Vert}{C}\}-N=]$ | **4** | 71/2 |
| $SN_3C_{24}H_{20}IP_2$ | $[\dot{-}SI\dot{-}N\dot{-}(P(C_6H_5)_2\dot{-}N)_2\dot{-}]\ = SN_3P_2(C_6H_5)_4I$ | **3** | 20/4, 27/8 |
| $SN_3C_{24}H_{20}I_3P_2$ | $[\dot{-}S\dot{-}N\dot{-}(P(C_6H_5)_2\dot{-}N)_2\dot{-}]^+I_3^-\ = SN_3P_2(C_6H_5)_4^+I_3^-$ | **3** | 20/4, 29/30 |
| $SN_3C_{24}H_{20}P_2^+$ | $[\dot{-}S\dot{-}N\dot{-}(P(C_6H_5)_2\dot{-}N)_2\dot{-}]^+\ = SN_3P_2(C_6H_5)_4^+$ | **3** | 20/4 |
| | Salts: | | |
| | $[\dot{-}S\dot{-}N\dot{-}(P(C_6H_5)_2\dot{-}N)_2\dot{-}]^+Br_3^-\ = SN_3P_2(C_6H_5)_4^+Br_3^-$ | **3** | 20/4, 29/30 |
| | $[\dot{-}S\dot{-}N\dot{-}(P(C_6H_5)_2\dot{-}N)_2\dot{-}]^+I_3^-\ = SN_3P_2(C_6H_5)_4^+I_3^-$ | **3** | 20/4, 29/30 |
| $SN_3C_{25}H_{19}O$ | $[=S=N-CO-\{\overset{\Vert}{C}-C(C_6H_5)=C(C_6H_5)-N(CH_2CH_2C_6H_5)-\overset{\Vert}{C}\}-N=]$ | **4** | 71/2 |
| $SN_3C_{25}H_{25}O_4$ | $[-SO-N(COC_6H_5)-CO-NC_6H_5-CH(C_6H_4N(C_2H_5)_2\text{-}4)-O-]$ | **4** | 104/5 |
| $SN_3C_{30}H_{25}P_2$ | $[\dot{-}S(C_6H_5)\dot{-}N\dot{-}(P(C_6H_5)_2\dot{-}N)_2\dot{-}]\ = SN_3P_2(C_6H_5)_5$ | **3** | 20/6, 28/9 |
| $SN_3Cl_4P_2^+$ | $[\dot{-}S\dot{-}N\dot{-}(PCl_2\dot{-}N)_2\dot{-}]^+\ = SN_3P_2Cl_4^+$ | **3** | 20/2 |
| | Salt: | | |
| | $[\dot{-}S\dot{-}N\dot{-}(PCl_2\dot{-}N)_2\dot{-}]^+SbCl_6^-\ = SN_3P_2Cl_4^+SbCl_6^-$ | **3** | 20/2, 29 |
| $SN_3Cl_5P_2$ | $[\dot{-}SCl\dot{-}N\dot{-}(PCl_2\dot{-}N)_2\dot{-}]\ = SN_3P_2Cl_5$ | **3** | 20/4 |
| $SN_3Cl_{10}P_2Sb$ | $[\dot{-}S\dot{-}N\dot{-}(PCl_2\dot{-}N)_2\dot{-}]^+SbCl_6^-\ = SN_3P_2Cl_4^+SbCl_6^-$ | **3** | 20/2, 29 |
| $SN_4BrC_4Cl$ | $[\dot{-}S\dot{-}N\dot{-}\{\overset{:\vert}{C}\dot{-}N\dot{-}CBr\dot{-}CCl\dot{-}N\dot{-}\overset{:\vert}{C}\}\dot{-}N\dot{-}]$ | **3** | 229, 231 |
| $SN_4Br_2C_4$ | $[\dot{-}S\dot{-}N\dot{-}\{\overset{:\vert}{C}\dot{-}N\dot{-}CBr\dot{-}CBr\dot{-}N\dot{-}\overset{:\vert}{C}\}\dot{-}N\dot{-}]$ | **3** | 229, 231 |

| | | |
|---|---|---|
| $SN_4C_{28}H_{24}O$ | $[{-}SO{-}N(C_6H_4CH_3\text{-}4){-}C(=NC_6H_5){-}C(=NC_6H_5){-}N(C_6H_4CH_3\text{-}4){-}]$ | **3** 273/5 |
| $SN_4C_{28}H_{30}P_2$ | $[∸S(N(C_2H_5)_2)∸N∸(P(C_6H_5)_2∸N)_2∸]$ $= SN_3P_2(C_6H_5)_4N(C_2H_5)_2$ | **3** 20/4, 28 |
| $SN_4C_{29}H_{30}P_2$ | $[∸S({-}NC_5H_{10})∸N∸(P(C_6H_5)_2∸N)_2∸]$ $= SN_3P_2(C_6H_5)_4NC_5H_{10}$ ($NC_5H_{10}$ = Piperidino) | **3** 20/4, 28 |
| $SN_4C_{30}H_{28}O$ | $[{-}SO{-}N(C_6H_4CH_3\text{-}4){-}C(=NC_6H_4CH_3\text{-}4){-}C(=NC_6H_4CH_3\text{-}4){-}N(C_6H_4CH_3\text{-}4){-}]$ | **3** 273/5 |
| $SN_4C_{34}H_{24}O$ | $[{-}SO{-}N(1\text{-}C_{10}H_7){-}C(=NC_6H_5){-}C(=NC_6H_5){-}N(1\text{-}C_{10}H_7){-}]$ ($1\text{-}C_{10}H_7$ = 1-Naphthalenyl) | **3** 273/5 |
| $SN_5C_3HO$ | 1,2,5-$SN_2C_2H$-3-$CON_3$ | **3** 174 |
| $SN_5C_3H_5O$ | 1,2,5-$SN_2C_2$-3-$CONHNH_2$-4-$NH_2$ | **3** 178/9 |
| $SN_5C_4ClH_4O$ | $[∸S∸N∸\{C{-}N=C(NH_2){-}NH{-}CO{-}C\}∸N∸]\cdot HCl$ | **3** 233, 237 |
| $SN_5C_4H_3$ | $[∸S∸N∸\{C∸N∸CH∸N∸C(NH_2)∸C\}∸N∸]$ | **3** 232, 235 |
| $SN_5C_4H_3O$ | $[∸S∸N∸\{C∸N∸C(NH_2)∸N∸C(OH)∸C\}∸N∸]$ | **3** 232/3, 237 |
| | $[∸S∸N∸\{C{-}N=C(NH_2){-}NH{-}CO{-}C\}∸N∸]$ | **3** 232/3, 237 |
| $SN_5C_5H_9$ | 1,2,5-$SN_2C_2$-3-$NH_2$-4-$N=CHN(CH_3)_2$ | **3** 148 |
| $SN_5C_7H_{15}O_2$ | $[{-}SO{-}N(C_3H_7\text{-}i){-}CO{-}N(NH_2){-}C(N(CH_3)_2)=N{-}]$ | **4** 23, 25/7 |
| $SN_5C_8ClH_{16}$ | $[∸S(N(C_3H_7\text{-}i)_2)∸N∸CCl∸N∸C(NH_2)∸N∸]$ | **4** 39/40, 45/9 |
| $SN_5C_8H_7O$ | $[{-}SO{-}N=C(NH_2){-}NH{-}\{C=N{-}\{1,2\text{-}C_6H_4\}{-}N\}{-}]$ | **4** 30/1 |
| $SN_5C_9ClH_{18}$ | $[∸S(N(C_3H_7\text{-}i)_2)∸N∸CCl∸N∸C(NHCH_3)∸N∸]$ | **4** 39/40, 45/9 |
| $SN_5C_{10}ClH_{14}$ | $[∸S∸N∸\{C∸N∸C(N(C_3H_7)_2)∸CCl∸N∸C\}∸N∸]$ | **3** 229, 231 |
| $SN_5C_{10}ClH_{18}O_2$ | $[{-}SO{-}N=C({-}N_4C_6H_{12}){-}C(CH_3)_2{-}O{-}]^+Cl^-$ ($N_4C_6H_{12}$ = 1,3,5,7-Tetraazatricyclo[3.3.1.1$^{3,7}$]decane) | **3** 286/7 |
| $SN_5C_{10}ClH_{20}$ | $[∸S(N(C_3H_7\text{-}i)_2)∸N∸CCl∸N∸C(NHC_2H_5)∸N∸]$ | **4** 39/40, 45/7 |
| | $[∸S(N(C_3H_7\text{-}i)_2)∸N∸CCl∸N∸C(N(CH_3)_2)∸N∸]$ | **4** 39/40, 45 |
| $SN_5C_{10}H_{18}O_2^+$ | $[{-}SO{-}N=C({-}N_4C_6H_{12}){-}C(CH_3)_2{-}O{-}]^+$ ($N_4C_6H_{12}$ = 1,3,5,7-Tetraazatricyclo[3.3.1.1$^{3,7}$]decane) | |
| | Salt: | |
| | $[{-}SO{-}N=C({-}N_4C_6H_{12}){-}C(CH_3)_2{-}O{-}]^+Cl^-$ | **3** 286/7 |
| $SN_5C_{10}H_{19}O_2$ | $[{-}SO{-}N(C_6H_{11}\text{-}c){-}CO{-}N(NH_2){-}C(N(CH_3)_2)=N{-}]$ | **4** 23, 25/7 |
| $SN_5C_{11}ClH_{20}$ | $[∸S(N(C_3H_7\text{-}i)_2)∸N∸CCl∸N∸C(NHCH_2CH=CH_2)∸N∸]$ | **4** 39/40, 45, 47/9 |
| $SN_5C_{11}ClH_{22}$ | $[∸S(N(C_3H_7\text{-}i)_2)∸N∸CCl∸N∸C(NHC_3H_7)∸N∸]$ | **4** 39/40, 45/9 |

| | | | |
|---|---|---|---|
| $SN_5C_{12}ClH_{24}$ | [∸S(N($C_3H_7$-i)$_2$)∸N∸CCl∸N∸C(NH$C_4H_9$-t)∸N∸] | **4** | 39/40, 45/9 |
| | [∸S(N($C_3H_7$-i)$_2$)∸N∸CCl∸N∸C(N($C_2H_5$)$_2$)∸N∸] | **4** | 39/40, 45 |
| $SN_5C_{12}H_{15}O_2$ | [–SO–NH–C(CONH$CH_2C_6H_5$)=N–C(N($CH_3$)$_2$)=N–] | **4** | 21/2 |
| $SN_5C_{13}ClH_{24}$ | [∸S(N($C_3H_7$-i)$_2$)∸N∸CCl∸N∸C(–N$C_5H_{10}$)∸N∸] (N$C_5H_{10}$ = Piperidino) | **4** | 39/40, 45, 47/9 |
| $SN_5C_{13}H_9O_7$ | [∸S∸N$CH_3$∸{1,2-$C_6H_4$}∸N∸]$^+$ 2,4,6-($NO_2$)$_3C_6H_2O^-$ | **3** | 246, 248/9 |
| $SN_5C_{14}ClH_{26}$ | [∸S(N($C_3H_7$-i)$_2$)∸N∸CCl∸N∸C(NH$C_6H_{11}$-c)∸N∸] | **4** | 39/40, 45/9 |
| $SN_5C_{14}H_{11}O_7$ | [∸S∸N$C_2H_5$∸{1,2-$C_6H_4$}∸N∸]$^+$ 2,4,6-($NO_2$)$_3C_6H_2O^-$ | **3** | 246, 249 |
| $SN_5C_{14}H_{17}O_3$ | [–SO–N(CONH$C_6H_4CH_3$-4)–C(=NH)–C(–N$C_4H_8$O)=N–] (N$C_4H_8$O = Morpholino) | **3** | 264 |
| $SN_5C_{15}ClH_{22}$ | [∸S(N($C_3H_7$-i)$_2$)∸N∸CCl∸N∸C(NH$CH_2C_6H_5$)∸N∸] | **4** | 39/40, 45, 47/9 |
| $SN_5C_{16}ClH_{30}$ | [∸S(N($C_3H_7$-i)$_2$)∸N∸CCl∸N∸C(N($C_2H_5$)($C_6H_{11}$-c))∸N∸] | **4** | 39/40, 45, 47/9 |
| $SN_5C_{16}ClH_{32}$ | [∸S(N($C_3H_7$-i)$_2$)∸N∸CCl∸N∸C(N($C_4H_9$-i)$_2$)∸N∸] | **4** | 39/40, 45, 47 |
| $SN_5C_{18}ClH_{36}$ | [∸S(N($C_4H_9$-i)$_2$)∸N∸CCl∸N∸C(N($C_4H_9$-i)$_2$)∸N∸] | **4** | 39/40, 45/8 |
| $SN_5C_{18}H_{17}O_3$ | [–S(=NC($CH_3$)$_2$CN)–N($C_6H_4CH_3$-4)–C($C_6H_4NO_2$-3)=N–O–] | **3** | 321/3 |
| $SN_5C_{20}ClH_{34}$ | [∸S(N($C_6H_{11}$-c)$_2$)∸N∸CCl∸N∸C(NH$C_6H_{11}$-c)∸N∸] | **4** | 39/40, 45 |
| $SN_5C_{20}H_{15}O_5$ | [–S(=N$C_6H_4NO_2$-4)–N($C_6H_4CH_3$-4)–C($C_6H_4NO_2$-3)=N–O–] | **3** | 321/3 |
| | [–S(=N$C_6H_4NO_2$-4)–N($C_6H_4CH_3$-4)–C($C_6H_4NO_2$-4)=N–O–] | **3** | 321/3 |
| $SN_5C_{20}H_{15}O_6$ | [–S(=N$C_6H_4NO_2$-4)–N($C_6H_4OCH_3$-4)–C($C_6H_4NO_2$-4)=N–O–] | **3** | 321/3 |
| $SN_5C_{22}ClH_{28}$ | [∸S(N($C_3H_7$-i)$_2$)∸N∸CCl∸N∸C(N($CH_2C_6H_5$)$_2$)∸N∸] | **4** | 39/40, 45, 47 |
| $SN_5C_{23}ClH_{28}$ | [∸S(–N$C_5H_8$($CH_3$)$_2$-2,6)∸N∸CCl∸N∸C(N($CH_2C_6H_5$)$_2$)∸N∸] (N$C_5H_8$($CH_3$)$_2$-2,6 = 2,6-Dimethylpiperidino) | **4** | 39/40, 45 |
| $SN_6C_4F_5H_{12}P_3Si_2$ | [=S=N–Si($CH_3$)$_2$–N(–$P_3N_3F_5$)–Si($CH_3$)$_2$–N=] ($P_3N_3F_5$ = Pentafluoro-cyclotri-$\lambda^5$-phosphazen-2-yl) | **3** | 8/9 |
| $SN_6C_4H_4$ | [∸S∸N∸{C∸C($NH_2$)∸N∸N∸C($NH_2$)∸C}∸N∸] | **3** | 227, 229 |
| | [∸S∸N∸{C–C(=NH)–NH–NH–C(=NH)–C}∸N∸] | **3** | 227, 229 |
| | [∸S∸N∸{C∸N∸C($NH_2$)∸N∸C($NH_2$)∸C}∸N∸] | **3** | 232, 236 |
| $SN_6C_4H_6O_2$ | 1,2,5-$SN_2C_2$-3-CONH$NH_2$-4-CONH$NH_2$ | **3** | 188 |
| $SN_6C_6$ | [∸S∸N∸{C∸N∸C(CN)∸C(CN)∸N∸C}∸N∸] | **3** | 229, 232 |

| Formula | Compound | Vol. Page |
|---|---|---|
| $SN_6C_{10}ClH_{19}O$ | $[-SO-NH-C(CH_3)_2-C(-N_4C_6H_{12})=N-]^+Cl^-$<br>($N_4C_6H_{12}$ = 1,3,5,7-Tetraazatricyclo[3.3.1.1$^{3,7}$]decane) | **3** 260 |
| $SN_6C_{10}H_{19}O^+$ | $[-SO-NH-C(CH_3)_2-C(-N_4C_6H_{12})=N-]^+$<br>($N_4C_6H_{12}$ = 1,3,5,7-Tetraazatricyclo[3.3.1.1$^{3,7}$]decane)<br>Salt:<br>$[-SO-NH-C(CH_3)_2-C(-N_4C_6H_{12})=N-]^+Cl^-$ | <br><br><br>**3** 260 |
| $SN_6C_{16}H_{28}$ | [∸S∸N∸{C∸N∸C(N(C$_3$H$_7$)$_2$)∸ C(N(C$_3$H$_7$)$_2$)∸N∸C}∸N∸] | **3** 229, 231 |
| $SN_6C_{19}H_{12}O_7$ | $[-S(=NC_6H_4NO_2\text{-}3)-N(C_6H_4NO_2\text{-}3)-C(C_6H_4NO_2\text{-}4)=N-O-]$ | **3** 321/3 |
| $SN_6C_{22}H_{30}$ | [∸S(N(C$_3$H$_7$-i)$_2$)∸N∸C(NHC$_6$H$_4$CH$_3$-4)∸N∸C(NHC$_6$H$_4$CH$_3$-4)∸N∸] | **4** 39, 52 |
| $SN_6C_{22}H_{30}O_2$ | [∸S(N(C$_3$H$_7$-i)$_2$)∸N∸C(NHC$_6$H$_4$OCH$_3$-4)∸N∸C(NHC$_6$H$_4$OCH$_3$-4)∸N∸] | **4** 39, 52 |
| $SN_6C_{28}H_{38}$ | [∸S(N(C$_6$H$_{11}$-c)$_2$)∸N∸C(NHC$_6$H$_4$CH$_3$-4)∸N∸C(NHC$_6$H$_4$CH$_3$-4)∸N∸] | **4** 39, 52 |
| $SN_6C_{28}H_{38}O_2$ | [∸S(N(C$_6$H$_{11}$-c)$_2$)∸N∸C(NHC$_6$H$_4$OCH$_3$-4)∸N∸C(NHC$_6$H$_4$OCH$_3$-4)∸N∸] | **4** 39, 52 |
| $SN_6C_{30}H_{30}O$ | $[-SO-N(C_6H_4N(CH_3)_2\text{-}4)-C(=NC_6H_5)-C(=NC_6H_5)-N(C_6H_4N(CH_3)_2\text{-}4)-]$ | **3** 273/5 |
| $SN_6F_6P_4$ | [∸S∸N∸{2,8-P$_4$N$_4$F$_6$}∸N∸] = SN$_6$P$_4$F$_6$ | **3** 41/2 |
| $SN_7C_4F_7H_{12}P_4Si_2$ | $[=S=N-Si(CH_3)_2-N(-P_4N_4F_7)-Si(CH_3)_2-N=]$<br>($P_4N_4F_7$ = Heptafluoro-cyclotetra-$\lambda^5$-phosphazen-2-yl) | **3** 9 |
| $SN_8C_{30}H_{32}O_5$ | $[-SO-N(C_6H_3(NH_2)OCH_3\text{-}4)-C(=NC_6H_3(NH_2)OCH_3\text{-}4)-$<br>$C(=NC_6H_3(NH_2)OCH_3\text{-}4)-N(C_6H_3(NH_2)OCH_3\text{-}4)-]$ | **3** 273/5 |

**$S_2$**

| Formula | Compound | Vol. Page |
|---|---|---|
| $S_2NC_7Cl_6H_{16}O_2Sb$ | $[-SO-N(-CH_2-)_5-S(CH_3)_2-O-]^+SbCl_6^-$ | **3** 1 |
| $S_2NC_7H_{16}O_2^+$ | $[-SO-N(-CH_2-)_5-S(CH_3)_2-O-]^+$<br>Salt:<br>$[-SO-N(-CH_2-)_5-S(CH_3)_2-O-]^+SbCl_6^-$ | **3** 1<br><br>**3** 1 |
| $S_2NC_{11}F_6H_{11}O_2$ | $[-S(CF_3)_2-N=C(C_6H_5)-O-S(CH_3)_2-O-]$ | **4** 84/5 |
| $S_2NC_{11}F_6H_{11}O_4$ | $[-S(CF_3)_2-N=C(C_6H_5)-O-S(OCH_3)_2-O-]$ | **4** 84/5 |
| $S_2NC_{12}F_6H_{13}O_2$ | $[-S(CF_3)_2-N=C(CH_2C_6H_5)-O-S(CH_3)_2-O-]$ | **4** 84/5 |
| $S_2NC_{12}F_6H_{13}O_4$ | $[-S(CF_3)_2-N=C(CH_2C_6H_5)-O-S(OCH_3)_2-O-]$ | **4** 84/5 |

| | | | |
|---|---|---|---|
| $S_2N_3C_2H_3$ | $[\dot{-}(S\dot{-}N)_2\dot{-}C(CH_3)\dot{-}N\dot{-}] = S_2N_3C(CH_3)$ | **4** | 1/2 |
| $S_2N_3C_2H_6P$ | $[\dot{-}(S\dot{-}N)_2\dot{-}P(CH_3)_2\dot{-}N\dot{-}] = S_2N_3P(CH_3)_2$ | **3** | 14/9 |
| $S_2N_3C_3ClH_6O_2$ | $[-SO-N{=}C(SC({=}NH)NH_2\cdot HCl)-CH_2-O-]$ | **3** | 286/7 |
| $S_2N_3C_4FH_8O_3$ | $[1,2,5\text{-}SN_2C_2H\text{-}2\text{-}CH_3\text{-}3\text{-}NHCH_3]^+FSO_3^-$ | **3** | 243/4 |
| $S_2N_3C_4F_{10}P$ | $[\dot{-}(S\dot{-}N)_2\dot{-}P(C_2F_5)_2\dot{-}N\dot{-}] = S_2N_3P(C_2F_5)_2$ | **3** | 14, 17/9 |
| $S_2N_3C_4H_7O_2$ | $[-SO-NCH_3-CO-NH-C(SCH_3){=}N-]$ | **4** | 23/4, 26/7 |
| $S_2N_3C_5ClH_{10}O_2$ | $[-SO-N{=}C(SC({=}NH)NH_2\cdot HCl)-C(CH_3)_2-O-]$ | **3** | 286/7 |
| $S_2N_3C_5H_6O_3P$ | $1,2,5\text{-}SN_2C_2\text{-}3\text{-}OP(S)(OCH_3)_2\text{-}4\text{-}CN$ | **3** | 169 |
| $S_2N_3C_5H_7O$ | $1,2,5\text{-}SN_2C_2H\text{-}3\text{-}OC(S)N(CH_3)_2$ | **3** | 159 |
| | $1,2,5\text{-}SN_2C_2H\text{-}3\text{-}SC(O)N(CH_3)_2$ | **3** | 162 |
| $S_2N_3C_5H_9O_2$ | $[-SO-NCH_3-CO-NCH_3-C(SCH_3){=}N-]$ | **4** | 23/4, 26/7 |
| $S_2N_3C_5H_{11}O_3$ | $[-SO-NCH_3-N{=}C(CH_3)-CH_2-N(SO_2CH_3)-]$ | **4** | 14/7 |
| $S_2N_3C_6FH_{12}O_3$ | $[1,2,5\text{-}SN_2C_2H\text{-}2\text{-}CH_3\text{-}3\text{-}NHC_3H_7]^+FSO_3^-$ | **3** | 243/4 |
| | $[1,2,5\text{-}SN_2C_2H\text{-}2\text{-}C_3H_7\text{-}3\text{-}NHCH_3]^+FSO_3^-$ | **3** | 243/4 |
| $S_2N_3C_6H_{11}O$ | $[-SO-N{=}C(SCH_3)-C(NHC_3H_7\text{-}i){=}N-]$ | **3** | 250/1, 254/6 |
| $S_2N_3C_6H_{11}O_2$ | $[-SO-N(C_3H_7\text{-}i)-CO-NH-C(SCH_3){=}N-]$ | **4** | 23/4, 26/7 |
| $S_2N_3C_7Cl_2H_5$ | $[\dot{-}(SCl\dot{-}N)_2\dot{-}C(C_6H_5)\dot{-}N\dot{-}] = S_2N_3CCl_2(C_6H_5)$ | **4** | 6, 11 |
| $S_2N_3C_7H_5$ | $[\dot{-}(S\dot{-}N)_2\dot{-}C(C_6H_5)\dot{-}N\dot{-}] = S_2N_3C(C_6H_5)$ | **4** | 1/2 |
| | Adduct: | | |
| | $[\dot{-}(S\dot{-}N)_2\dot{-}C(C_6H_5)\dot{-}N\dot{-}]\cdot C_7H_8 = S_2N_3C(C_6H_5)\cdot C_7H_8$ ($C_7H_8$ = Norbornadiene) | **4** | 4/5 |
| $S_2N_3C_7H_{11}O$ | $[-SO-N{=}C(SCH_3)-C(-NC_4H_8){=}N-]$ ($NC_4H_8$ = 1-Pyrrolidinyl) | **3** | 250/1, 254/6 |
| $S_2N_3C_7H_{11}O_2$ | $[-SO-N{=}C(SCH_3)-C(-NC_4H_8O){=}N-]$ ($NC_4H_8O$ = Morpholino) | **3** | 250/1, 254/6 |
| $S_2N_3C_7H_{13}O_2$ | $[-SO-N(C_3H_7\text{-}i)-CO-NCH_3-C(SCH_3){=}N-]$ | **4** | 23/4, 26/7 |
| $S_2N_3C_8ClH_4O$ | $[{=}S{=}N-C({=}NCOC_6H_4Cl\text{-}3)-S-N{=}] = S_2N_2C{=}NCOC_6H_4Cl\text{-}3$ | **3** | 68/70 |
| | $[{=}S{=}N-C({=}NCOC_6H_4Cl\text{-}4)-S-N{=}] = S_2N_2C{=}NCOC_6H_4Cl\text{-}4$ | **3** | 68/70 |
| $S_2N_3C_8FH_{16}O_3$ | $[1,2,5\text{-}SN_2C_2H\text{-}2\text{-}C_3H_7\text{-}3\text{-}NHC_3H_7]^+FSO_3^-$ | **3** | 243/4 |

| | | |
|---|---|---|
| $S_2N_3C_{12}H_{17}O_3$ | $[-SO-NCH_3-N{=}C(CH_3)-CH(CH_3)-N(SO_2C_6H_4CH_3\text{-}4)-]$ | **4** 14/7 |
| $S_2N_3C_{12}H_{17}O_4$ | $[-SO-NC_2H_5-CO-\{1,2\text{-}C_6H_3(SO_2NHC_2H_5\text{-}4)\}-NCH_3-]$ | **4** 78/80 |
| $S_2N_3C_{14}H_{13}$ | $[\dot{-}(S\dot{-}N)_2\dot{-}C(C_6H_5)\dot{-}N\dot{-}]\cdot C_7H_8 = S_2N_3C(C_6H_5)\cdot C_7H_8$ ($C_7H_8$ = Norbornadiene) | **4** 4/5 |
| $S_2N_3C_{14}H_{19}O_3$ | $[-SO-NCH_3-N{=}\{1,2\text{-}C_6H_9\}-N(SO_2C_6H_4CH_3\text{-}4)-]$ | **4** 14/7 |
| $S_2N_3C_{14}H_{19}O_8$ | $[-SO-N(2\text{-}C_5H_6O(OOCCH_3)_3)-CS-N{=}C(CH_3)-NH-]$ ($2\text{-}C_5H_9O$ = Tetrahydro-2H-pyran-2-yl) | **4** 27/8 |
| $S_2N_3C_{14}H_{21}O_3$ | $[-SO-N(SO_2C_6H_4CH_3\text{-}4)-C({=}NC_3H_7\text{-}i)-N(C_3H_7\text{-}i)-]$ | **3** 60/2 |
| $S_2N_3C_{14}H_{25}O_3$ | $[-SO-N(SO_2CH_3)-C({=}NC_6H_{11}\text{-}c)-N(C_6H_{11}\text{-}c)-]$ | **3** 60/2 |
| $S_2N_3C_{15}H_{19}O_{10}$ | $[SO-N(2\text{-}C_5H_5O(OOCCH_3)_4)-CS-N{=}CH-NH-]$ ($2\text{-}C_5H_9O$ = Tetrahydro-2H-pyran-2-yl) | **4** 27/8 |
| $S_2N_3C_{16}ClH_{22}O_4$ | $[-S({=}NSO_2C_6H_4Cl\text{-}4)-N(C_4H_9\text{-}t)-CO-CO-N(C_4H_9\text{-}t)-]$ | **3** 276/7 |
| $S_2N_3C_{16}H_{21}O_{10}$ | $[-SO-N(2\text{-}C_5H_5O(OOCCH_3)_4)-CS-N{=}C(CH_3)-NH-]$ ($2\text{-}C_5H_9O$ = Tetrahydro-2H-pyran-2-yl) | **4** 27/8 |
| $S_2N_3C_{16}H_{25}O_3$ | $[-SO-N(SO_2C_6H_4CH_3\text{-}4)-C({=}NC_4H_9)-NC_4H_9-]$ | **3** 60/2 |
| $S_2N_3C_{17}H_{25}O_4$ | $[-S({=}NSO_2C_6H_4CH_3\text{-}4)-N(C_4H_9\text{-}t)-CO-CO-N(C_4H_9\text{-}t)-]$ | **3** 276/7 |
| $S_2N_3C_{17}H_{25}O_5$ | $[-S({=}NSO_2C_6H_4OCH_3\text{-}4)-N(C_4H_9\text{-}t)-CO-CO-N(C_4H_9\text{-}t)-]$ | **3** 276/7 |
| $S_2N_3C_{18}H_{27}O_4$ | $[-S({=}NSO_2C_6H_4C_2H_5\text{-}4)-N(C_4H_9\text{-}t)-CO-CO-N(C_4H_9\text{-}t)-]$ | **3** 276/7 |
| $S_2N_3C_{19}H_{21}O_3$ | $[-SO-NC_6H_5-N{=}\{1,2\text{-}C_6H_9\}-N(SO_2C_6H_4CH_3\text{-}4)-]$ | **4** 14, 16/7 |
| $S_2N_3C_{20}H_{17}O_3$ | $[-SO-N(SO_2C_6H_4CH_3\text{-}4)-C({=}NC_6H_5)-NC_6H_5-]$ | **3** 60/2 |
| | $[-S({=}NSO_2C_6H_4CH_3\text{-}4)-NC_6H_5-C(C_6H_5){=}N-O-]$ | **3** 321/3 |
| $S_2N_3C_{20}H_{29}O_3$ | $[-SO-N(SO_2C_6H_4CH_3\text{-}4)-C({=}NC_6H_{11}\text{-}c)-N(C_6H_{11}\text{-}c)-]$ | **3** 60/2 |
| $S_2N_3C_{21}H_{29}O_5$ | $[-S({=}NSO_2C_6H_4OCH_3\text{-}4)-N(C_6H_{11}\text{-}c)-CO-CO-N(C_6H_{11}\text{-}c)-]$ | **3** 276/8 |
| $S_2N_3C_{21}H_{35}O_5$ | $[\dot{-}S(NH_2)\dot{-}N(C(CH_3)_2C_2H_5)-CO-CO-N(C(CH_3)_2C_2H_5)\dot{-}]^+ 2,4,6\text{-}(CH_3)_3C_6H_2SO_3^-$ | **3** 282/3 |
| $S_2N_3C_{25}H_{43}O_5$ | $[\dot{-}S(NH_2)\dot{-}N(C_4H_9\text{-}t)-CO-CO-N(C_4H_9\text{-}t)\dot{-}]^+ 2,4,6\text{-}(i\text{-}C_3H_7)_3C_6H_2SO_3^-$ | **3** 282/3 |
| $S_2N_3C_{28}H_{23}O_8$ | $[-SO-N(2\text{-}C_4H_4O(OOCC_6H_5)_2(CH_2OOCC_6H_5))-CS-N{=}CH-NH-]$ ($2\text{-}C_4H_7O$ = Tetrahydro-2-furanyl) | **4** 27/8 |
| $S_2N_3C_{29}H_{25}O_8$ | $[-SO-N(2\text{-}C_4H_4O(OOCC_6H_5)_2(CH_2OOCC_6H_5))-CS-N{=}C(CH_3)-NH-]$ ($2\text{-}C_4H_7O$ = Tetrahydro-2-furanyl) | **4** 27/8 |
| $S_2N_3F_2P$ | $[\dot{-}(S\dot{-}N)_2\dot{-}PF_2\dot{-}N\dot{-}] = S_2N_3PF_2$ | **3** 14, 17/9 |

| Formula | Compound | Vol. | Page |
|---|---|---|---|
| $S_2N_3H_2P$ | $[\dot{-}(S\dot{-}N)_2\dot{-}PH_2\dot{-}N\dot{-}]\ = S_2N_3PH_2$ | **3** | 14, 18 |
| $S_2N_3H_3Pb$ | $[-(S\dot{-}N)_2-Pb-]\cdot NH_3\ = S_2N_2Pb\cdot NH_3$ | **3** | 58/9 |
| $S_2N_4As_2C_2H_6$ | $[(=S=N-AsCH_3-N=)_2]\ = S_2N_4As_2(CH_3)_2$ | **3** | 49/50 |
| $S_2N_4As_2C_8H_{18}$ | $[(=S=N-As(C_4H_9\text{-t})-N=)_2]\ = S_2N_4As_2(C_4H_9\text{-t})_2$ | **3** | 49/51 |
| $S_2N_4As_2C_{12}H_{10}$ | $[(=S=N-AsC_6H_5-N=)_2]\ = S_2N_4As_2(C_6H_5)_2$ | **3** | 49/52 |
| $S_2N_4As_2C_{18}H_{22}$ | $[(=S=N-As(C_6H_2(CH_3)_3\text{-}2,4,6)-N=)_2]\ = S_2N_4As_2(C_6H_2(CH_3)_3\text{-}2,4,6)_2$ | **3** | 49/53 |
| $S_2N_4As_2F_{12}Se_4$ | $[\dot{-}S\dot{-}N\dot{-}Se\dot{-}Se\dot{-}N\dot{-}]_2^{2+}(AsF_6^-)_2\ = [SN_2Se_2^+AsF_6^-]_2$ | **3** | 2/5 |
| $S_2N_4BC_5Cl_5H_{10}$ | $[\dot{-}SCl\dot{-}N\dot{-}S\dot{-}N\dot{-}C(N(C_2H_5)_2)\dot{-}N\dot{-}]^+BCl_4^-\ = S_2N_3CCl(N(C_2H_5)_2)^+BCl_4^-$ | **4** | 6 |
| $S_2N_4BrC_{24}H_{20}P_2^+$ | $[\dot{-}S\dot{-}N\dot{-}SBr\dot{-}N\dot{-}(P(C_6H_5)_2\dot{-}N)_2\dot{-}]^+\ = S_2N_4P_2(C_6H_5)_4Br^+$ [5,7 P] | **3** | 34 |
| $S_2N_4Br_2C_6H_2$ | 1,2,5-$SN_2C_2H$-3-CBr=CBr-3-$HC_2N_2S$-1,2,5 | **3** | 128 |
| $S_2N_4Br_2C_{12}CuH_8$ | $2[\dot{-}S\dot{-}N\dot{-}\{1,2\text{-}C_6H_4\}\dot{-}N\dot{-}]\cdot CuBr_2\ = (SN_2C_6H_4)_2\cdot CuBr_2$ | **3** | 218 |
| $S_2N_4Br_2C_{18}H_{10}$ | 4-$C_6H_5$-1,2,5-$SN_2C_2$-3-CBr=CBr-3-$C_2N_2S$-1,2,5-$C_6H_5$-4 | **3** | 130/1 |
| $S_2N_4Br_2C_{24}H_{20}P_2$ | $[\dot{-}(SBr\dot{-}N)_2\dot{-}(P(C_6H_5)_2\dot{-}N)_2\dot{-}]\ = S_2N_4P_2(C_6H_5)_4Br_2$ [5,7 P] | **3** | 34 |
| | $[(\dot{-}SBr\dot{-}N\dot{-}P(C_6H_5)_2\dot{-}N\dot{-})_2]\ = S_2N_4P_2(C_6H_5)_4Br_2$ [3,7 P] | **3** | 37/8 |
| $S_2N_4Br_2Se_4$ | $[\dot{-}S\dot{-}N\dot{-}Se\dot{-}Se\dot{-}N\dot{-}]_2^{2+}(Br^-)_2\ = [SN_2Se_2^+Br^-]_2$ | **3** | 3/4 |
| $S_2N_4CH_2$ | $[\dot{-}(S\dot{-}N)_2\dot{-}C(NH_2)\dot{-}N\dot{-}]\ = S_2N_3C(NH_2)$ | **4** | 1/2 |
| $S_2N_4C_2$ | $[\dot{-}S\dot{-}N\dot{-}\{\overset{:|}{C}\dot{-}N\dot{-}S\dot{-}N\dot{-}\overset{:|}{C}\}\dot{-}N\dot{-}]\ = S_2N_4C_2$ | **3** | 189/90 |
| $S_2N_4C_2^{\bullet-}$ | $[\dot{-}S\dot{-}N\dot{-}\{\overset{:|}{C}\dot{-}N\dot{-}S\dot{-}N\dot{-}\overset{:|}{C}\}\dot{-}N\dot{-}]^{\bullet-}\ = S_2N_4C_2^{\bullet-}$ | **3** | 191 |
| $S_2N_4C_2H_2$ | $[(\dot{-}S\dot{-}N\dot{-}CH\dot{-}N\dot{-})_2]\ = S_2N_4C_2H_2$ [3,7 C] | **4** | 134/5 |
| $S_2N_4C_3ClH_6^+$ | $[\dot{-}SCl\dot{-}N\dot{-}S\dot{-}N\dot{-}C(N(CH_3)_2)\dot{-}N\dot{-}]^+ = S_2N_3CCl(N(CH_3)_2)^+$<br>Salt:<br>$[\dot{-}SCl\dot{-}N\dot{-}S\dot{-}N\dot{-}C(N(CH_3)_2)\dot{-}N\dot{-}]^+SbCl_6^-\ = S_2N_3CCl(N(CH_3)_2)^+SbCl_6^-$ | **4** | 6 |
| $S_2N_4C_3Cl_2H_6$ | $[\dot{-}(SCl\dot{-}N)_2\dot{-}C(N(CH_3)_2)\dot{-}N\dot{-}]\ = S_2N_3CCl_2(N(CH_3)_2)$ | **4** | 6/9 |
| $S_2N_4C_3Cl_7H_6Sb$ | $[\dot{-}SCl\dot{-}N\dot{-}S\dot{-}N\dot{-}C(N(CH_3)_2)\dot{-}N\dot{-}]^+SbCl_6^-\ = S_2N_3CCl(N(CH_3)_2)^+SbCl_6^-$ | **4** | 6 |
| $S_2N_4C_4Cl_8Ti$ | (1,2,5-$SN_2C_2$-3-Cl-4-Cl$)_2\cdot TiCl_4$ | **3** | 140 |

| | | |
|---|---|---|
| S$_2$N$_4$C$_6$H$_{12}$O$_2$ | [–SO–N(C$_3$H$_7$-i)–CO–N(NH$_2$)–C(SCH$_3$)=N–] | **4** 23, 25/7 |
| S$_2$N$_4$C$_6$H$_{13}$OP | 1,2,5-SN$_2$C$_2$H-3-OP(S)(N(CH$_3$)$_2$)$_2$ | **3** 160 |
| S$_2$N$_4$C$_6$H$_{18}$O$_2$Si$_2$ | [=S=N–Si(CH$_3$)$_2$–N(SO$_2$N(CH$_3$)$_2$)–Si(CH$_3$)$_2$–N=] | **3** 7/8 |
| S$_2$N$_4$C$_7$ClH$_{14}^+$ | [∸SCl∸N∸S∸N∸C(N(C$_3$H$_7$-i)$_2$)∸N∸]$^+$ = S$_2$N$_3$CCl(N(C$_3$H$_7$-i)$_2$)$^+$ | |
| | Salt: | |
| | ([∸SCl∸N∸S∸N∸C(N(C$_3$H$_7$-i)$_2$)∸N∸]$^+$)$_2$SnCl$_6^{2-}$ = (S$_2$N$_3$CCl(N(C$_3$H$_7$-i)$_2$)$^+$)$_2$SnCl$_6^{2-}$ | **4** 6 |
| S$_2$N$_4$C$_7$Cl$_2$H$_{14}$ | [∸(SCl∸N)$_2$∸C(N(C$_3$H$_7$-i)$_2$)∸N∸] = S$_2$N$_3$CCl$_2$(N(C$_3$H$_7$-i)$_2$) | **4** 6, 10 |
| S$_2$N$_4$C$_7$H$_{14}$ | [∸(S∸N)$_2$∸C(N(C$_3$H$_7$-i)$_2$)∸N∸] = S$_2$N$_3$C(N(C$_3$H$_7$-i)$_2$) | |
| | Adduct: | |
| | [∸(S∸N)$_2$∸C(N(C$_3$H$_7$-i)$_2$)∸N∸]·C$_7$H$_8$ = (S$_2$N$_3$C(N(C$_3$H$_7$-i)$_2$)·C$_7$H$_8$ (C$_7$H$_8$ = Norbornadiene) | **4** 5 |
| (S$_2$N$_4$C$_7$H$_{14}$)$_n$ | ([∸(S∸N)$_2$∸C(N(C$_3$H$_7$-i)$_2$)∸N∸])$_n$ = (S$_2$N$_3$C(N(C$_3$H$_7$-i)$_2$))$_n$ | **4** 5/6 |
| S$_2$N$_4$C$_7$H$_{15}$OP | 1,2,5-SN$_2$C$_2$-3-OP(S)(N(CH$_3$)$_2$)$_2$-4-CH$_3$ | **3** 160 |
| S$_2$N$_4$C$_8$H$_4$O$_3$ | [=S=N–C(=NCOC$_6$H$_4$NO$_2$-4)–S–N=] = S$_2$N$_2$C=NCOC$_6$H$_4$NO$_2$-4 | **3** 68/70 |
| S$_2$N$_4$C$_8$H$_6$O$_4$ | 1,2,5-SN$_2$C$_2$H-3-NHSO$_2$C$_6$H$_4$NO$_2$-4′ | **3** 143 |
| S$_2$N$_4$C$_8$H$_8$O$_2$ | 1,2,5-SN$_2$C$_2$H-3-NHSO$_2$C$_6$H$_4$NH$_2$-4′ | **3** 143 |
| S$_2$N$_4$C$_9$ClH$_{17}$ | [∸S(N(C$_3$H$_7$-i)$_2$)∸N∸CCl∸N∸C(SCH$_3$)∸N∸] | **4** 39/40, 43/4 |
| S$_2$N$_4$C$_9$H$_{16}$O$_2$ | [–SO–N(C$_6$H$_{11}$-c)–CO–N(NH$_2$)–C(SCH$_3$)=N–] | **4** 23, 25/7 |
| S$_2$N$_4$C$_{10}$ClH$_{11}$O$_4$ | [–S(=NSO$_2$C$_6$H$_4$Cl-4)–N=C(OCH$_3$)–NH–C(OCH$_3$)=N–] | **4** 32/3 |
| | [–S(=NSO$_2$C$_6$H$_4$Cl-4)–NH–C(OCH$_3$)=N–C(OCH$_3$)=N–] | **4** 32/3 |
| S$_2$N$_4$C$_{10}$ClH$_{19}$ | [∸S(N(C$_3$H$_7$-i)$_2$)∸N∸CCl∸N∸C(SC$_2$H$_5$)∸N∸] | **4** 39/40, 43/4 |
| S$_2$N$_4$C$_{10}$H$_4$ | [∸N∸S∸N∸]$_2${1,8,4,5-C$_{10}$H$_4$} = S$_2$N$_4$C$_{10}$H$_4$ | **4** 68/70 |
| S$_2$N$_4$C$_{10}$H$_{12}$O$_4$ | [–S(=NSO$_2$C$_6$H$_5$)–N=C(OCH$_3$)–NH–C(OCH$_3$)=N–] | **4** 32/3 |
| | [–S(=NSO$_2$C$_6$H$_5$)–NH–C(OCH$_3$)=N–C(OCH$_3$)=N–] | **4** 32/3 |
| S$_2$N$_4$C$_{10}$H$_{18}$ | [(∸S∸N∸C(C$_4$H$_9$-t)∸N∸)$_2$] = S$_2$N$_4$C$_2$(C$_4$H$_9$-t)$_2$ [3,7C] | **4** 136, 139/43 |
| S$_2$N$_4$C$_{11}$ClH$_{13}$O$_4$ | [–S(=NSO$_2$C$_6$H$_4$Cl-4)–NCH$_3$–C(OCH$_3$)=N–C(OCH$_3$)=N–] | **4** 32/3, 35 |
| | [–S(=NSO$_2$C$_6$H$_4$Cl-4)–N=C(OCH$_3$)–NCH$_3$–C(OCH$_3$)=N–] | **4** 32/3, 35 |
| S$_2$N$_4$C$_{11}$H$_{14}$O$_4$ | [–S(=NSO$_2$C$_6$H$_4$CH$_3$-4)–NH–C(OCH$_3$)=N–C(OCH$_3$)=N–] | **4** 32/4 |
| | [–S(=NSO$_2$C$_6$H$_4$CH$_3$-4)–N=C(OCH$_3$)–NH–C(OCH$_3$)=N–] | **4** 32/4 |

| Formula | Structure | Vol. | Page |
|---|---|---|---|
| $S_2N_5C_{12}H_{36}PSi_4$ | [=S=N–P(=NSi(CH$_3$)$_3$)(N(Si(CH$_3$)$_3$)$_2$)–N(Si(CH$_3$)$_3$)–S–N=]<br>=S$_2$N$_3$P(=NSi(CH$_3$)$_3$)(N(Si(CH$_3$)$_3$)$_2$)(Si(CH$_3$)$_3$) | **3** | 19 |
| $S_2N_5C_{17}Cl_3H_{14}O_3$ | [–S(=NSO$_2$C$_6$H$_4$CH$_3$-4)–N=C(OCH$_2$CCl$_3$)–{N–{1,2-C$_6$H$_4$}–NH–C}=N–] | **4** | 36/7 |
| $S_2N_5C_{17}H_{15}O_3$ | [–S(=NSO$_2$C$_6$H$_4$CH$_3$-4)–NH–CO–{N–N=C(C$_6$H$_5$)–CH=C}–NH–] | **4** | 37/8 |
| $S_2N_5C_{20}ClH_{14}O_3$ | [–S(=NSO$_2$C$_6$H$_4$Cl-4)–N=C(OC$_6$H$_5$)–{N–{1,2-C$_6$H$_4$}–NH–C}=N–] | **4** | 36 |
| $S_2N_5C_{20}H_{15}O_3$ | [–S(=NSO$_2$C$_6$H$_5$)–N=C(OC$_6$H$_5$)–{N–{1,2-C$_6$H$_4$}–NH–C}=N–] | **4** | 36 |
| $S_2N_5C_{21}H_{17}O_3$ | [–S(=NSO$_2$C$_6$H$_4$CH$_3$-4)–N=C(OC$_6$H$_5$)–{N–{1,2-C$_6$H$_4$}–NH–C}=N–] | **4** | 36/7 |
| $S_2N_5C_{23}H_{19}O_3$ | [–S(=NSO$_2$C$_6$H$_4$CH$_3$-4)–NC$_6$H$_5$–CO–{N–N=C(C$_6$H$_5$)–CH=C}–NH–] | **4** | 37/8 |
| $S_2N_6BrC_6H_{12}^+$ | [∸SBr∸N∸C(N(CH$_3$)$_2$)∸N∸S∸N∸C(N(CH$_3$)$_2$)∸N∸]$^+$ = S$_2$N$_4$C$_2$Br(N(CH$_3$)$_2$)$_2^+$ [3,7C]<br>Salt:<br>[∸SBr∸N∸C(N(CH$_3$)$_2$)∸N∸S∸N∸C(N(CH$_3$)$_2$)∸N∸]$^+$Br$_3^-$<br>= S$_2$N$_4$C$_2$Br(N(CH$_3$)$_2$)$_2^+$Br$_3^-$ [3,7C] | **4** | 143/4 |
| $S_2N_6Br_4C_6H_{12}$ | [∸SBr∸N∸C(N(CH$_3$)$_2$)∸N∸S∸N∸C(N(CH$_3$)$_2$)∸N∸]$^+$Br$_3^-$ = S$_2$N$_4$C$_2$Br(N(CH$_3$)$_2$)$_2^+$Br$_3^-$ [3,7C] | **4** | 143/4 |
| $S_2N_6Br_6C_{48}H_{40}P_4$ | [(∸S∸N∸(P(C$_6$H$_5$)$_2$∸N)$_2$∸)$_2$]$^{2+}$[Br$_3^-$]$_2$ = S$_2$N$_6$P$_4$(C$_6$H$_5$)$_8^{2+}$(Br$_3^-$)$_2$ | **3** | 43/7 |
| $S_2N_6C_2H_4$ | [(∸S∸N∸C(NH$_2$)∸N∸)$_2$] = S$_2$N$_4$C$_2$(NH$_2$)$_2$ [3,7C] | **4** | 135 |
| $S_2N_6C_4H_4$ | ([∸S∸N∸CH∸N∸CH∸N∸])$_2$ = (SN$_3$C$_2$H$_2$)$_2$ | **4** | 19 |
| $S_2N_6C_6ClH_{12}^+$ | [∸SCl∸N∸C(N(CH$_3$)$_2$)∸N∸S∸N∸C(N(CH$_3$)$_2$)∸N∸]$^+$ = S$_2$N$_4$C$_2$Cl(N(CH$_3$)$_2$)$_2^+$ [3,7C]<br>Salt:<br>[∸SCl∸N∸C(N(CH$_3$)$_2$)∸N∸S∸N∸C(N(CH$_3$)$_2$)∸N∸]$^+$Cl$_3^-$<br>= S$_2$N$_4$C$_2$Cl(N(CH$_3$)$_2$)$_2^+$Cl$_3^-$ [3,7C] | **4** | 143/4 |
| $S_2N_6C_6Cl_2H_{12}$ | [∸(SCl∸N)$_2$∸(C(N(CH$_3$)$_2$)∸N)$_2$∸] = S$_2$N$_4$C$_2$Cl$_2$(N(CH$_3$)$_2$)$_2$ [5,7C] | **4** | 133/4 |
| | [(∸SCl∸N∸C(N(CH$_3$)$_2$)∸N∸)$_2$] = S$_2$N$_4$C$_2$Cl$_2$(N(CH$_3$)$_2$)$_2$ [3,7C] | **4** | 145 |
| $S_2N_6C_6Cl_4H_{12}$ | [∸SCl∸N∸C(N(CH$_3$)$_2$)∸N∸S∸N∸C(N(CH$_3$)$_2$)∸N∸]$^+$Cl$_3^-$ = S$_2$N$_4$C$_2$Cl(N(CH$_3$)$_2$)$_2^+$Cl$_3^-$ [3,7C] | **4** | 143/4 |
| $S_2N_6C_6H_{12}$ | [(∸S∸N∸C(N(CH$_3$)$_2$)∸N∸)$_2$] = S$_2$N$_4$C$_2$(N(CH$_3$)$_2$)$_2$ [3,7C] | **4** | 136/43 |
| $S_2N_6C_{14}ClH_{11}O_2$ | [–S(=NSO$_2$C$_6$H$_4$Cl-4)–N=C(NH$_2$)–NH–{C=N–{1,2-C$_6$H$_4$}–N}–] | **4** | 38/9 |
| $S_2N_6C_{14}H_{12}O_2$ | [–S(=NSO$_2$C$_6$H$_5$)–N=C(NH$_2$)–NH–{C=N–{1,2-C$_6$H$_4$}–N}–] | **4** | 38/9 |
| $S_2N_6C_{15}H_{14}O_2$ | [–S(=NSO$_2$C$_6$H$_4$CH$_3$-4)–N=C(NH$_2$)–NH–{C=N–{1,2-C$_6$H$_4$}–N}–] | **4** | 38/9 |

| | | | |
|---|---|---|---|
| $S_2N_6C_{26}H_{20}$ | $[(∸S∸N∸C(N(C_6H_5)_2)∸N∸)_2]$ = $S_2N_4C_2(N(C_6H_5)_2)_2$ [3,7C] | **4** | 136, 138/40, 142 |
| $S_2N_6C_{28}H_{20}$ | $([∸S∸N∸C(C_6H_5)∸N∸C(C_6H_5)∸N∸])_2$ = $(SN_3C_2(C_6H_5)_2)_2$ | **4** | 20/1 |
| $S_2N_6C_{48}H_{40}I_6P_4$ | $[(∸S∸N∸(P(C_6H_5)_2∸N)_2∸)_2]^{2+}[I_3^-]_2$ = $S_2N_6P_4(C_6H_5)_8^{2+}(I_3^-)_2$ | **3** | 43/7 |
| $S_2N_6C_{48}H_{40}P_4$ | $[∸S∸N∸(P(C_6H_5)_2∸N)_2∸S∸\{N∸(P(C_6H_5)_2∸N)_2\}∸]$ = $S_2N_6P_4(C_6H_5)_8$ | **3** | 43/6 |
| $S_2N_6C_{48}H_{40}P_4^{2+}$ | $[(∸S∸N∸(P(C_6H_5)_2∸N)_2∸)_2]^{2+}$ = $S_2N_6P_4(C_6H_5)_8^{2+}$ | **3** | 43/6 |
| | Salts: | | |
| | $[(∸S∸N∸(P(C_6H_5)_2∸N)_2∸)_2]^{2+}[Br_3^-]_2$ = $S_2N_6P_4(C_6H_5)_8^{2+}(Br_3^-)_2$ | **3** | 43/7 |
| | $[(∸S∸N∸(P(C_6H_5)_2∸N)_2∸)_2]^{2+}[I_3^-]_2$ = $S_2N_6P_4(C_6H_5)_8^{2+}(I_3^-)_2$ | **3** | 43/7 |
| $S_2N_7C_4H_3$ | 1,2,5-$SN_2C_2H$-3-N=N–NH-3-$HC_2N_2S$-1,2,5 | **3** | 143 |
| $S_2N_8C_{52}H_{52}P_4$ | $[(∸S(N(CH_3)_2)∸N∸(P(C_6H_5)_2∸N)_2∸)_2]$ = $S_2N_6P_4(C_6H_5)_8(N(CH_3)_2)_2$ | **3** | 43/7 |
| $S_2N_8C_{56}H_{60}P_4$ | $[(∸S(N(C_2H_5)_2)∸N∸(P(C_6H_5)_2∸N)_2∸)_2]$ = $S_2N_6P_4(C_6H_5)_8(N(C_2H_5)_2)_2$ | **3** | 43, 47 |
| $S_2N_8C_{58}H_{60}P_4$ | $[(∸S(–NC_5H_{10})∸N∸(P(C_6H_5)_2∸N)_2∸)_2]$ = $S_2N_6P_4(C_6H_5)_8(NC_5H_{10})_2$ ($NC_5H_{10}$ = Piperidino) | **3** | 43, 45, 48 |

## $S_3$

| | | | |
|---|---|---|---|
| $S_3N_2BrC^+$ | $[∸S∸N∸C(SBr)∸S∸N∸]^+$ = $S_2N_2C–SBr^+$ | **3** | 73 |
| | Salts: | | |
| | $[∸S∸N∸C(SBr)∸S∸N∸]^+Br^-$ = $S_2N_2C–SBr^+Br^-$ | **3** | 72 |
| | $[∸S∸N∸C(SBr)∸S∸N∸]^+Br_3^-$ = $S_2N_2C–SBr^+Br_3^-$ | **3** | 72/4 |
| $S_3N_2BrC_2H$ | [=S=N–S–CBr=CH–S–N=] = $S_3N_2C_2HBr$ | **4** | 112/3, 117 |
| $S_3N_2Br_2C$ | $[∸S∸N∸C(SBr)∸S∸N∸]^+Br^-$ = $S_2N_2C–SBr^+Br^-$ | **3** | 72 |
| $S_3N_2Br_2C_2$ | [=S=N–S–CBr=CBr–S–N=] = $S_3N_2C_2Br_2$ | **4** | 112/3, 117 |
| $S_3N_2Br_2C_6H_{12}O_5$ | $[–SO–N(SO_2CH_3)–CH(CH_2Br)–CH(CH_2Br)–N(SO_2CH_3)–]$ | **3** | 266/7 |
| $S_3N_2Br_4C$ | $[∸S∸N∸C(SBr)∸S∸N∸]^+Br_3^-$ = $S_2N_2C–SBr^+Br_3^-$ | **3** | 72/4 |
| $S_3N_2CO_4$ | $[=S=N–CO–S–N=]\cdot SO_3$ = $S_2N_2CO\cdot SO_3$ | **3** | 67 |
| $S_3N_2C_2HI$ | [=S=N–S–CI=CH–S–N=] = $S_3N_2C_2HI$ | **4** | 112/3, 117 |
| $S_3N_2C_2H_2$ | [=S=N–S–CH=CH–S–N=] = $S_3N_2C_2H_2$ | **4** | 112, 115/7 |

| Formula | Structure | Vol. | Pages |
|---|---|---|---|
| $S_3N_3C_{17}H_{17}O_4$ | [–SO–N(CH$_2$C$_6$H$_5$)–C(=NSO$_2$C$_6$H$_4$CH$_3$-4)–S–N(COCH$_3$)–] | **3** | 76/7, 79 |
| $S_3N_3C_{18}H_{21}O_3$ | [–SO–N(CH$_2$C$_6$H$_5$)–C(=NSO$_2$C$_6$H$_4$CH$_3$-4)–S–NC$_3$H$_7$–] | **3** | 76/7 |
| $S_3N_3C_{18}H_{23}O_3Si$ | [–SO–N(CH$_2$C$_6$H$_5$)–C(=NSO$_2$C$_6$H$_4$CH$_3$-4)–S–N(Si(CH$_3$)$_3$)–] | **3** | 76, 78, 80 |
| $S_3N_3C_{20}H_{17}O_7$ | [–SO–N(SO$_2$C$_6$H$_4$CH$_3$-4)–{1,2-C$_6$H$_3$NO$_2$-4}–N(SO$_2$C$_6$H$_4$CH$_3$-4)–] | **3** | 274 |
| $S_3N_3C_{21}ClH_{18}O_3$ | [–SO–N(CH$_2$C$_6$H$_5$)–C(=NSO$_2$C$_6$H$_4$CH$_3$-4)–S–N(C$_6$H$_4$Cl-4)–] | **3** | 76/7 |
| $S_3N_3C_{22}H_{21}O_3$ | [–SO–N(CH$_2$C$_6$H$_5$)–C(=NSO$_2$C$_6$H$_4$CH$_3$-4)–S–N(C$_6$H$_4$CH$_3$-4)–] | **3** | 76, 78 |
| $S_3N_3Se^+$ | [∸(S∸N)$_2$∸S∸Se∸N∸]$^+$ | **3** | 5 |
| | [∸(S∸N)$_2$∸Se∸N∸S∸]$^+$ | **3** | 5 |
| $S_3N_4C_2F_3^+$ | [∸(S∸N)$_3$∸C(CF$_3$)∸N∸]$^+$ = S$_3$N$_4$C(CF$_3$)$^+$ | **4** | 126 |
| | Salt: | | |
| | [∸(S∸N)$_3$∸C(CF$_3$)∸N∸]$^+$S$_3$N$_3$O$_4^-$ = S$_3$N$_4$C(CF$_3$)$^+$S$_3$N$_3$O$_4^-$ | **4** | 126 |
| | (S$_3$N$_3$O$_4^-$ = 1,3,5,2,4,6-Trithia(5-S$^{IV}$)triazine 1,1,3,3-tetraoxide ion(1−)) | | |
| $S_3N_4C_2O_4$ | [=S=N–S–C(NO$_2$)=C(NO$_2$)–S–N=] = S$_3$N$_2$C$_2$(NO$_2$)$_2$ | **4** | 112/3, 117 |
| $S_3N_4C_4H_2$ | [∸S∸N∸{C∸C(SH)∸N∸N∸C(SH)∸C}∸N∸] | **3** | 227/9 |
| | [–S–N–{C–CS–NH–NH–CS–C}∸N∸] | **3** | 227/9 |
| $S_3N_4C_5ClH_9$ | [∸S∸N∸SCl∸N∸S∸N∸C(C$_4$H$_9$-t)∸N∸] = S$_3$N$_4$CCl(C$_4$H$_9$-t) | **4** | 127/9 |
| $S_3N_4C_6H_6$ | [∸S∸N∸{C∸C(SCH$_3$)∸N∸N∸C(SCH$_3$)∸C}∸N∸] | **3** | 227, 229 |
| $S_3N_4C_{10}H_{14}$ | [∸S∸N∸{C∸N∸C(SC$_3$H$_7$)∸C(SC$_3$H$_7$)∸N∸C}∸N∸] | **3** | 229, 231/2 |
| $S_3N_4C_{10}H_{20}$ | [∸S(N(C$_3$H$_7$-i)$_2$)∸N∸C(SCH$_3$)∸N∸C(SCH$_3$)∸N∸] | **4** | 39, 50/1 |
| $S_3N_4C_{12}H_{24}$ | [∸S(N(C$_3$H$_7$-i)$_2$)∸N∸C(SC$_2$H$_5$)∸N∸C(SC$_2$H$_5$)∸N∸] | **4** | 39, 50/1 |
| $S_3N_4C_{14}H_{28}$ | [∸S(N(C$_3$H$_7$-i)$_2$)∸N∸C(SC$_3$H$_7$-i)∸N∸C(SC$_3$H$_7$-i)∸N∸] | **4** | 39, 50/1 |
| $S_3N_4C_{16}H_{10}$ | (4-C$_6$H$_5$-1,2,5-SN$_2$C$_2$-3)–N=(4′-C$_2$NS$_2$-1′,3′2′-C$_6$H$_5$-5′) | **3** | 144/5 |
| | (4-C$_2$NS$_2$-1,3,2-C$_6$H$_5$-5 = 5-Phenyl-1,3,2-dithiazol-4-ylidene) | | |
| $S_3N_4C_{16}H_{28}$ | [∸S(N(C$_6$H$_{11}$-c)$_2$)∸N∸C(SCH$_3$)∸N∸C(SCH$_3$)∸N∸] | **4** | 39, 50/1 |
| $S_3N_4C_{16}H_{32}$ | [∸S(N(C$_3$H$_7$-i)$_2$)∸N∸C(SC$_4$H$_9$)∸N∸C(SC$_4$H$_9$)∸N∸] | **4** | 39, 50/2 |
| $S_3N_4C_{20}H_{24}$ | [∸S(N(C$_3$H$_7$-i)$_2$)∸N∸C(SC$_6$H$_5$)∸N∸C(SC$_6$H$_5$)∸N∸] | **4** | 39, 50/1 |

| | | | |
|---|---|---|---|
| $S_3N_5C_{13}ClH_{24}$ | [∸S(N(C$_3$H$_7$-i)$_2$)∸N∸CCl∸N∸C(SC(S)N(C$_2$H$_5$)$_2$)∸N∸] | **4** | 39/40, 44 |
| $S_3N_5C_{17}H_{15}O_2$ | [–S(=NSO$_2$C$_6$H$_4$CH$_3$-4)–NH–CS–{N–N=C(C$_6$H$_5$)–CH=C}–NH–] | **4** | 37/8 |
| $S_3N_5C_{19}ClH_{32}$ | [∸S(N(C$_6$H$_{11}$-c)$_2$)∸N∸CCl∸N∸C(SC(S)N(C$_2$H$_5$)$_2$)∸N∸] | **4** | 39/40, 44 |
| $S_3N_5C_{20}H_{21}O_2$ | [–S(=NSO$_2$C$_6$H$_4$CH$_3$-4)–NC$_3$H$_7$–CS–{N–N=C(C$_6$H$_5$)–CH=C}–NH–] | **4** | 37/8 |
| $S_3N_5C_{22}H_{21}O_4$ | [–S(=NSO$_2$C$_6$H$_4$CH$_3$-4)–N(SO$_2$C$_6$H$_4$CH$_3$-4)–C(NH$_2$)=N–C(C$_6$H$_5$)=N–] | **4** | 35/6 |
| | [–S(=NSO$_2$C$_6$H$_4$CH$_3$-4)–N(SO$_2$C$_6$H$_4$CH$_3$-4)–C(=NH)–N=C(C$_6$H$_5$)–NH–] | **4** | 35 |
| $S_3N_5C_{25}H_{20}P$ | [∸S∸N∸S(N=P(C$_6$H$_5$)$_3$)∸N∸S∸N∸C(C$_6$H$_5$)∸N∸] = S$_3$N$_4$C(N=P(C$_6$H$_5$)$_3$)(C$_6$H$_5$) | **4** | 127/30 |
| $S_3N_5F_2P$ | [∸{1,3-S$_3$N$_3$}∸N∸PF$_2$∸N∸] = S$_3$N$_5$PF$_2$ | **3** | 39/40 |
| $S_3N_6C_2H_6Si_2$ | [=S=N–SiCH$_3$–N=S=N–SiCH$_3$–N=] (Si–N=S=N–Si bridge) = S$_3$N$_6$Si$_2$(CH$_3$)$_2$ | **3** | 11/2 |
| $S_3N_6C_3H_6$ | [–{1,3-S$_3$N$_3$}–N=C(N(CH$_3$)$_2$)–N=] = S$_3$N$_5$C(N(CH$_3$)$_2$) | **4** | 146/8 |
| $S_3N_6C_5H_{10}$ | [–{1,3-S$_3$N$_3$}–N=C(N(C$_2$H$_5$)$_2$)–N=] = S$_3$N$_5$C(N(C$_2$H$_5$)$_2$) | **4** | 146/8 |
| $S_3N_6C_5H_{12}O$ | [–{1,3-S$_3$N$_3$(N(CH$_3$)$_2$-3′)}–NCH$_3$–CO–NCH$_3$–] | **4** | 153/4 |
| $S_3N_6C_6$ | [∸N∸S∸N∸]$_3${C$_6$} = S$_3$N$_6$C$_6$ | **3** | 223/4 |
| $S_3N_6C_7H_{14}$ | [={1,3-S$_3$N$_3$}–N=C(N(C$_3$H$_7$-i)$_2$)–N=] = S$_3$N$_5$C(N(C$_3$H$_7$-i)$_2$) | **4** | 146/9 |
| $S_3N_6C_7H_{16}O$ | [–{1,3-S$_3$N$_3$(N(C$_2$H$_5$)$_2$-3′)}–NCH$_3$–CO–NCH$_3$–] | **4** | 153/4 |
| $S_3N_6C_{18}Cl_3H_{12}Rh$ | [([∸S∸N∸{1,2-C$_6$H$_4$}∸N∸])$_3$RhCl$_3$] = [(SN$_2$C$_6$H$_4$)$_3$·RhCl$_3$] | **3** | 218 |
| $S_3N_6C_{18}H_{22}O_4$ | [–S(=NSO$_2$C$_6$H$_4$CH$_3$-4)–N(SO$_2$C$_6$H$_4$CH$_3$-4)–C(NH$_2$)=N–C(N(CH$_3$)$_2$)=N–] | **4** | 35 |
| $S_3N_6C_{25}H_{28}O_4$ | [–S(=NSO$_2$C$_6$H$_4$CH$_3$-4)–N(SO$_2$C$_6$H$_4$CH$_3$-4)–C(NH$_2$)=N–C(N(C$_2$H$_5$)(CH$_2$C$_6$H$_5$))=N–] | **4** | 35 |

**$S_4$**

| | | | |
|---|---|---|---|
| $S_4N_2C_7H_8$ | [=S=N–S–{2,3-C$_7$H$_8$}–S–S–N=] | **4** | 132/3 |
| $S_4N_3C_{22}H_{21}O_5$ | [–SO–N(CH$_2$C$_6$H$_5$)–C(=NSO$_2$C$_6$H$_4$CH$_3$-4]–S–N(SO$_2$C$_6$H$_4$CH$_3$-4)–] | **3** | 76, 78/9 |
| $S_4N_4C_2$ | [∸S∸N∸S∸N∸{C∸S∸N∸S∸C}∸N∸] = S$_4$N$_4$C$_2$ | **4** | 122/4 |
| $S_4N_4C_2Cl_4O_2Sn$ | 2[=S=N–CO–S–N=]·SnCl$_4$ = 2S$_2$N$_2$CO·SnCl$_4$ | **3** | 67 |

## Monocyclic Systems

**4**

**SNCO**

{$S^{IV}NCO$}

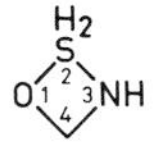

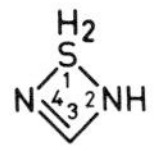

**$SN_2C$**

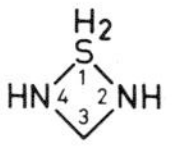

{$S^{IV}N_2C$}

**4**
**$SN_2C$**
$\{S^{IV}N_2C\}$

1,2,4-Thiadiazetidine, 1,1-dihydro-
(=1λ⁴,2,4-Thiadiazetidine)
Derivatives (cont.)

| | |
|---|---|
| (S: 1) =O; (N: 2) $-SO_2C_6H_5$; (N: 4) $-C_6H_5$; (C: 3) $-C_6H_5$, $-C_6H_5$ | **3** 60 |
| (S: 1) =O; (N: 2) $-SO_2C_6H_4CH_3$-4; (N: 4) $-CH_3$; (C: 3) $=NCH_3$ | **3** 60/2 |
| (S: 1) =O; (N: 2) $-SO_2C_6H_4CH_3$-4; (N: 4) $-C_3H_7$-i; (C: 3) $=NC_3H_7$-i | **3** 60/2 |
| (S: 1) =O; (N: 2) $-SO_2C_6H_4CH_3$-4; (N: 4) $-C_4H_9$; (C: 3) $=NC_4H_9$ | **3** 60/2 |
| (S: 1) =O; (N: 2) $-SO_2C_6H_4CH_3$-4; (N: 4) $-C_6H_{11}$-c; (C: 3) $=NC_6H_{11}$-c | **3** 60/2 |
| (S: 1) =O; (N: 2) $-SO_2C_6H_4CH_3$-4; (N: 4) $-C_6H_5$; (C: 3) $=NC_6H_5$ | **3** 60/2 |
| (S: 1) =O; (N: 2) $-SO_2C_6H_4CH_3$-4; (N: 4) $-C_6H_4CH_3$-4; (C: 3) $-C_6H_4CH_3$-4, $-C_6H_4CH_3$-4 | **3** 60 |

**$SN_2P$**
$\{S^{IV}N_2P^{V}\}$

1,2,4,3-Thiadiazaphosphetidine, 1,1,3,3-tetrahydro-
(=1λ⁴,2,4,3λ⁵-Thiadiazaphosphetidine)
Derivatives:

| | |
|---|---|
| (S: 1) $=NC_4H_9$-t; (N: 2 and 4) $-C_4H_9$-t; (P: 3) $-N(C_4H_9\text{-t})(Si(CH_3)_3)$, =S | **3** 13/4 |
| (S: 1) =O; (N: 2 and 4) $-C_4H_9$-t; (P: 3) –Cl, =S | **3** 13 |

**$S_2NO$**
$\{S_2^{IV}NO\}$

1,2,4,3-Oxadithiazetidine, 2,2,4,4-tetrahydro-
(=1,2λ⁴,4λ⁴,3-Oxadithiazetidine)
Derivatives:
(S: 2) =O; (S: 4) $-CH_3$, $-CH_3$; (N: 3) $-(CH_2)_5-$
Ion(1+)
Salt: $SbCl_6^-$ ... **3** 1
(S: 2) =O; (S: 4) $-N(CH_3)_2$, $-CH_3$; (N: 3) $-CH_3$, $-CH_3$
Ion(1+)
Salt: $BF_4^-$ ... **3** 1

**5**
**$SNC_2O$**
$\{S^{IV}NC_2O\}$

5H-1,2,3-Oxathiazole, 2,2-dihydro-
(=5H-1,2λ⁴,3-Oxathiazole)
Derivatives:

| | |
|---|---|
| (S: 2) =O; (C: 4) –Cl | **3** 286/7 |
| (S: 2) =O; (C: 4) –Cl; (C: 5) $-CH_3$, $-CH_3$ | **3** 286/7 |
| (S: 2) =O; (C: 4) $-N(CH_3)_2\cdot HCl$ | **3** 286/7 |

**5**
**$SNC_2O$**
$\{S^{IV}NC_2O\}$

1,2,3-Oxathiazolidine, 2,2-dihydro-
(=1,2$\lambda^4$,3-Oxathiazolidine)

Derivatives (cont.)

(S: 2) =O; (N: 3) –$C_6H_4Cl$-4; (C: 4) =$NCONHC_6H_5$; (C: 5) –$CH_3$, –$CH_3$ ........ **3** 291/2
(S: 2) =O; (N: 3) –$C_6H_3Cl_2$-3,4; (C: 4) =O; (C: 5) –$CH_3$ ........ **3** 288/90
(S: 2) =O; (N: 3) –(1-$C_{10}H_7$) (1-$C_{10}H_7$ = 1-Naphthalenyl) ........ **3** 294/5, 297

**$SN_2CO$**
{$S^{IV}N_2CO$}

3H-1,3,2,4-Oxathiadiazole, 2,5-dihydro-
(=1,3λ[4],2,4-Oxathiadiazole, 2,5-dihydro-)
Derivative:
(S: 3) –F; (N: 2) –$CF_3$; (C: 5) –$CF_3$, –$CF_3$ ........ **3** 313

3H-1,2,3,4-Oxathiadiazole, 2,2-dihydro-
(=3H-1,2λ[4],3,4-Oxathiadiazole)
Derivatives:
(S: 2) =O; (N: 3) –(2-$C_5H_4N$); (C: 5) –$CH_3$ (2-$C_5H_4N$ = 2-Pyridinyl) ........ **3** 309/10, 312
(S: 2) =O; (N: 3) –(2-$C_5H_4N$); (C: 5) –$C_6H_5$ (2-$C_5H_4N$ = 2-Pyridinyl) ........ **3** 309/10, 312
(S: 2) =O; (N: 3) –(2-$C_5H_3N$-5-Cl); (C: 5) –(2-$C_5H_4N$) ........ **3** 309, 311/2
(2-$C_5H_3N$-5-Cl = 5-Chloro-2-pyridinyl; 2-$C_5H_4N$ = 2-Pyridinyl)
(S: 2) =O; (N: 3) –(2-$C_5H_3N$-5-Cl); (C: 5) –(4-$C_5H_4N$) ........ **3** 309, 311/2
(2-$C_5H_3N$-5-Cl = 5-Chloro-2-pyridinyl; 4-$C_5H_4N$ = 4-Pyridinyl)
(S: 2) =O; (N: 3) –(2-($C_5H_3N$-6-Cl); (C: 5) –$C_6H_4NO_2$-2 ........ **3** 309, 311/2
(2-$C_5H_3N$-6-Cl = 6-Chloro-2-pyridinyl)
(S: 2) =O; (N: 3) –(1-$C_9H_5N$-3-Cl); (C: 5) –$C_6H_4NO_2$-2 ........ **3** 309, 311/2
(1-$C_9H_5N$-3-Cl = 3-Chloro-1-isoquinolinyl)
(S: 2) =O; (N: 3) –CHO; (C:5) –$C_6H_5$ ........ **3** 309/10, 312
(S: 2) =O; (N: 3) –$COCH_3$; (C: 5) –$CH_3$ ........ **3** 309/10, 312
(S: 2) =O; (N: 3) –$COCH_3$; (C: 5) –$C_6H_5$ ........ **3** 309/10, 312
(S: 2) =O; (N: 3) –$COC_6H_5$; (C: 5) –$C_6H_5$ ........ **3** 309/10, 312
(S: 2) =O; (N: 3) –$COOC_2H_5$; (C: 5) –$C_6H_5$ ........ **3** 309/10, 312
(S: 2) =O; (N: 3) –$COOCH_2C_6H_5$; (C: 5) –$C_6H_5$ ........ **3** 309/10, 312
(S: 2) =O; (N: 3) –$Si(CH_3)_3$; (C: 5) –$Si(CH_3)_3$ ........ **3** 309, 311/2

3H-1,2,3,5-Oxathiadiazole, 2,2-dihydro-
(=3H-1,2λ[4],3,5-Oxathiadiazole)
Derivatives:
(S: 2) =$NC(CH_3)_2CN$; (N: 3) –$C_6H_4CH_3$-4; (C: 4) –$C_6H_4NO_2$-3 ........ **3** 321/3
(S: 2) =$NC_6H_5$; (N: 3) –$C_6H_5$; (C: 4) –$C_6H_4NO_2$-4 ........ **3** 321/3
(S: 2) =$NC_6H_4NO_2$-3; (N: 3) –$C_6H_4NO_2$-3; (C: 4) –$C_6H_4NO_2$-4 ........ **3** 321/3
(S: 2) =$NC_6H_4NO_2$-4; (N: 3) –$C_6H_4CH_3$-4; (C: 4) –$C_6H_4NO_2$-3 ........ **3** 321/3

**5**
**$SN_2CO$**
{$S^{IV}N_2CO$}

3H-1,2,3,5-Oxathiadiazole, 2,2-dihydro-
(=3H-1,2λ[4],3,5-Oxathiadiazole)
Derivatives (cont.)



1,2,3,5-Oxathiadiazolidine, 2,2-dihydro-
(=1,2λ[4],3,5-Oxathiadiazolidine)
Derivatives:

**$SN_2C_2$**
{$S^{IV}N_2C_2$}

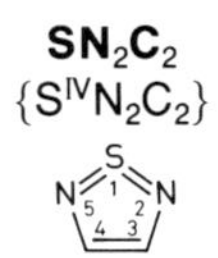

**5**
$SN_2C_2$
$\{S^{IV}N_2C_2\}$

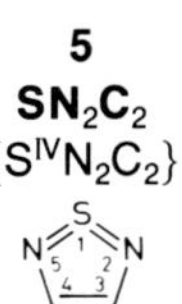

**5**
**$SN_2C_2$**
$\{S^{IV}N_2C_2\}$

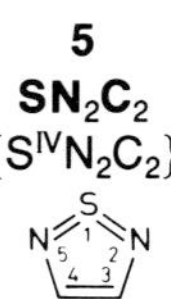

1,2,5-Thia($S^{IV}$)diazole
(=1λ[4],2,5-Thiadiazole)

**5**
**$SN_2C_2$**
$\{S^{IV}N_2C_2\}$

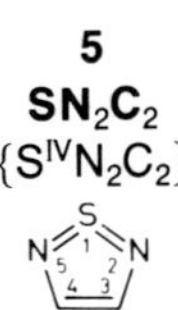

1,2,5-Thia($S^{IV}$)diazole
(=1$\lambda^4$,2,5-Thiadiazole)

**5**
**$SN_2C_2$**
$\{S^{IV}N_2C_2\}$

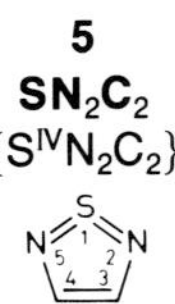

**5**
**$SN_2C_2$**
$\{S^{IV}N_2C_2\}$

**5**
**$SN_2C_2$**
$\{S^{IV}N_2C_2\}$

**5**
**$SN_2C_2$**
$\{S^{IV}N_2C_2\}$

1,2,5-Thiadiazolidine, 1,1-dihydro-
(=1λ$^4$,2,5-Thiadiazolidine)

Derivatives (cont.)

(S: 1) =O; (N: 2) $-CH_2C_6H_5$; (N: 5) $-C_4H_9$-t; (C: 3) =O ........ **3** 267/9
(S: 1) =O; (N: 2 and 5) $-C_6H_{11}$-c; (C: 3 and 4) =O ........ **3** 270, 272
(S: 1) =O; (N: 2) $-C_6H_5$; (C: 3) =NH; (C: 4) $-CH_3$, $-CH_3$ ........ **3** 273
(S: 1) =O; (N: 2) $-C_6H_5$; (N: 5) $-C_4H_9$-t; (C: 3) =O; (C: 4) $-C_6H_5$ ........ **3** 267/8
(S: 1) =O; (N: 2 and 5) $-C_6H_5$; (C: 3 and 4) $=NC_6H_5$ ........ **3** 273/5
(S: 1) =O; (N: 2 and 5) $-C_6H_5$; (C: 3) =O; (C: 4) $-C_6H_5$, $C_6H_5$ ........ **3** 267, 269
(S: 1) =O; (N: 2) $-C_6H_4CH_3$-2; (N: 5) $-C_4H_9$-t; (C: 3) =O ........ **3** 267/9
(S: 1) =O; (N: 2) $-C_6H_4CH_3$-2; (N: 5) $-C_6H_5$; (C: 3) =O ........ **3** 267/9
(S: 1) =O; (N: 2 and 5) $-C_6H_4CH_3$-4; (C: 3 and 4) $=NC_6H_5$ ........ **3** 273/5
(S: 1) =O; (N: 2 and 5) $-C_6H_4CH_3$-4; (C: 3 and 4) $=NC_6H_4CH_3$-4 ........ **3** 273/5
(S: 1) =O; (N: 2) $-C_6H_4Cl$-2; (C: 3) =NH; (C: 4) $-CH_3$, $-CH_3$ ........ **3** 273
(S: 1) =O; (N: 2) $-C_6H_4Cl$-3; (C: 3) =NH; (C: 4) $-CH_3$, $-CH_3$ ........ **3** 273
(S: 1) =O; (N: 2 and 5) $-C_6H_4N(CH_3)_2$-4; (C: 3 and 4) $=NC_6H_5$ ........ **3** 273/5
(S: 1) =O; (N: 2) $-C_6H_4NO_2$-3; (C: 3) =NH; (C: 4) $-CH_3$, $CH_3$ ........ **3** 273
(S: 1) =O; (N: 2 and 5) $-C_6H_3(NH_2)OCH_3$-4; (C: 3 and 4) $=NC_6H_3(NH_2)OCH_3$-4 ........ **3** 273/5
(S: 1) =O; (N: 2 and 5) $-(1-C_{10}H_7)$; (C: 3 and 4) $=NC_6H_5$ ........ **3** 273/5
($1-C_{10}H_7$ = 1-Naphthalenyl)
(S: 1) =O; (N: 2 and 5) $-COC_6H_5$; (C: 3 and 4) =O ........ **3** 270, 272
(S: 1) =O; (N: 2 and 5) $-SO_2CH_3$; (C: 3 and 4) $-CH_2Br$ ........ **3** 266/7
(S: 1) =O; (N: 2 and 5) $-SO_2CH_3$; (C: 3 and 4) $-CH_2Cl$ ........ **3** 266/7
(S: 1) =O; (N: 2 and 5) $-SO_2CH_3$; (C: 3 and 4) $-SO_3CH_3$ ........ **3** 266/7
(S: 1) =O; (N: 2 and 5) $-SO_2C_6H_4CH_3$-4 ........ **3** 264/6

**$SN_2Se_2$**
$\{S^{IV}N_2Se^{II}_2\}$

1,3,4,2,5-Thia(1-$S^{IV}$)diselenadiazole
(=1λ$^4$,3,4,2,5-Thiadiselenadiazole)
*(for the dimeric ion and salts see $\{S^{IV}N_2Se^{IV}_2-Se^{IV}_4-S^{IV}N_2Se^{IV}_2\}$)*

**$SN_2Si_2$**
$\{S^{IV}N_2Si_2\}$

1-Thia-2,5-diaza-3,4-disilacyclopentane, 1,1-dihydro-
(=1λ$^4$-Thia-2,5-diaza-3,4-disilacyclopentane)

Derivatives:

(S: 1) =O; (N: 2 and 5) $-CH_3$; (Si: 3 and 4) $-CH_3$, $-CH_3$ ........ **3** 6/7
(S: 1) =O; (N: 2 and 5) $-Si(CH_3)_3$; (Si: 3 and 4) $-CH_3$, $-CH_3$ ........ **3** 6/7

1,2,3,5-Thiatriazole, 1,1,2,3-tetrahydro-
(=1λ[4],2,3,5-Thiatriazole, 2,3-dihydro-)

Derivatives:

(S: 1) =O; (N: 2) $-CH_3$; (C: 4) $-CONHCH_2C_6H_5$ ..... **3** 80
(S: 1) =O; (N: 2) $-C_6H_5$; (C: 4) $-CONHCH_2C_6H_5$ ..... **3** 80

1,2,3,5-Thiatriazole, 1,1,2,5-tetrahydro-
(=1λ[4],2,3,5-Thiatriazole, 2,5-dihydro-)

Derivatives:

(S: 1) =O; (N: 2 and 5) $-C_6H_5$; (C: 4) $-C_6H_5$ ..... **3** 81/2
(S: 1) =O; (N: 2) $-C_6H_5$; (N: 5) $-C_6H_4Cl$-4; (C: 4) $-COOCH_3$ ..... **3** 81/2
(S: 1) =O; (N: 2) $-C_6H_4CF_3$-3; (N: 5) $-C_3H_7$-i; (C: 4) $-COOCH_3$ ..... **3** 81/2
(S: 1) =O; (N: 2) $-C_6H_4CF_3$-3; (N: 5) $-C_6H_5$; (C: 4) $-COOCH_3$ ..... **3** 81/2
(S: 1) =O; (N: 2) $-C_6H_4Cl$-3; (N: 5) $-C_3H_7$-i; (C: 4) $-COOCH_3$ ..... **3** 81/2
(S: 1) =O; (N: 2) $-C_6H_4Cl$-4; (N: 5) $-C_3H_7$-i; (C: 4) $-COOCH_3$ ..... **3** 81/2
(S: 1) =O; (N: 2) $-C_6H_4Cl$-4; (N: 5) $-C_6H_3(CF_3)_2$-3,5; (C: 4) $-COOCH_3$ ..... **3** 81/2
(S: 1) =O; (N: 2) $-C_6H_3Cl_2$-3,4; (N: 5) $-CH_3$; (C: 4) $-COOCH_3$ ..... **3** 81/2
(S: 1) =O; (N: 2) $-C_6H_3Cl_2$-3,4; (N: 5) $-C_3H_7$-i; (C: 4) $-COOCH_3$ ..... **3** 81/2
(S: 1) =O; (N: 2) $-COC_6H_5$; (N: 5) $-C_6H_5$; (C: 4) $-C_6H_5$ ..... **3** 81/2

1,2,3,5-Thiatriazolidine, 1,1-dihydro-
(=1λ[4],2,3,5-Thiatriazolidine)

Derivatives:

(S: 1) =O; (N: 2 and 3) $-CH_3$; (N: 5) $-C_4H_9$-t; (C: 4) =O ..... **3** 81, 83, 85
(S: 1) =O; (N: 2 and 3) $-CH_3$; (N: 5) $-C_6H_5$; (C: 4) =O ..... **3** 81, 83, 85
(S: 1) =O; (N: 2 and 3) $-CH_3$; (N: 5) $-C_6H_5$; (C: 4) =S ..... **3** 81, 84/5
(S: 1) =O; (N: 2 and 3) $-CH_3$; (N: 5) $-C_6H_4Br$-4; (C: 4) =O ..... **3** 81, 83, 85
(S: 1) =O; (N: 2 and 3) $-CH_3$; (N: 5) $-C_6H_4CF_3$-3; (C: 4) =O ..... **3** 81, 83, 85
(S: 1) =O; (N: 2 and 3) $-CH_3$; (N: 5) $-C_6H_4Cl$-4; (C: 4) =O ..... **3** 81, 83, 85
(S: 1) =O; (N: 2 and 3) $-CH_3$; (N: 5) $-C_6H_3Cl_2$-3,4; (C: 4) =O ..... **3** 81, 83, 85
(S: 1) =O; (N: 2 and 3) $-CH_3$; (N: 5) $-C_6H_2Cl_3$-2,4,5; (C: 4) =O ..... **3** 81, 83, 85
(S: 1) =O; (N: 2 and 3) $-CH_3$; (N: 5) $-C_6H_4F$-2; (C: 4) =O ..... **3** 81, 83, 85
(S: 1) =O; (N: 2 and 3) $-CH_3$; (N: 5) $-C_6H_4F$-4; (C: 4) =O ..... **3** 81, 83, 85
(S: 1) =O; (N: 2 and 3) $-CH_3$; (N: 5) $-C_6H_4F$-4; (C: 4) =S ..... **3** 81, 84/5
(S: 1) =O; (N: 2 and 3) $-CH_3$; (N: 5) $-C_6H_4I$-4; (C: 4) =O ..... **3** 81, 83, 85
(S: 1) =O; (N: 2 and 5) $-CH_3$; (N: 3) $-C_6H_5$; (C: 4) =O ..... **3** 81, 83, 85
(S: 1) =O; (N: 2 and 5) $-CH_3$; (N: 3) $-C_6H_3Cl_2$-3,4; (C: 4) =O ..... **3** 81, 84/5

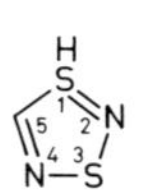

**5**
**$S_2N_2Pb$**
$\{S^{IV}S^{II}N_2Pb^{II}\}$

5H-1,3,2,4,5-Dithia(3-$S^{IV}$)diazaplumbole, 5,5-didehydro- ..... **3** 58
(=1,3λ[4],2,4,5λ[2]-Dithiadiazaplumbole)
Adduct:
$S_2N_2Pb \cdot NH_3$ ..... **3** 58/9

**$S_2N_2Sn$**
$\{S^{IV}S^{II}N_2Sn^{IV}\}$

5H-1,3,2,4,5-Dithia(3-$S^{IV}$)diazastannole
(=5H-1,3λ[4],2,4,5-Dithiadiazastannole)
Derivatives, dimeric:
(Sn: 5) $-CH_3$, $-CH_3$ ..... **3** 54/5
(Sn: 5) –Cl, –Cl ..... **3** 54

**6**
**$SNC_3O$**
$\{S^{IV}NC_3O\}$

1,2,3-Oxathiazine, 2,2,3,4-tetrahydro-
(=1,2λ[4],3-Oxathiazine, 3,4-dihydro-)
Derivatives:
(S: 2) =O; (N: 3) $-C_3H_7$; (C: 4) =O; (C: 5 and 6) $-C_6H_5$ ..... **4** 95/7
(S: 2) =O; (N: 3) $-C_6H_{11}$-c; (C: 4) =O; (C: 6) $-C_6H_5$ ..... **4** 95/7
(S: 2) =O; (N: 3) $-C_6H_{11}$-c; (C: 4) =O; (C: 5 and 6) $-C_6H_5$ ..... **4** 95/7
(S: 2) =O; (N: 3) $-C_6H_5$; (C: 4) =O; (C: 6) $-C_6H_5$ ..... **4** 95/6, 98, 100
(S: 2) =O; (N: 3) $-C_6H_5$; (C: 4) =O; (C: 5 and 6) $-C_6H_5$ ..... **4** 95/6, 98
(S: 2) =O; (N: 3) $-C_6H_4CH_3$-4; (C: 4) =O; (C: 6) $-C_6H_5$ ..... **4** 95/6, 99
(S: 2) =O; (N: 3) $-C_6H_4CH_3$-4; (C: 4) =O; (C: 5 and 6) $-C_6H_5$ ..... **4** 95/6, 99/100
(S: 2) =O; (N: 3) $-C_6H_3(CH_3)_2$-2,6; (C: 4) =O; (C: 5 and 6) $-C_6H_5$ ..... **4** 95/6, 99
(S: 2) =O; (N: 3) $-C_6H_4Cl$-4; (C: 4) =O; (C: 6) $-C_6H_5$ ..... **4** 95/6, 98, 100
(S: 2) =O; (N: 3) $-C_6H_4NO_2$-4; (C: 4) =O; (C: 5 and 6) $-C_6H_5$ ..... **4** 95/6, 98
(S: 2) =O; (N: 3) $-C_6H_4OCH_3$-4; (C: 4) =O; (C: 5 and 6) $-C_6H_5$ ..... **4** 95/6, 98
(S: 2) =O; (N: 3) $-C_6H_4OCH_3$-4; (C: 4) =O; (C: 5 and 6) $-C_6H_4OCH_3$-4 ..... **4** 95/6, 98
(S: 2) =O; (N: 3) $-N(CH_3)(C_6H_5)$; (C: 4) =O; (C: 5 and 6) $-C_6H_5$ ..... **4** 95/7
(S: 2) =O; (N: 3) $-SO_2C_6H_4CH_3$-4; (C: 4) =O; (C: 6) $-C_6H_5$ ..... **4** 95/7

1,2,3-Oxathiazine, 2,2,3,4,5,6-hexahydro-
(=1,2λ[4],3-Oxathiazine, 3,4,5,6-tetrahydro-)
Derivatives:
(S: 2) =O; (N: 3) $-CH_3$; (C: 4) $-CH_3$ ..... **4** 86, 88/91

(S: 2) =O; (N: 3) $-C_3H_7$-i . . . . . . **4** 86, 88/91
(S: 2) =O; (N: 3) $-C_4H_9$-t . . . . . . **4** 86, 88/91
(S: 2) =O; (N: 3) $-C_4H_9$-t; (C: 4) $-CH_3$ . . . . . . **4** 86/92
(S: 2) =O; (N: 3) $-CH_2C_6H_5$ . . . . . . **4** 86/91
(S: 2) =O; (N: 3) $-C_6H_5$ . . . . . . **4** 86/91, 93/4
(S: 2) =O; (N: 3) $-C_6H_5$; (C: 4) $-CH_3$ . . . . . . **4** 86/91
(S: 2) =O; (N: 3) $-C_6H_5$; (C: 6) $-CH_3$ . . . . . . **4** 86/8, 90/1
(S: 2) =O; (N: 3) $-C_6H_5$; (C: 4 and 6) $-CH_3$ . . . . . . **4** 86/91
(S: 2) =O; (N: 3) $-C_6H_5$; (C: 4 and 6) $-C_4H_9$-t . . . . . . **4** 86/9, 91
(S: 2) =O; (N: 3) $-C_6H_4Cl$-4 . . . . . . **4** 86/7
(S: 2) =O; (N: 3) $-C_6H_4OCH_3$-4 . . . . . . **4** 86/7

**$SN_2CP_2$**
$\{S^{IV}N_2CP^{III}P^{V}\}$

1H-1,2,6,3,5-Thiadiazadiphosphorine, 3,4,5,5-tetrahydro-
(=1λ⁴,2,6,3,5λ⁵-Thiadiazadiphosphorine, 3,4-dihydro-)
Derivative:
(S: 1) $-CH_3$; (P: 3 and 5) $-C_6H_5$, $-C_6H_5$
Ion (1+) . . . . . . **4** 108
Salt: $Br^-$ . . . . . . **4** 108

$\{S^{IV}N_2CP^{V}_2\}$

1H-1,2,6,3,5-Thiadiazadiphosphorine, 3,3,5,5-tetrahydro-
(=1λ⁴,2,6,3λ⁵,5λ⁵-Thiadiazadiphosphorine)
Derivative:
(S: 1) $-CH_3$; (P: 3 and 5) $-C_6H_5$, $-C_6H_5$ . . . . . . **4** 106/8

**$SN_2C_2O$**
$\{S^{IV}N_2C_2O\}$

1,2,3,5-Oxathiadiazine, 2,2,3,4,5,6-hexahydro-
(=1,2λ⁴,3,5-Oxathiadiazine, 3,4,5,6-tetrahydro-)
Derivatives:
(S: 2) =O; (N: 3) $-COCCl_3$; (N: 5) $-C_6H_5$; (C: 4) =O; (C: 6) $-C_6H_4N(CH_3)_2$-4 . . . . . . **4** 104/5
(S: 2) =O; (N: 3) $-COCCl_3$; (N: 5) $-C_6H_5$; (C: 4) =O; (C: 6) $-C_6H_4NO_2$-4 . . . . . . **4** 104/5
(S: 2) =O; (N: 3) $-COCCl_3$; (N: 5) $-C_6H_4OCH_3$-4; (C: 4) =O; (C: 6) $-C_6H_5$ . . . . . . **4** 104/5
(S: 2) =O; (N: 3) $-COCCl_3$; (N: 5) $-C_6H_4OCH_3$-4; (C: 4) =O; (C: 6) $-C_6H_4N(C_2H_5)_2$-4 . . . . . . **4** 104/5
(S: 2) =O; (N: 3) $-COC_6H_5$; (N: 5) $-C_6H_5$; (C: 4) =O; (C: 6) $-C_6H_4N(CH_3)_2$-4 . . . . . . **4** 104/5
(S: 2) =O; (N: 3) $-COC_6H_5$; (N: 5) $-C_6H_5$; (C: 4) =O; (C: 6) $-C_6H_4N(C_2H_5)_2$-4 . . . . . . **4** 104/5

**6**
**$SN_2C_2O$**
$\{S^{IV}N_2C_2O\}$

1,4,3,5-Oxathiadiazine, 2,3,4,4-tetrahydro-
(=1,4λ[4],3,5-Oxathiadiazine, 2,3-dihydro-)
Derivatives:

| | |
|---|---|
| (S: 4) =O; (N: 3) $-C_6H_{11}$-c; (C: 2) $=NC_6H_{11}$-c; (C: 6) $-CH_3$ | **4** 104, 106 |
| (S: 4) =O; (N: 3) $-C_6H_{11}$-c; (C: 2) $=NC_6H_{11}$-c; (C: 6) $-C_6H_5$ | **4** 104, 106 |
| (S: 4) =O; (N: 3) $-C_6H_5$; (C: 2) $=NC_6H_5$; (C: 6) $-C_6H_5$ | **4** 104, 106 |

**$SN_2C_3$**
$\{S^{IV}N_2C_3\}$

3H-1,2,6-Thia($S^{IV}$)diazine, 4,5-dihydro- ........ **4** 61/3
(=3H-1λ[4],2,6-Thiadiazine, 4,5-dihydro-)
Derivatives:

| | |
|---|---|
| (C: 4) $-C_6H_4Cl$-4 | **4** 61/4 |
| (C: 4) $-OH$ | **4** 61/3 |
| (C: 4) $-OOCCH_3$ | **4** 61/3 |

2H-1,2,6-Thiadiazine, 1,1-dihydro-
(=2H-1λ[4],2,6-Thiadiazine)
Derivatives:

| | |
|---|---|
| (S: 1) =O; (N: 2) $-C_6H_{11}$-c; (C: 3 and 5) $-C_6H_5$; (C: 4) $-CH_3$ | **4** 73/5 |
| (S: 1) =O; (N: 2) $-C_6H_{11}$-c; (C: 3) $-C_6H_5$; (C: 4) $-CH_3$; (C: 5) $-C_6H_4CH_3$-4 | **4** 73/5 |
| (S: 1) =O; (N: 2) $-C_6H_5$; (C: 3 and 5) $-C_6H_5$ | **4** 73/5 |
| (S: 1) =O; (N: 2) $-C_6H_5$; (C: 3) $-C_6H_5$; (C: 5) $-C_6H_4CH_3$-4 | **4** 73/5 |
| (S: 1) =O; (N: 2) $-C_6H_5$; (C: 3 and 5) $-C_6H_5$; (C: 4) $-Br$ | **4** 73/4 |
| (S: 1) =O; (N: 2) $-C_6H_5$; (C: 3 and 5) $-C_6H_5$; (C: 4) $-CH_3$ | **4** 73/5 |
| (S: 1) =O; (N: 2) $-C_6H_5$; (C: 3) $-C_6H_5$; (C: 4) $-CH_3$; (C: 5) $-C_6H_4CH_3$-4 | **4** 73/5 |
| (S: 1) =O; (N: 2) $-C_6H_5$; (C: 3) $-C_6H_4Cl$-4; (C: 4) $-CH_3$; (C: 5) $-C_6H_4CH_3$-4 | **4** 73/5 |
| (S: 1) =O; (N: 2) $-C_6H_4CH_3$-2; (C: 3 and 5) $-C_6H_5$; (C: 4) $-CH_3$ | **4** 73/5 |
| (S: 1) =O; (N: 2) $-C_6H_4CH_3$-4; (C: 3) $-C_6H_5$; (C: 5) $-C_6H_{11}$-c | **4** 73/5 |
| (S: 1) =O; (N: 2) $-C_6H_4CH_3$-4; (C: 3) $-C_6H_5$; (C: 5) $-C_6H_4CH_3$-4 | **4** 73/5 |
| (S: 1) =O; (N: 2) $-C_6H_4CH_3$-4; (C: 3) $-C_6H_5$; (C: 4) $-Br$; (C: 5) $-C_6H_{11}$-c | **4** 73/4 |
| (S: 1) =O; (N: 2) $-C_6H_4CH_3$-4; (C: 3 and 5) $-C_6H_5$; (C: 4) $-Br$ | **4** 73/4 |
| (S: 1) =O; (N: 2) $-C_6H_4CH_3$-4; (C: 3 and 5) $-C_6H_5$; (C: 4) $-CH_3$ | **4** 73/5 |
| (S: 1) =O; (N: 2) $-C_6H_4CH_3$-4; (C: 3) $-C_6H_5$; (C: 4) $-Br$-; (C: 5) $-C_6H_4CH_3$-4 | **4** 73/4 |
| (S: 1) =O; (N: 2) $-C_6H_4CH_3$-4; (C: 3) $-C_6H_5$; (C: 4) $-CH_3$; (C: 5) $-C_6H_4CH_3$-4 | **4** 73/5 |

**6**
**$SN_3C_2$**
$\{S^{IV}N_3C_2\}$

1H-1,2,4,6-Thiatriazine
(=1λ[4],2,4,6-Thiatriazine)

Derivatives (cont.)

| Compound | Vol. | Pages |
|---|---|---|
| (S: 1) $-CH_3$; (C: 3 and 5) $-C_6H_4Cl$-4 | **4** | 58/9 |
| (S: 1) $-C_2H_5$; (C: 3 and 5) $-C_6H_5$ | **4** | 58/60 |
| (S: 1) $-C_3H_7$-i; (C: 3 and 5) $-C_6H_5$ | **4** | 58/60 |
| (S: 1) $-C_4H_9$; (C: 3 and 5) $-C_6H_5$ | **4** | 58/60 |
| (S: 1) $-C_6H_5$; (C: 3 and 5) $-C_6H_5$ | **4** | 58/9 |
| (S: 1) $-C_6H_5$; (C: 3) $-C_6H_5$; (C: 5) $-C_6H_4Cl$-4 | **4** | 58/9 |
| (S: 1) $-C_6H_5$; (C: 3 and 5) $-C_6H_4Cl$-4 | **4** | 58/9 |
| (S: 1) $-C_6H_5$; (C: 3) $-N(C_6H_5)_2$; (C: 5) $-C_6H_5$ | **4** | 58, 60 |
| (S: 1) $-C_6H_4NO_2$-3; (C: 3 and 5) $-C_6H_5$ | **4** | 58/9 |
| (S: 1) $-C_6H_4NO_2$-4; (C: 3 and 5) $-C_6H_5$ | **4** | 58/60 |
| (S: 1) $-Cl$; (C: 3 and 5) $-CCl_3$ | **4** | 53, 55/7 |
| (S: 1) $-Cl$; (C: 3) $-CCl_3$; (C: 5) $-CF_3$ | **4** | 53, 55/6 |
| (S: 1) $-Cl$; (C: 3) $-CCl_3$; (C: 5) $-C_6H_5$ | **4** | 53, 55/7 |
| (S: 1) $-Cl$; (C: 3 and 5) $-C_6H_5$ | **4** | 53, 55/7 |
| (S: 1) $-Cl$; (C: 3) $-Cl$; (C: 5) $-CCl_3$ | **4** | 53/4 |
| (S: 1) $-Cl$; (C: 3) $-Cl$; (C: 5) $-C_6H_5$ | **4** | 53/5 |
| (S: 1) $-Cl$; (C: 3 and 5) $-Cl$ | **4** | 53/4 |
| (S: 1) $-I$; (C: 3 and 5) $-C_6H_5$ | **4** | 57/8 |
| (S: 1) $-N(CH_3)_2$; (C: 3) $-CCl_3$; (C: 5) $-C_6H_5$ | **4** | 39, 52/3 |
| (S: 1) $-N(C_2H_5)_2$; (C: 3) $-CCl_3$; (C: 5) $-C_6H_5$ | **4** | 39, 52/3 |
| (S: 1) $-N(C_3H_7\text{-i})_2$; (C: 3) $-Cl$; (C: 5) $-C_6H_5$ | **4** | 39/40, 49 |
| (S: 1) $-N(C_3H_7\text{-i})_2$; (C: 3 and 5) $-Cl$ | **4** | 39/43 |
| (S: 1) $-N(C_3H_7\text{-i})_2$; (C: 3) $-Cl$; (C: 5) $-NH_2$ | **4** | 39/40, 45/9 |
| (S: 1) $-N(C_3H_7\text{-i})_2$; (C: 3) $-Cl$; (C: 5) $-NHCH_3$ | **4** | 39/40, 45/9 |
| (S: 1) $-N(C_3H_7\text{-i})_2$; (C: 3) $-Cl$; (C: 5) $-NHC_2H_5$ | **4** | 39/40, 45/7 |
| (S: 1) $-N(C_3H_7\text{-i})_2$; (C: 3) $-Cl$; (C: 5) $-NHC_3H_7$ | **4** | 39/40, 45/9 |
| (S: 1) $-N(C_3H_7\text{-i})_2$; (C: 3) $-Cl$; (C: 5) $-NHC_4H_9$-t | **4** | 39/40, 45/9 |
| (S: 1) $-N(C_3H_7\text{-i})_2$; (C: 3) $-Cl$; (C: 5) $-NHC_6H_{11}$-c | **4** | 39/40, 45/9 |
| (S: 1) $-N(C_3H_7\text{-i})_2$; (C: 3) $-Cl$; (C: 5) $-NHCH_2CH{=}CH_2$ | **4** | 39/40, 45, 47/9 |
| (S: 1) $-N(C_3H_7\text{-i})_2$; (C: 3) $-Cl$; (C: 5) $-NHCH_2C_6H_5$ | **4** | 39/40, 45, 47/9 |
| (S: 1) $-N(C_3H_7\text{-i})_2$; (C: 3) $-Cl$; (C: 5) $-N(CH_3)_2$ | **4** | 39/40, 45 |
| (S: 1) $-N(C_3H_7\text{-i})_2$; (C: 3) $-Cl$; (C: 5) $-N(C_2H_5)_2$ | **4** | 39/40, 45 |

**6**
**$SN_3C_2$**
$\{S^{IV}N_3C_2\}$

1H-1,2,4,6-Thiatriazine, 2,5-dihydro-
(=1λ[4],2,4,6-Thiatriazine, 2,5-dihydro-)
Derivatives:
(S: 1) $-C_6H_5$; (C: 3) $-C_6H_5$; (C: 5) $-CF_3$, $-CF_3$ ..... **4** 60/1
(S: 1) $-C_6H_5$; (N: 2) $-C_6H_5$; (C: 3) $-C_6H_5$; (C: 5) $-CF_3$, $-CF_3$ ..... **4** 60/1

2H-1,2,4,6-Thiatriazine, 1,1-dihydro-
(=2H-1λ[4],2,4,6-Thiatriazine)
Derivatives:
(S: 1) $=NSO_2C_6H_5$; (C: 3 and 5) $-OCH_3$ ..... **4** 32/3
(S: 1) $=NSO_2C_6H_4CH_3$-4; (C: 3 and 5) $-OCH_3$ ..... **4** 32/4
(S: 1) $=NSO_2C_6H_4CH_3$-4; (N: 2) $-SO_2C_6H_4CH_3$-4; (C: 3) $-NH_2$; (C: 5) $-C_6H_5$ ..... **4** 35/6
(S: 1) $=NSO_2C_6H_4CH_3$-4; (N: 2) $-SO_2C_6H_4CH_3$-4; (C: 3) $-NH_2$; (C: 5) $-N(CH_3)_2$ ..... **4** 35
(S: 1) $=NSO_2C_6H_4CH_3$-4; (N: 2) $-SO_2C_6H_4CH_3$-4; (C: 3) $-NH_2$; (C: 5) $-N(C_2H_5)(CH_2C_6H_5)$ ..... **4** 35
(S: 1) $=NSO_2C_6H_4Cl$-4; (C: 3 and 5) $-OCH_3$ ..... **4** 32/3
(S: 1) $=NSO_2C_6H_4Cl$-4; (N: 2) $-CH_3$; (C: 3 and 5) $-OCH_3$ ..... **4** 32/3, 35
(S: 1) =O; (C: 3 and 5) $-CCl_3$ ..... **4** 21/2
(S: 1) =O; (C: 3) $-C_6H_5$; (C: 5) $-CCl_3$ ..... **4** 21/2
(S: 1) =O; (C: 3) $-CONHCH_2C_6H_5$; (C: 5) $-C_6H_5$ ..... **4** 21/2
(S: 1) =O; (C: 3) $-CONHCH_2C_6H_5$; (C: 5) $-N(CH_3)_2$ ..... **4** 21/2
(S: 1) =O; (C: 3) $-CONHCH_2C_6H_5$; (C: 5) $-SCH_2C_6H_5$ ..... **4** 21/2
(S: 1) =O; (N: 2) $-SO_2C_6H_4CH_3$-4; (C: 3) $-NH_2$; (C: 5) $-N(CH_3)_2$ ..... **4** 22/3

4H-1,2,4,6-Thiatriazine, 1,1-dihydro-
(=4H-1λ[4],2,4,6-Thiatriazine)
Derivatives:
(S: 1) $=NSO_2C_6H_5$; (C: 3 and 5) $-OCH_3$ ..... **4** 32/3
(S: 1) $=NSO_2C_6H_4CH_3$-4; (C: 3 and 5) $-OCH_3$ ..... **4** 32/4
(S: 1) $=NSO_2C_6H_4Cl$-4; (C: 3 and 5) $-OCH_3$ ..... **4** 32/3
(S: 1) $=NSO_2C_6H_4Cl$-4; (N: 4) $-CH_3$; (C: 3 and 5) $-OCH_3$ ..... **4** 32/3, 35
(S: 1) =O; (N: 4) $-CH_3$; (C: 3 and 5) $-OC_6H_4CH_3$ ..... **4** 23

2H-1,2,4,6-Thiatriazine, 1,1,3,4-tetrahydro-
(=2H-1λ[4],2,4,6-Thiatriazine, 3,4-dihydro-)
Derivatives:
(S: 1) =O; (N: 2) $-CH_3$; (C: 3) =O; (C: 5) $-SCH_3$ ..... **4** 23/4, 26/7
(S: 1) =O; (N: 2 and 4) $-CH_3$; (C: 3) =O; (C: 5) $-N(CH_3)_2$ ..... **4** 23, 25/7
(S: 1) =O; (N: 2 and 4) $-CH_3$; (C: 3) =O; (C: 5) $-SCH_3$ ..... **4** 23/4, 26/7

(S: 1) =O; (N: 2) –$C_3H_7$-i; (C: 3) =O; (C: 5) –$SCH_3$ .......... **4** 23/4, 26/7
(S: 1) =O; (N: 2) –$C_3H_7$-i; (N: 4) –$CH_3$; (C: 3) =O; (C: 5) –$N(CH_3)_2$ .......... **4** 23, 25/7
(S: 1) =O; (N: 2) –$C_3H_7$-i; (N: 4) –$CH_3$; (C: 3) =O; (C: 5) –$SCH_3$ .......... **4** 23/4, 26/7
(S: 1) =O; (N: 2) –$C_3H_7$-i; (N: 4) –$NH_2$; (C: 3) =O; (C: 5) –$N(CH_3)_2$ .......... **4** 23, 25/7
(S: 1) =O; (N: 2) –$C_3H_7$-i; (N: 4) –$NH_2$; (C: 3) =O; (C: 5) –$SCH_3$ .......... **4** 23, 25/7
(S: 1) =O; (N: 2) –$C_6H_{11}$-c; (C: 3) =O; (C: 5) –$SCH_3$ .......... **4** 23/4, 26/7
(S: 1) =O; (N: 2) –$C_6H_{11}$-c; (N: 4) –$CH_3$; (C: 3) =O; (C: 5) –$N(CH_3)_2$ .......... **4** 23, 25/7
(S: 1) =O; (N: 2) –$C_6H_{11}$-c; (N: 4) –$CH_3$; (C: 3) =O; (C: 5) –$SCH_3$ .......... **4** 23/4, 26/7
(S: 1) =O; (N: 2) –$C_6H_{11}$-c; (N: 4) –$NH_2$; (C: 3) =O; (C: 5) –$N(CH_3)_2$ .......... **4** 23, 25/7
(S: 1) =O; (N: 2) –$C_6H_{11}$-c; (N: 4) –$NH_2$; (C: 3) =O; (C: 5) –$SCH_3$ .......... **4** 23, 25/7
(S: 1) =O; (N: 2) –$C_6H_5$; (C: 3) =O; (C: 5) –$SCH_3$ .......... **4** 23/4, 26/7
(S: 1) =O; (N: 2) –$C_6H_5$; (N: 4) –$CH_3$; (C: 3) =O; (C: 5) –$SCH_3$ .......... **4** 23/4, 26/7

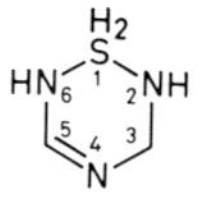

2H-1,2,4,6-Thiatriazine, 1,1,3,6-tetrahydro-
(= 2H-1$\lambda^4$,2,4,6-Thiatriazine, 3,6-dihydro-)
Derivatives:
(S: 1) =$NSO_2C_6H_4CH_3$-4; (N: 2) –$SO_2C_6H_4CH_3$-4; (C: 3) =NH; (C: 5) –$C_6H_5$ .......... **4** 35
(S: 1) =O; (N: 2) –(2-$C_4H_4O(OOCC_6H_5)_2(CH_2OOCC_6H_5)$); (C: 3) =S .......... **4** 27/8
(2-$C_4H_7O$ = Tetrahydro-2-furanyl)
(S: 1) =O; (N: 2) –(2-$C_4H_4O(OOCC_6H_5)_2(CH_2OOCC_6H_5)$); (C: 3) =S; (C: 5) –$CH_3$ .......... **4** 27/8
(2-$C_4H_7O$ = Tetrahydro-2-furanyl)
(S: 1) =O; (N: 2) –(2-$C_5H_6O(OOCCH_3)_3$); (C: 3) =S; (C: 5) –$CH_3$ .......... **4** 27/8
(2-$C_5H_9O$ = Tetrahydro-2H-pyran-2-yl)
(S: 1) =O; (N: 2) –(2-$C_5H_5O(OOCCH_3)_4$); (C: 3) =S .......... **4** 27/8
(2-$C_5H_9O$ = Tetrahydro-2H-pyran-2-yl)
(S: 1) =O; (N: 2) –(2-$C_5H_5O(OOCCH_3)_4$); (C: 3) =S; (C: 5) –$CH_3$ .......... **4** 27/8
(2-$C_5H_9O$ = Tetrahydro-2H-pyran-2-yl)

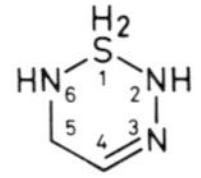

2H-1,2,3,6-Thiatriazine, 1,1,5,6-tetrahydro-
(= 2H-1$\lambda^4$,2,3,6-Thiatriazine, 5,6-dihydro-)
Derivatives:
(S: 1) =O; (N: 2) –$CH_3$; (N: 6) –$COC_6H_5$; (C: 4) –$CH_3$ .......... **4** 14, 16/7
(S: 1) =O; (N: 2) –$CH_3$; (N: 6) –$SO_2CH_3$; (C: 4) –$CH_3$ .......... **4** 14/7
(S: 1) =O; (N: 2) –$CH_3$; (N: 6) –$SO_2C_6H_4CH_3$-4; (C: 4) –$CH_3$ .......... **4** 14/7
(S: 1) =O; (N: 2) –$CH_3$; (N: 6) –$SO_2C_6H_4CH_3$-4; (C: 4 and 5) –$CH_3$ .......... **4** 14/7
(S: 1) =O; (N: 2) –$CH_3$; (N: 6) –$SO_2C_6H_4CH_3$-4; (C: 4) –$CH_3$; (C: 5) –$CONHC_6H_5$ .......... **4** 14/7
(S: 1) =O; (N: 2) –$C_6H_5$; (N: 6) –$SO_2C_6H_4CH_3$-4; (C: 4) –$CH_3$; (C: 5) –$CONHC_6H_5$ .......... **4** 14, 16/7

**6**
**$SN_3P_2$**
$\{S^{IV}N_3P_2^{V}\}$

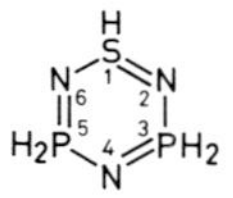

1H-1,2,4,6,3,5-Thiatriazadiphosphorine, 3,3,5,5-tetrahydro-
(= 1λ⁴,2,4,6,3λ⁵,5λ⁵-Thiatriazadiphosphorine)

Derivatives:

(P: 3 and 5) $-C_6H_5$, $-C_6H_5$

Ion(1+) .......... **3** 20/4

Salts: $Br_3^-$ .......... **3** 20/4, 29/30

$I_3^-$ .......... **3** 20/4, 29/30

(P: 3 and 5) –Cl, –Cl

Ion(1+) .......... **3** 20/2

Salt: $SbCl_6^-$ .......... **3** 20/2, 29

(S: 1) –Br; (P: 3 and 5) $-C_6H_5$, $-C_6H_5$ .......... **3** 20/4, 27

(S: 1) $-C_6H_5$; (P: 3 and 5) $-C_6H_5$, $-C_6H_5$ .......... **3** 20/6, 28/9

(S: 1) –Cl; (P: 3 and 5) $-C_6H_5$, $-C_6H_5$ .......... **3** 20/6

(S: 1) –Cl; (P: 3 and 5) –Cl, $-C_6H_5$ .......... **3** 20/5

(S: 1) –Cl; (P: 3 and 5) –Cl, –Cl .......... **3** 20/4

(S: 1) –I; (P: 3 and 5) $-C_6H_5$, $-C_6H_5$ .......... **3** 20/4, 27/8

(S: 1) $-N(CH_3)_2$; (P: 3 and 5) $-C_6H_5$, $-C_6H_5$ .......... **3** 20/4, 28

(S: 1) $-N(C_2H_5)_2$; (P: 3 and 5) $-C_6H_5$, $-C_6H_5$ .......... **3** 20/4, 28

(S: 1) $-NC_5H_{10}$; (P: 3 and 5) $-C_6H_5$, $-C_6H_5$ ($NC_5H_{10}$ = Piperidino) .......... **3** 20/4, 28

**$SN_3Si_2$**
$\{S^{IV}N_3Si_2\}$

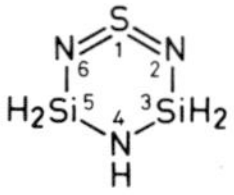

1-Thia($S^{IV}$)-2,4,6-triaza-3,5-disila-1,6-cyclohexadiene
(= 1λ⁴-Thia-2,4,6-triaza-3,5-disila-1,6-cyclohexadiene)

Derivatives:

(N: 4) $-CH_3$; (Si: 3 and 5) $-CH_3$, $-CH_3$ .......... **3** 7

(N: 4) $-P_3N_3F_5$; (Si: 3 and 5) $-CH_3$, $-CH_3$ .......... **3** 8/9
($P_3N_3F_5$ = Pentafluoro-cyclotri-λ⁵-phosphazen-2-yl)

(N: 4) $-P_4N_4F_7$; (Si: 3 and 5) $-CH_3$, $-CH_3$ .......... **3** 9
($P_4N_4F_7$ = Heptafluoro-cyclotetra-λ⁵-phosphazen-2-yl)

(N: 4) $-SO_2N(CH_3)_2$; (Si: 3 and 5) $-CH_3$, $-CH_3$ .......... **3** 7/8

(N: 4) $-Sn(CH_3)_3$; (Si: 3 and 5) $-CH_3$, $-CH_3$ .......... **3** 8

(Si: 3 and 4) $-CH_3$, $-CH_3$ .......... **3** 7

**$S_2N_4As_2$**
$\{S_2^{IV}N_4As_2^{III}\}$

3H,7H-1,5,2,4,6,8,3,7-Dithia(1,5-$S^{IV}$)tetrazadiarsocine
(=3H,7H-1$\lambda^4$,5$\lambda^4$,2,4,6,8,3,7-Dithiatetrazadiarsocine)
Derivatives:
(As: 3 and 7) $-CH_3$ ........ **3** 49/50
(As: 3 and 7) $-C_4H_9$-t ........ **3** 49/51
(As: 3 and 7) $-C_6H_5$ ........ **3** 49/52
(As: 3 and 7) $-C_6H_2(CH_3)_3$-2,4,6 ........ **3** 49/53

**$S_2N_4C_2$**
$\{S_2^{IV}N_4C_2\}$

1H,3H-1,3,2,4,6,8-Dithiatetrazocine
(=1$\lambda^4$,3$\lambda^4$,2,4,6,8-Dithiatetrazocine)
Derivative:
(S: 1 and 3) –Cl; (C: 5 and 7) $-N(CH_3)_2$ ........ **4** 133/4

1H,5H-1,5,2,4,6,8-Dithiatetrazocine
(=1$\lambda^4$,5$\lambda^4$,2,4,6,8-Dithiatetrazocine)
Derivatives:
(S: 1) –Br; (C: 3 and 7) $-N(CH_3)_2$
Ion(1+)
Salt: $Br_3^-$ ........ **4** 143/4
(S: 1) –Cl; (C: 3 and 7) $-N(CH_3)_2$
Ion(1+)
Salt: $Cl_3^-$ ........ **4** 143/4
(S: 1 and 5) –Cl; (C: 3 and 7) $-N(CH_3)_2$ ........ **4** 145

$\{S^{IV}S^{II}N_4C_2\}$

1,5,2,4,6,8-Dithia(5-$S^{IV}$)tetrazocine ........ **4** 134/5
(=1,5$\lambda^4$,2,4,6,8-Dithiatetrazocine)
Derivatives:
(C: 3 and 7) $-C_4H_9$-t ........ **4** 136, 139/43
(C: 3 and 7) $-C_6H_5$ ........ **4** 136/43
(C: 3 and 7) $-C_6H_4COOC_2H_5$-4 ........ **4** 136/40
(C: 3 and 7) $-C_6H_4OCH_3$-4 ........ **4** 136/40
(C: 3 and 7) $-NH_2$ ........ **4** 135

**8**

**$S_2N_4C_2$**
$\{S^{IV}S^{II}N_4C_2\}$

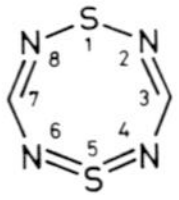

1,5,2,4,6,8-Dithia(5-$S^{IV}$)tetrazocine
(=1,5$\lambda^4$,2,4,6,8-Dithiatetrazocine)
Derivatives (cont.)
(C: 3 and 7) –$N(CH_3)_2$ . . . . . . **4** 136/43
(C: 3 and 7) –$N(C_6H_5)_2$ . . . . . . **4** 136, 138/40, 142

**$S_2N_4P_2$**
$\{S_2^{IV}N_4P_2^{V}\}$

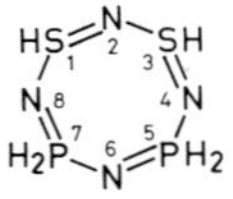

1H,3H,-1,3,2,4,6,8,5,7-Dithiatetrazadiphosphocine, 5,5,7,7-tetrahydro-
(=1$\lambda^4$,3$\lambda^4$,2,4,6,8,5$\lambda^5$,7$\lambda^5$-Dithiatetrazadiphosphocine)
Derivatives:
(S: 3) –Br; (P: 5 and 7) –$C_6H_5$, –$C_6H_5$
Ion(1+) . . . . . . **3** 34
(S: 3) –I; (P: 5 and 7) –$C_6H_5$, –$C_6H_5$
Ion(1+) . . . . . . **3** 34
(S: 1 and 3) –Br; (P: 5 and 7) –$C_6H_5$, –$C_6H_5$ . . . . . . **3** 34
(S: 1 and 3) –Cl; (P: 5 and 7) –$C_6H_5$, –$C_6H_5$ . . . . . . **3** 34
(S: 1 and 3) –I; (P: 5 and 7) –$C_6H_5$, –$C_6H_5$ . . . . . . **3** 34

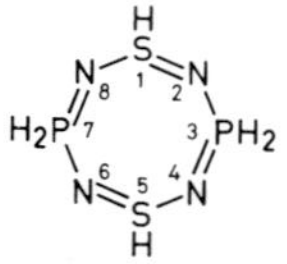

1H,5H-1,5,2,4,6,8,3,7-Dithiatetrazadiphosphocine, 3,3,7,7-tetrahydro-
(=1$\lambda^4$,5$\lambda^4$,2,4,6,8,3$\lambda^5$,7$\lambda^5$-Dithiatetrazadiphosphocine)
Derivatives:
(S: 1 and 5) –Br; (P: 3 and 7) –$C_6H_5$, –$C_6H_5$ . . . . . . **3** 37/8
(S: 1 and 5) –Cl; (P: 3 and 7) –$C_6H_5$, –$C_6H_5$ . . . . . . **3** 37

$\{S^{IV}S^{II}N_4P_2^{V}\}$

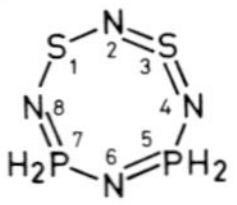

1,3,2,4,6,8,5,7-Dithia(3-$S^{IV}$)tetrazadiphosphocine, 5,5,7,7-tetrahydro- . . . . . . **3** 30, 32
(=1,3$\lambda^4$,2,4,6,8,5$\lambda^5$,7$\lambda^5$-Dithiatetrazadiphosphocine)
Derivatives:
(P: 5 and 7) –$CF_3$, –$CF_3$ . . . . . . **3** 30/3
(P: 5 and 7) –$CH_3$, –$CH_3$ . . . . . . **3** 30/2
(P: 5 and 7) –$C_6H_5$, –$C_6H_5$ . . . . . . **3** 30/3

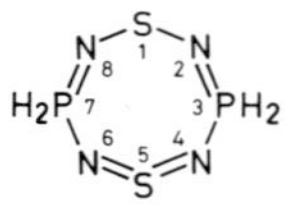

1,5,2,4,6,8,3,7-Dithia(5-$S^{IV}$)tetrazadiphosphocine, 3,3,7,7-tetrahydro-
(=1,5$\lambda^4$,2,4,6,8,3$\lambda^5$,7$\lambda^5$-Dithiatetrazadiphosphocine)
Derivatives:
(P: 3 and 7) –$CH_3$, –$CH_3$ . . . . . . **3** 34/6
(P: 3 and 7) –$C_6H_5$, –$C_6H_5$ . . . . . . **3** 34/7

**12**
**$S_2N_6P_4$**
$\{S_2^{IV}N_6P_4^{V}\}$

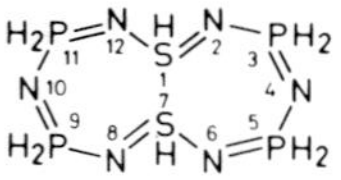

1,7-Dithia(1,7-$S^{IV}$)-2,4,6,8,10,12-hexaaza-3,5,9,11-tetraphosphacyclododeca-1,3,5,7,9,11-hexaene, 3,3,5,5,9,9,11,11-octahydro-
(=1λ⁴,7λ⁴-Dithia-2,4,6,8,10,12-hexaaza-3λ⁵,5λ⁵,9λ⁵,11λ⁵-tetraphospha-cyclododeca-1,3,5,7,9,11-hexaene)
Derivatives:
(P: 3,5,9, and 11) $-C_6H_5$, $-C_6H_5$
Ion (2+) (cont.)
Salts: $Br_3^-$ . . . . . . **3** 43/7
$I_3^-$ . . . . . . **3** 43/7
(S: 1 and 7) $-N(CH_3)_2$; (P: 3,5,9, and 11) $-C_6H_5$, $-C_6H_5$ . . . . . . **3** 43/7
(S: 1 and 7) $-N(C_2H_5)_2$; (P: 3,5,9, and 11) $-C_6H_5$, $-C_6H_5$ . . . . . . **3** 43, 47
(S: 1 and 7) $-NC_5H_{10}$; (P: 3,5,9, and 11) $-C_6H_5$, $-C_6H_5$ ($NC_5H_{10}$ = Piperidino) . . . . . . **3** 43, 45, 48

## Bicyclic Systems

**3,5**
(5,3)
**$SNC_2O-NC_2$ ($SNC_3O$)**
$\{S^{IV}NC_2O-NC_2\}$

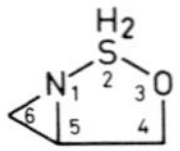

3-Oxa-2-thia-1-azabicyclo[3.1.0]hexane, 2,2-dihydro-
(=3-Oxa-2λ⁴-thia-1-azabicyclo[3.1.0]hexane)
Derivatives:
(S: 2) =O; (C: 4) $-CH_3$; (C: 5) $-C_6H_5$ . . . . . . **3** 303/5
(S: 2) =O; (C: 4) $-C_3H_7$, $-CH_3$; (C: 5) $-C_6H_5$ . . . . . . **3** 303/4, 306
(S: 2) =O; (C: 4) $-C_3H_7$, $-CH_3$; (C: 5) $-C_6H_5$; (C: 6) $-CH_3$ . . . . . . **3** 303/4, 306
(S: 2) =O; (C: 4) $-C_6H_5$ . . . . . . **3** 303/5
(S: 2) =O; (C: 4) $-C_6H_5$; (C: 5) $-C_2H_5$ . . . . . . **3** 303/5
(S: 2) =O; (C: 4 and 5) $-C_6H_5$ . . . . . . **3** 303/5
(S: 2) =O; (C: 4) $-C_6H_5$, $-CH_3$; (C: 5) $-C_2H_5$ . . . . . . **3** 303/5
(S: 2) =O; (C: 4) $-C_6H_5$, $-CH_3$; (C: 5) $-C_6H_5$ . . . . . . **3** 303/4, 306
(S: 2) =O; (C: 4) $-C_6H_5$, $-C_2H_5$; (C: 5) $-C_6H_5$ . . . . . . **3** 303/4, 306
(S: 2) =O; (C: 4) $-C_6H_5$, $-C_6H_5$; (C: 5) $-C_2H_5$ . . . . . . **3** 303/4, 306
(S: 2) =O; (C: 4) $-C_6H_5$, $-C_6H_5$; (C: 5) $-C_6H_5$ . . . . . . **3** 303/4, 306

**5,5**
**$SN_2C_2-C_5$ ($SN_2C_5$)**
$\{S^{IV}N_2C_2-C_5\}$

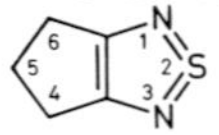

4H-Cyclopenta[1,2,5]thiadiazole-2-$S^{IV}$, 5,6-dihydro- . . . . . . **3** 191
(=4H-Cyclopenta[1λ⁴,2,5]thiadiazole, 5,6-dihydro-)

**$SN_2C_2$–$NC_4$ ($SN_3C_4$)**
{$S^{IV}N_2C_2$–$NC_4$}

5H-Pyrrolo[3,4-c][1,2,5]thiadiazole-2-$S^{IV}$, 4,6-dihydro-
(= 5H-Pyrrolo[3,4-c][1λ⁴,2,5]thiadiazole, 4,6-dihydro-)
Derivative:
(N: 5) –$C_6H_5$; (C: 4 and 6) =O ........ **3** 224

**$SN_2C_2$–$N_2C_2Se$ ($SN_4C_2Se$)**
{$S^{IV}N_2C_2$–$N_2C_2Se^{IV}$}

[1,2,5]Selenadiazolo[3,4-c][1,2,5]thiadiazole- 2-$S^{IV}$-5-$Se^{IV}$ ........ **3** 240
(= [1λ⁴,2,5]Selenadiazolo[3,4-c][1λ⁴,2,5]thiadiazole)

**$SN_2C_2$–$SC_4$ ($S_2N_2C_4$)**
{$S^{IV}N_2C_2$–$S^{II}C_4$}

Thieno[3,4-c][1,2,5]thiadiazole-2-$S^{IV}$ ........ **3** 239
(= Thieno[3,4-c][1λ⁴,2,5]thiadiazole)
Derivative:
(C: 4 and 6) –$C_6H_5$ ........ **3** 239/40

Thieno[3,4-c][1,2,5]thiadiazole-2-$S^{IV}$, 4,6-dihydro- ........ **3** 237/8
(= Thieno[3,4-c][1λ⁴,2,5]thiadiazole, 4,6-dihydro-)

{$S^{IV}N_2C_2$–$S^{IV}C_4$}

Thieno[3,4-c][1,2,5]thiadiazole-2-$S^{IV}$, 4,5,5,6-tetrahydro-
(= Thieno[3,4-c][1λ⁴,2,5]thiadiazole, 4,5,5,6-tetrahydro-)
Derivative:
(S: 5) =O ........ **3** 238/9

**$SN_2C_2$–$SN_2C_2$ ($S_2N_4C_2$)**
{$S^{IV}N_2C_2$–$S^{IV}N_2C_2$}

[1,2,5]Thiadiazolo[3,4-c][1,2,5]thiadiazole-2,5-$S^{IV}$ ........ **3** 189/90
(= [1λ⁴,2,5]Thiadiazolo[3,4-c][1λ⁴,2,5]thiadiazole)
Radical ion(1−) ........ **3** 191

**$S_2N_2C$–$S_3N_2$ ($S_3N_4C$)**
{$S_2^{IV}N_2C$-$S_3^{IV}N_2$}
see
*$S_3N_2$–$S_2N_2C$ ($S_3N_4C$)*
*{$S_3^{IV}N_2$–$S_2^{IV}N_2C$}*

**5,6**
**$SN_2C_2$–$C_6$ ($SN_2C_6$)**
{$S^{IV}N_2C_2$–$C_6$}

**5,6**
**$SN_2C_2$–$N_2C_4$ ($SN_4C_4$)**
{$S^{IV}N_2C_2$–$N_2C_4$}

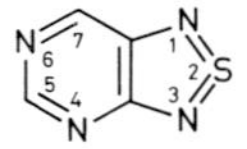

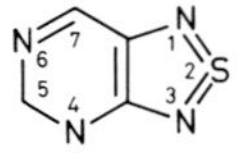

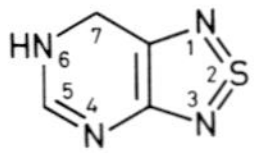

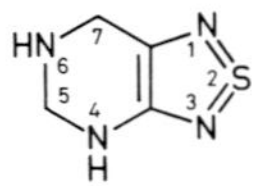

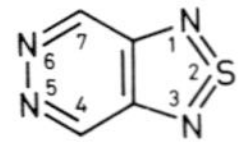

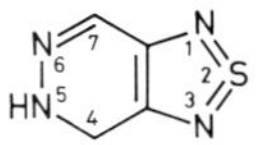

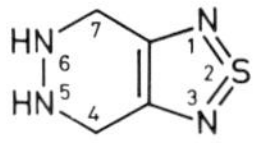

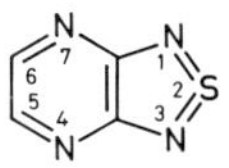

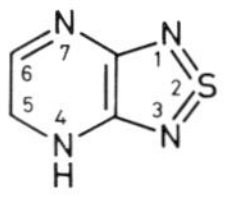

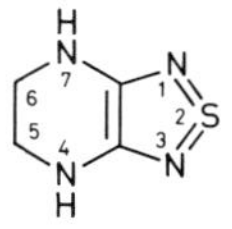

**5,6**
**$SN_3C–NC_5$ ($SN_3C_5$)**
{$S^{IV}N_3C–NC_5$}

2H-[1,2,3,5]Thiatriazolo[5,4-a]pyridine, 3,3-dihydro-
(=2H-[1$\lambda^4$,2,3,5]Thiatriazolo[5,4-a]pyridine)
Derivative:
(S: 3) =O ........ **3** 85/6

1H-[1,2,3,5]Thiatriazolo[5,4-a]pyridine, 3,3-dihydro-
(=1H-[1$\lambda^4$,2,3,5]Thiatriazolo[5,4-a]pyridine)
Derivatives:
(S: 3) =O ........ **3** 85/6
(S: 3) =O; (N: 1) $-CH_3$ ........ **3** 86/7

[1,2,3,5]Thiatriazolo[5,4-a]pyridin-4-ium, 1,2,3,3-tetrahydro-
(=[1$\lambda^4$,2,3,5]Thiatriazolo[5,4-a]pyridin-4-ium, 1,2-dihydro-)
Derivative:
(S: 3) =O ........ **3** 86

**$SN_3C–N_2C_4$ ($SN_4C_4$)**
{$S^{IV}N_3C–N_2C_4$}

2H-[1,2,3,5]Thiatriazolo[5,4-c]pyrimidine, 1,3,3,5,6,7-hexahydro-
(=2H-[1$\lambda^4$,2,3,5]Thiatriazolo[5,4-c]pyrimidine, 1,5,6,7-tetrahydro-)
Derivatives:
(S: 3) =O; (N: 6) $-CH_3$; (C: 5 and 7) =O ........ **3** 90
(S: 3) =O; (N: 1 and 6) $-CH_3$; (C: 5 and 7) =O ........ **3** 90

**5,6**
(6,5)
**$SN_2C_3–C_4O$ ($SN_2C_5O$)**
{$S^{IV}N_2C_3–C_4O$}

4H-Furo[2,3-c][1,2,6]thiadiazine-2-$S^{IV}$
(=4H-Furo[2,3-c][1$\lambda^4$,2,6]thiadiazine)
Derivatives:
(C: 4) =O; (C: 5 and 6) $-CH_3$ ........ **4** 70/1
(C: 4) =O; (C: 5 and 6) $-C_6H_5$ ........ **4** 70/1

1H-Furo[2,3-c][1,2,6]thiadiazine, 2,2,3,4-tetrahydro-
(=1H-Furo[2,3-c][1$\lambda^4$,2,6]thiadiazine, 3,4-dihydro-)
Derivative:
(S: 2) =O; (C: 4) =O; (C: 5 and 6) $-CH_3$ ........ **4** 82/3

**$SN_2C_3$–$NC_4$ ($SN_3C_5$)**
{$S^{IV}N_2C_3$–$NC_4$}

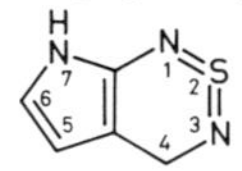

Pyrrolo[2,3-c][1,2,6]thiadiazine-2-$S^{IV}$, 4,7-dihydro-
(= Pyrrolo[2,3-c][1λ$^4$,2,6]thiadiazine, 4,7-dihydro-)
Derivatives:
(N: 7) –$CH_3$; (C: 5 and 6) –$CH_3$ . . . . . . **4** 71
(N: 7) –$CH_3$; (C: 5 and 6) –$C_6H_5$ . . . . . . **4** 71/2
(N: 7) –$C_3H_7$-i; (C: 5 and 6) –$C_6H_5$ . . . . . . **4** 71/2
(N: 7) –$C_6H_{13}$; (C: 5 and 6) –$C_6H_5$ . . . . . . **4** 71/2
(N: 7) –$C_6H_{11}$-c; (C: 5 and 6) –$C_6H_5$ . . . . . . **4** 71/2
(N: 7) –$CH_2C_6H_5$; (C: 5 and 6) –$C_6H_5$ . . . . . . **4** 71/2
(N: 7) –$CH_2CH_2C_6H_5$; (C: 5 and 6) –$C_6H_5$ . . . . . . **4** 71/2
(N: 7) –$(CH_2)_3N(CH_3)_2 \cdot HCl$; (C: 5 and 6) –$C_6H_5$ . . . . . . **4** 71/2
(N: 7) –$C_6H_5$; (C: 5 and 6) –$C_6H_5$ . . . . . . **4** 71/2

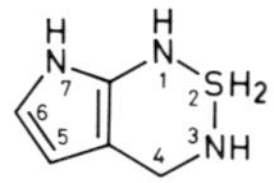

Pyrrolo[2,3-c][1,2,6]thiadiazine, 1,2,2,3,4,7-hexahydro-
(= Pyrrolo[2,3-c][1λ$^4$,2,6]thiadiazine, 1,3,4,7-tetrahydro-)
Derivatives:
(S: 2) =O; (N: 7) –$C_6H_{13}$; (C: 4) =O; (C: 5 and 6) –$CH_3$ . . . . . . **4** 82/3
(S: 2) =O; (N: 7) –$C_6H_{11}$-c; (C: 4) =O; (C: 5 and 6) –$CH_3$ . . . . . . **4** 82/3
(S: 2) =O; (N: 7) –$C_6H_4Cl$-4; (C: 4) =O; (C: 5 and 6) –$CH_3$ . . . . . . **4** 82/3

**$SN_2C_3$–$SC_4$ ($S_2N_2C_5$)**
{$S^{IV}N_2C_3$–$S^{II}C_4$}

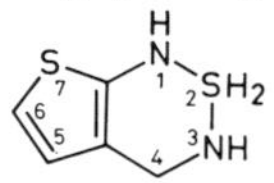

1H-Thieno[2,3-c][1,2,6]thiadiazine, 2,2,3,4-tetrahydro-
(= 1H-Thieno[2,3-c][1λ$^4$,2,6]thiadiazine, 3,4-dihydro-)
Derivative:
(S: 2) =O; (C: 4) =O; (C: 5 and 6) –$CH_3$ . . . . . . **4** 82/3

**$SN_3C_2$–$N_2C_3$ ($SN_4C_4$)**
{$S^{IV}N_3C_2$–$N_2C_3$}

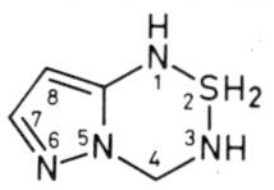

2H-Pyrazolo[5,1-c][1,2,4,6]thiatriazine, 1,2,3,4-tetrahydro-
(= 2H-Pyrazolo[5,1-c][1λ$^4$,2,4,6]thiatriazine, 1,3,4-trihydro-)
Derivatives:
(S: 2) =$NSO_2C_6H_4CH_3$-4; (C: 4) =O; (C: 7) –$C_6H_5$ . . . . . . **4** 37/8
(S: 2) =$NSO_2C_6H_4CH_3$-4; (C: 4) =S; (C: 7) –$C_6H_5$ . . . . . . **4** 37/8
(S: 2) =$NSO_2C_6H_4CH_3$-4; (N: 3) –$C_3H_7$; (C: 4) =S; (C: 7) –$C_6H_5$ . . . . . . **4** 37/8
(S: 2) =$NSO_2C_6H_4CH_3$-4; (N: 3) –$C_6H_5$; (C: 4) =O; (C: 7) –$C_6H_5$ . . . . . . **4** 37/8

**5,7**
**$SN_3C–NC_6$ ($SN_3C_6$)**
{$S^{IV}N_3C–NC_6$}

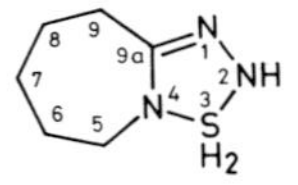

[1,2,3,5]Thiatriazolo[5,4-a]azepine, 2,3,3,5,6,7,8,9-octahydro-
(= [1λ[4],2,3,5]Thiatriazolo[5,4-a]azepine, 2,5,6,7,8,9-hexahydro-)
Derivatives:
(S: 3) =O; (N: 2) –$COC_6H_5$ . . . . . . **3** 87
(S: 3) =O; (N: 2) –$COC_6H_4Cl$-4 . . . . . . **3** 87
(S: 3) =O; (N: 2) –$COC_6H_4OCH_3$-4 . . . . . . **3** 87
(S: 3) =O; (N: 2) –CO(4-$C_5H_4N$) (4-$C_5H_4N$ = 4-Pyridinyl) . . . . . . **3** 87

**5,7**
(7,5)
**$S_2N_3C_2–S_2NC_2$ ($S_4N_4C_2$)**
{$S^{IV}S^{II}N_3C_2–S^{IV}S^{II}NC_2$}

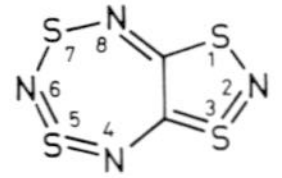

[1,3,2]Dithiazolo[4,5-e][1,3,2,4,7]dithiatriazepine-3,5-$S^{IV}$ . . . . . . **4** 122/4
(= [1,3λ[4],2]Dithiazolo[4,5-e][1,3λ[4],2,4,7]dithiatriazepine)

**5,13**
**$SN_2CO–NC_{12}$ ($SN_2C_{12}O$)**
{$S^{IV}N_2CO–NC_{12}$}

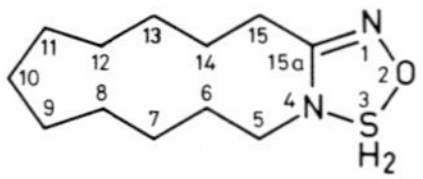

5H-[1,2,3,5]Oxathiadiazolo[3,4-a]azacyclotridecine,
3,3,6,7,8,9,10,11,12,13,14,15-dodecahydro-
(= 5H-[1,2λ[4],3,5]Oxathiadiazolo[3,4-a]azacyclotridecine,
6,7,8,9,10,11,12,13,14,15-decahydro-)
Derivative:
(S: 3) =O . . . . . . **3** 320/1

**6,6**
**$SNC_3O–C_6$ ($SNC_7O$)**
{$S^{IV}NC_3O-C_6$}

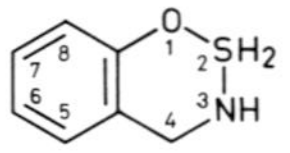

1,2,3-Benzoxathiazine, 2,2,3,4-tetrahydro-
(= 1,2λ[4],3-Benzoxathiazine, 3,4-dihydro-)
Derivatives:
(S: 2) =O; (C: 4) =O . . . . . . **4** 95/6, 99/100
(S: 2) =O; (N: 3) –$CH_3$ . . . . . . **4** 86/9
(S: 2) =O; (N: 3) –$CH_3$; (C: 4) –$CH_3$ . . . . . . **4** 86/9
(S: 2) =O; (N: 3) –$C_3H_7$-i . . . . . . **4** 86/9
(S: 2) =O; (N: 3) –$C_3H_7$-i; (C: 4) –$CH_3$ . . . . . . **4** 86/9

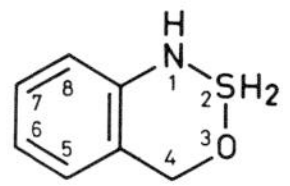

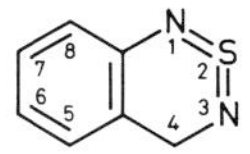

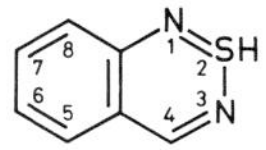

**6,6**
**$SN_2C_3$–$C_6$ ($SN_2C_7$)**
{$S^{IV}N_2C_3$–$C_6$}

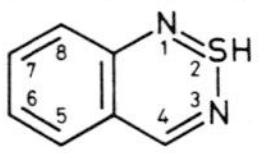

2H-2,1,3-Benzothiadiazine
(= 2λ[4],1,3-Benzothiadiazine)
Derivatives (cont.)
(C: 4) –$OC_2H_5$; (C: 6 and 8) –Cl
Ion(1+) .......... **4** 65
(C: 4) –$OC_2H_5$; (C: 6) –$NO_2$
Ion(1+) .......... **4** 65

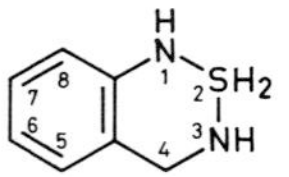

1H-2,1,3-Benzothiadiazine, 2,2,3,4-tetrahydro-
(=1H-2λ[4],1,3-Benzothiadiazine, 3,4-dihydro-)
Derivatives:
(S: 2) =O; (C: 4) =O .......... **4** 78/80
(S: 2) =O; (C: 4) =S .......... **4** 81
(S: 2) =O; (N: 1) –$CH_3$; (N: 3) –$C_2H_5$; (C: 4) =O; (C: 6) –Cl .......... **4** 78/80
(S: 2) =O; (N: 1) –$CH_3$; (N: 3) –$C_2H_5$; (C: 4) =O; (C: 6) –$SO_2NHC_2H_5$ .......... **4** 78/80
(S: 2) =O; (N: 1) –$CH_3$; (N: 3) –$C_5H_9$-c; (C: 4) =O; (C: 6) –Cl .......... **4** 78/80
(S: 2) =O; (N: 1) –$CH_3$; (N: 3) –$CH_2CH_2Cl$; (C: 4) =O; (C: 6) –Cl .......... **4** 78/80
(S: 2) =O; (N: 1) –$CH_3$; (N: 3) –$CH_2CH_2Cl$; (C: 4) =O; (C: 6) –$NO_2$ .......... **4** 78/80
(S: 2) =O; (N: 1) –$CH_3$; (N: 3) –$CH_2CH_2OC_2H_5$; (C: 4) =O; (C: 6) –Cl .......... **4** 78/80
(S: 2) =O; (N: 1) –$CH_3$; (N: 3) –$CH_2C_6H_4Cl$-2; (C: 4) =O .......... **4** 78/80
(S: 2) =O; (N: 1) –$CH_3$; (N: 3) –$CH_2C_6H_4Cl$-2; (C: 4) =O; (C: 6) –Cl .......... **4** 78/80
(S: 2) =O; (N: 1) –$CH_3$; (N: 3) –$CH_2CH_2C_6H_5$; (C: 4) =O; (C: 6) –Cl .......... **4** 78/80
(S: 2) =O; (N: 1 and 3) –$C_2H_5$; (C: 4) =O; (C: 6) –Cl .......... **4** 78/80
(S: 2) =O; (N: 3) –$C_6H_5$; (C: 4) –$C_6H_5$ .......... **4** 78

**$SN_2C_3$–$NC_5$ ($SN_2C_7$)**
{$S^{IV}N_2C_3$–$NC_5$}

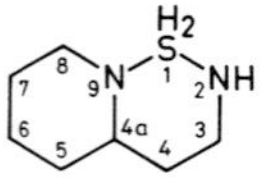

Pyrido[1,2-b][1,2,6]thiadiazine, 1,1,2,3,4,4a,5,6,7,8-decahydro-
(= Pyrido[1,2-b][1λ[4],2,6]thiadiazine, 2,3,4,4a,5,6,7,8-octahydro-)
Derivative:
(S: 1) =O; (N: 2) –$CH_2C_6H_5$ .......... **4** 80

**$SN_3C_2$–$C_6$ ($SN_3C_6$)**
{$S^{IV}N_3C_2$–$C_6$}

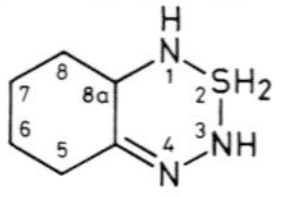

1H-2,1,3,4-Benzothiatriazine, 2,2,3,5,6,7,8,8a-octahydro-
(=1H-2λ[4],1,3,4-Benzothiatriazine, 3,5,6,7,8,8a-hexahydro-)
Derivatives:
(S: 2) =O; (N: 1) –$C_6H_5$; (N: 3) –$CH_3$ .......... **4** 14, 16/7
(S: 2) =O; (N: 1) –$COC_6H_5$; (N: 3) –$CH_3$ .......... **4** 14, 16/7

**6,6**
**$S_3N_3$–$S_2N_3C$ ($S_3N_5C$)**
{$S_3^{IV}N_3$–$S_2^{IV}N_3C$}

1,3,5-Trithia(1,3,5-$S^{IV}$)-2,4,6,8,9-pentaazabicyclo[3.3.1]nona-1,3,5(9)-triene
(=1λ⁴,3λ⁴,5λ⁴-Trithia-2,4,6,8,9-pentaazabicyclo[3.3.1]nona-1,3,5(9)-triene)

Derivatives

(N: 6 and 8) $-CH_3$; (C: 7) =O

Ion(1+) (cont.)

| | |
|---|---|
| Salts: $AsF_6^-$ | **4** 150/3 |
| $Cl^-$ | **4** 150/3 |
| $SbCl_6^-$ | **4** 150/2 |
| $SnCl_5^-$ | **4** 150/2 |
| $TiCl_5^-$ | **4** 150/2 |
| (N: 6) $-CH_3$; (N: 8) $-C_6H_5$; (C: 7) =O | |
| Ion(1+) | **4** 150 |
| Salts: $Cl^-$ | **4** 150/3 |
| $SbCl_6^-$ | **4** 150/2 |
| (S: 3) $-N(CH_3)_2$; (N: 6 and 8) $-CH_3$; (C: 7) =O | **4** 153/4 |
| (S: 3) $-N(C_2H_5)_2$; (N: 6 and 8) $-CH_3$; (C: 7) =O | **4** 153/4 |

**$S_3N_3$–$S_2N_3P$ ($S_3N_5P$)**
{$S_3^{IV}N_3$–$S_2^{IV}N_3P^V$}

1,3,5-Trithia(1,3,5-$S^{IV}$)-2,4,6,8,9-pentaaza-7-phosphabicyclo[3.3.1]-nona-1(9),2,3,5,7-pentaene, 7,7-dihydro-
(=1λ⁴,3λ⁴,5λ⁴-Trithia-2,4,6,8,9-pentaaza-7λ⁵-phosphabicyclo[3.3.1]-nona-1(9),2,3,5,7-pentaene)

Derivatives:

| | |
|---|---|
| (P: 7) $-CH_3$, $-CH_3$ | **3** 39/41 |
| (P: 7) $-C_6H_5$, $-C_6H_5$ | **3** 39/41 |
| (P: 7) –F, $-C_6H_5$ | **3** 39/41 |
| (P: 7) –F, –F | **3** 39/40 |

**6,7**
(7,6)
**$S_3N_2C_2$–$C_6$ ($S_3N_2C_6$)**
{$S^{IV}S_2^{II}N_2C_2$–$C_6$}

1,3,5,2,4-Benzotrithiadiazepine-3-$S^{IV}$ .......... **4** 115/6, 118/9
(=1,3λ⁴,5,2,4-Benzotrithiadiazepine)

**6,8**
**$SN_3P_2$–$N_4P_4$($SN_6P_4$)**
$\{S^{IV}N_3P^{V}_2–N_4P^{V}_4\}$

9-Thia(9-$S^{IV}$)-2,4,6,8,10,11-hexaaza-1,3,5,7-tetraphosphabicyclo[5.3.1]undeca-1,3,5,7(11),8,9-hexaene, 1,3,3,5,5,7-hexahydro-
(=9$\lambda^4$-Thia-2,4,6,8,10,11-hexaaza-1$\lambda^5$,3$\lambda^5$,5$\lambda^5$,7$\lambda^5$-tetraphosphabicyclo[5.3.1]-undeca-1,3,5,7(11),8,9-hexaene)
Derivative:
(P: 1 and 7) –F; (P: 3 and 5) –F, –F ........ **3** 41/2

**7,7**
**$S_2N_3P_2$–$S_2N_3P_2$ ($S_2N_6P_4$)**
$\{S^{IV}_2N_3P^{V}_2–S^{IV}_2N_3P^{V}_2\}$

[1,2,3,5,7,4,6]Dithiatriazadiphosphepino[1,2-a][1,2,3,5,7,4,6]dithiatriaza-diphosphepine-6,12-$S^{IV}$, 2,2,4,4,8,8,10,10-octahydro-
(=[1$\lambda^4$,2$\lambda^4$,3,5,7,4$\lambda^5$,6$\lambda^5$]Dithiatriazadiphosphepino[1,2-a][1$\lambda^4$,2$\lambda^4$,3,5,7,4$\lambda^5$,6$\lambda^5$]-dithiatriazadiphosphepine)
Derivative:
(P: 2,4,8, and 10) –$C_6H_5$, –$C_6H_5$ ........ **3** 43/6

**8,8**
**$S_2N_4Si_2$–$S_2N_4Si_2$ ($S_3N_6Si_2$)**
$\{S^{IV}_2N_4Si_2–S^{IV}_2N_4Si_2\}$

3,7,10-Trithia(3,7,10-$S^{IV}$)-2,4,6,8,9,11-hexaaza-1,5-disilabicyclo[3.3.3]undeca-2,3,6,7,9,10-hexaene
(=3$\lambda^4$,7$\lambda^4$,10$\lambda^4$-Trithia-2,4,6,8,9,11-hexaaza-1,5-disilabicyclo[3.3.3]-undeca-2,3,6,7,9,10-hexaene)
Derivative:
(Si: 1 and 5) –$CH_3$ ........ **3** 11/2

## Tricyclic Systems

**4,5,5**
(5,4,5)
**$SN_2Se_2$–$Se_4$–$SN_2Se_2$**
**($S_2N_4Se$)**
$\{S^{IV}N_2Se^{IV}_2–Se^{IV}_4–S^{IV}N_2Se^{IV}_2\}$

2H,7H-Tetraseleneto[1,2-c:3,4-c′]bis[1,3,4,2,5]thiadiselenadiazole-4,5,9,10-$Se^{IV}$
(=Tetraseleneto[1,2-c:3,4-c′]bis[1$\lambda^4$,3$\lambda^4$,4$\lambda^4$,2,5]thiadiselenadiazole)
Ion(2+) ........ **3** 2/3

**4,5,5**
(5,4,5)
**$SN_2Se_2$–$Se_4$–$SN_2Se_2$**
**($S_2N_4Se$)**
$\{S^{IV}N_2Se^{IV}_2–Se^{IV}_4–S^{IV}N_2Se^{IV}_2\}$

2H,7H-Tetraseleneto[1,2-c:3,4-c′]bis[1,3,4,2,5]thiadiselenadiazole-4,5,9,10-$Se^{IV}$
(= Tetraseleneto[1,2-c:3,4-c′]bis[1λ$^4$,3λ$^4$,4λ$^4$,2,5]thiadiselenadiazole)
Ion(2+) (cont.)
Salts: $AsF_6^-$ .......... **3** 2/4
$Br^-$ .......... **3** 3/4
$Cl^-$ .......... **3** 3/4
$SbF_6^-$ .......... **3** 2/4

**5,5,6**
(5,6,5)
**$SN_2C_2$–$C_6$–$SN_2C_2$ ($S_2N_4C_6$)**
$\{S^{IV}N_2C_2–C_6–S^{IV}N_2C_2\}$

Benzo[1,2-c:3,4-c′]bis[1,2,5]thiadiazole-2,7-$S^{IV}$ .......... **3** 222/3
(= Benzo[1,2-c:3,4-c′]bis[1λ$^4$,2,5]thiadiazole)
Derivative:
(C: 4) $-NO_2$ .......... **3** 222/3

**5,5,6**
**$SN_3C$–$N_2C_3$–$C_6$ ($SN_3C_8$)**
$\{S^{IV}N_3C–N_2C_3–C_6\}$

5H-[1,2,3,5]Thiatriazolo[2,3-a]indazole, 1,1,2,3-tetrahydro-
(= 5H-[1λ$^4$,2,3,5]Thiatriazolo[2,3-a]indazole, 2,3-dihydro-)
Derivative:
(S: 1) =O; (N: 2) $-CH_3$; (C: 3 and 5) =O .......... **3** 91

**5,5,8**
(8,5,5)
**$S_4N_2C_2$–$C_5$–$C_5$ ($S_4N_2C_7$)**
$\{S^{IV}S^{II}_3N_2C_2–C_5–C_5\}$

7,10-Methano-1,2,4,6,3,5-benzotetrathiadiazocine-4-$S^{IV}$, 6a,7,10,10a-tetrahydro- .......... **4** 132/3
(= 7,10-Methano-1,2,4λ$^4$,6,3,5-benzotetrathiadiazocine, 6a,7,10,10a-tetrahydro-)

**5,6,6**
**$SN_3C$–$NC_5$–$C_6$ ($SN_3C_9$)**
$\{S^{IV}N_3C–NC_5–C_6\}$

2H-[1,2,3,5]Thiatriazolo[4,5-a]isoquinoline, 3,3-dihydro-
(= 2H-[1λ$^4$,2,3,5]Thiatriazolo[4,5-a]isoquinoline)
Derivatives:
(S: 3) =O; (N: 2) $-(1-C_9H_6N)$ ($1-C_9H_6N$ = 1-Isoquinolinyl) .......... **3** 89

(S: 3) =O; (N: 2) –$COCH_3$ .......... **3** 88/9
(S: 3) =O; (N: 2) –$COCH_3$; (C: 5) –Cl .......... **3** 88/9
(S: 3) =O; (N: 2) –$COC{\equiv}CC_6H_5$ .......... **3** 88/9
(S: 3) =O; (N: 2) –$COC_6H_5$ .......... **3** 88/9
(S: 3) =O; (N: 2) –$COC_6H_4NO_2$-2 .......... **3** 88/9
(S: 3) =O; (N: 2) –$COC_6H_4NO_2$-2; (C: 5) –Cl .......... **3** 88/9

**5,6,6**

(6,6,5)

**$SNC_3O–C_6–C_3O_2$ ($SNC_8O_3$)**
{$S^{IV}NC_3O–C_6–C_3O_2$}

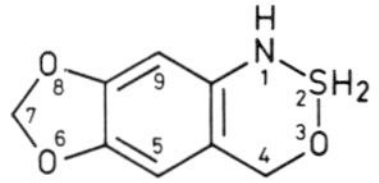

[1,3]Dioxolo[4,5-g]-3,2,1-benzoxathiazine, 1,2,2,4-tetrahydro-
(= [1,3]Dioxolo[4,5-g]-3,2λ$^4$,1-benzoxathiazine, 1,4-dihydro-)
Derivative:
(S: 2) =O; (C: 4) =O .......... **4** 100/1, 103

(6,5,6)

**$SN_2C_3–N_2C_3–NC_5$ ($SN_4C_8$)**
{$S^{IV}N_2C_3–N_2C_3–NC_5$}

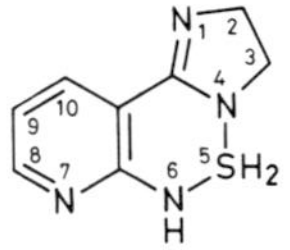

2H-Imidazo[1,2-b]pyrido[3,2-d][1,2,6]thiadiazine, 3,5,5,6-tetrahydro-
(= 2H-Imidazo[1,2-b]pyrido[3,2-d][1λ$^4$,2,6]thiadiazine, 3,6-dihydro-)
Derivative:
(S: 5) =O; (N: 6) –$CH_3$ .......... **4** 83/4

**$SN_2C_3–SC_4–C_6$ ($S_2N_2C_9$)**
{$S^{IV}N_2C_3–S^{II}C_4–C_6$}

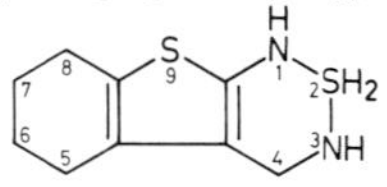

1H-[1]Benzothieno[2,3-c][1,2,6]thiadiazine, 2,2,3,4,5,6,7,8-octahydro-
(= 1H-[1]Benzothieno[2,3-c][1λ$^4$,2,6]thiadiazine, 3,4,5,6,7,8-hexahydro-)
Derivative:
(S: 2) =O; (C: 4) =O .......... **4** 82/3

**$SN_3C_2–N_2C_3–C_6$ ($SN_4C_8$)**
{$S^{IV}N_3C_2–N_2C_3–C_6$}

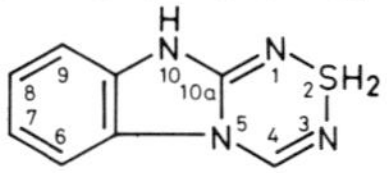

10H(or 1H)-[1,2,4,6]Thiatriazino[4,3-a]benzimidazole, 2,2-dihydro-
(= 10H(or 1H)-[1λ$^4$,2,4,6]Thiatriazino[4,3-a]benzimidazole)
Derivatives:
(S: 2) =$NSO_2C_6H_5$; (C: 4) –$OC_6H_5$ .......... **4** 36
(S: 2) =$NSO_2C_6H_4CH_3$-4; (C: 4) –$OCH_2CCl_3$ .......... **4** 36/7
(S: 2) =$NSO_2C_6H_4CH_3$-4; (C: 4) –$OC_6H_5$ .......... **4** 36/7
(S: 2) =$NSO_2C_6H_4Cl$-4; (C: 4) –$OC_6H_5$ .......... **4** 36
(S: 2) =O; (C: 4) –$OC_6H_5$ .......... **4** 29/30

**5,6,6**
(6,5,6)
**$SN_3C_2$–$N_2C_3$–$C_6$ ($SN_4C_8$)**
{$S^{IV}N_3C_2$–$N_2C_3$–$C_6$}

4H-[1,2,4,6]Thiatriazino[2,3-a]benzimidazole, 1,1-dihydro-
(= 4H-[1λ⁴,2,4,6]Thiatriazino[2,3-a]benzimidazole)
Derivatives:
(S: 1) =$NSO_2C_6H_5$; (C: 3) –$NH_2$ ........ **4** 38/9
(S: 1) =$NSO_2C_6H_4CH_3$-4; (C: 3) –$NH_2$ ........ **4** 38/9
(S: 1) =$NSO_2C_6H_4Cl$-4; (C: 3) –$NH_2$ ........ **4** 38/9
(S: 1) =O; (C: 3) –$NH_2$ ........ **4** 30/1

**5,6,7**
(7,6,5)
**$S_2N_3C_2$–$C_6$–$S_2NC_2$**
**($S_4N_4C_6$)**
{$S^{IV}S^{II}N_3C_2$–$C_6$–$S^{II}_2NC_2$}

[1,2,3]Dithiazolo[5,4-g]-2,4,1,3,5-benzodithiatriazepine-2-$S^{IV}$ ........ **4** 119/21
(= [1,2,3]Dithiazolo[5,4-g]-2λ⁴,4,1,3,5-benzodithiatriazepine)

**6,6,6**
**$SN_2C_3$–$C_6$–$C_6$ ($SN_2C_{10}$)**
{$S^{IV}N_2C_3$–$C_6$–$C_6$}

Naphtho[1,8-cd][1,2,6]thiadiazine-2-$S^{IV}$ ........ **4** 66/7
(= Naphtho[1,8-cd][1λ⁴,2,6]thiadiazine)
Radical ion(1–) ........ **4** 68

1H,3H-Naphtho[1,8-cd][1,2,6]thiadiazine, 2,2-dihydro-
(= 1H,3H-Naphtho[1,8-cd][1λ⁴,2,6]thiadiazine)
Derivative:
(S: 2) =O ........ **4** 81/2

**$SN_2C_3$–$NC_5$–$C_6$ ($SN_2C_{10}$)**
{$S^{IV}N_2C_3$–$NC_5$–$C_6$}

7H-Pyrido[3,2,1-ij][2,1,3]benzothiadiazine, 1,1,2,3,8,9-hexahydro-
(= 7H-Pyrido[3,2,1-ij][2λ⁴,1,3]benzothiadiazine, 2,3,8,9-tetrahydro-)
Derivative:
(S: 1) =O; (N: 2) –$CH_3$; (C: 3) =O ........ **4** 82

**$S_2N_3C–S_4N_2–S_2N_3C$**
**($S_4N_6C_2$)**
{$S_2^{IV}N_3C–S_4^{IV}N_2–S_2^{IV}N_3C$}

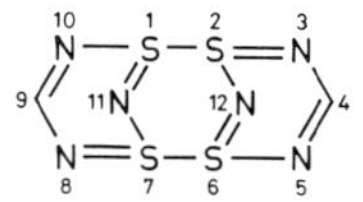

1,2,6,7-Tetrathia(1,2,6,7-$S^{IV}$)-3,5,8,10,11,12-hexaazatricyclo[5.3.1.$1^{2,6}$]-dodeca-1(11),2,4,6(12),7,9-hexaene ........ **4** 2/3
(=1$\lambda^4$,2$\lambda^4$,6$\lambda^4$,7$\lambda^4$-Tetrathia-3,5,8,10,11,12-hexaazatricyclo[5.3.1.$1^{2,6}$]-dodeca-1(11),2,4,6(12)7,9-hexaene)
Derivative:
(C: 4 and 9) $–C_6H_5$ ........ **4** 2/4

**$S_4N_2–S_2N_3C–S_2N_3C$**
**($S_4N_6C_2$)**
{$S_4^{IV}N_2–S_2^{IV}N_3C–S_2^{IV}N_3C$}
see
*$S_2N_3C–S_4N_2–S_2N_3C$($S_4N_6C_2$)*
*{$S_2^{IV}N_3C–S_4^{IV}N_2–S_2^{IV}N_3C$}*

## Tetracyclic Systems

**5,5,5,6**
(5,6,5,5)
**$SN_2C_2–C_6–SN_2C_2–SN_2C_2$**
**($S_3N_6C_6$)**
{$S^{IV}N_2C_2$-$C_6$-$S^{IV}N_2C_2$–$S^{IV}N_2C_2$}

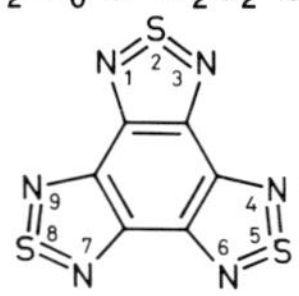

Benzo[1,2-c:3,4-c′:5,6-c″]tris[1,2,5]thiadiazole-2,5,8-$S^{IV}$ ........ **3** 223/4
(=Benzo[1,2-c:3,4-c′:5,6-c″]tris[1$\lambda^4$,2,5]thiadiazole)

**6,6,6,6**
**$SN_2C_3–C_6–C_6–SN_2C_3$**
**($S_2N_4C_{10}$)**
{$S^{IV}N_2C_3–C_6–C_6–S^{IV}N_2C_3$}

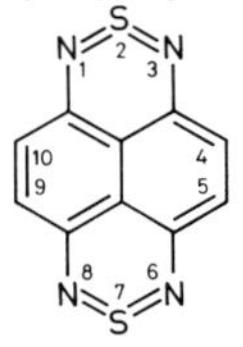

Naphtho[1,8-cd:4,5-c′d′]bis[1,2,6]thiadiazine-2,7-$S^{IV}$ ........ **4** 68/70
(=Naphtho[1,8-cd:4,5-c′d′]bis[1$\lambda^4$,2,6]thiadiazine)

**6,6,7,7**
**$SN_2C_3$–$SN_2C_3$–$NC_6$–$NC_6$**
**($SN_2C_{14}$)**
{$S^{IV}N_2C_3$–$S^{IV}N_2C_3$–$NC_6$–$NC_6$}

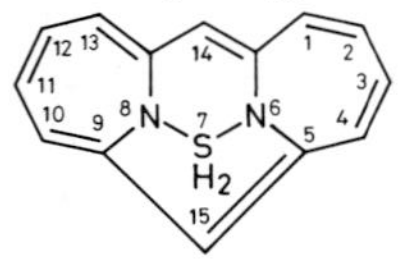

9,5-Metheno-5H-[1,2,6]thiadiazino[2,3-a:6,5-a′]bisazepine, 7,7-dihydro-
(=9,5-Metheno-5H-[1λ$^4$,2,6]thiadiazino[2,3-a:6,5-a′]bisazepine)
Derivative:
(S: 7) =O ........ **4** 154/5

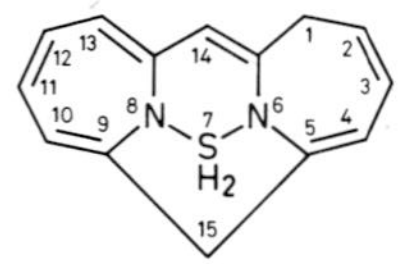

5,9-Methano-1H(or 14H)-[1,2,6]thiadiazino[2,3-a:6,5-a′]bisazepine, 7,7-dihydro-
(=5,9-Methano-1H(or 14H)-[1λ$^4$,2,6]thiadiazino[2,3-a:6,5-a′]bisazepine)
Derivative:
(S: 7) =O ........ **4** 154/5

## Hexacyclic System

**3,3,6,6,6,6**
(6,6,3,6,3,6)
**$SN_2C_3$–$SN_2C_3$–$NC_2$–$C_6$–$NC_2$–$C_6$ ($SN_2C_{14}$)**
{$S^{IV}N_2C_3$–$S^{IV}N_2C_3$–$NC_2$–$C_6$–$NC_2$–$C_6$}

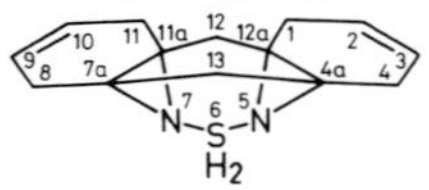

4a,7a-Methano-4aH,7aH,12H-bisbenz[2,3]azirino[1,2-b:2′,1′-e][1,2,6]thiadiazine, 1,4,6,6,8,11-hexahydro-
(=4a,7a-Methano-4aH,7aH,12H-bisbenz[2,3]azirino[1,2-b:2′,1′-e][1λ$^4$,2,6]thiadiazine, 1,4,8,11-tetrahydro-)
Derivative:
(S: 6) =O ........ **4** 154/5

## Table of Conversion Factors

Following the notation in Landolt-Börnstein [7], values which have been fixed by convention are indicated by a bold-face last digit. The conversion factor between calorie and Joule that is given here is based on the thermochemical calorie, $cal_{th\,ch}$, and is defined as 4.184**0** J/cal. However, for the conversion of the "Internationale Tafelkalorie", $cal_{IT}$, into Joule, the factor 4.1868 J/cal is to be used [1, p. 147]. For the conversion factor for the British thermal unit, the Steam Table Btu, $BTU_{ST}$, is used [1, p. 95].

| **Force** | N | dyn | kp |
|---|---|---|---|
| 1 N (Newton) | 1 | $10^5$ | 0.1019716 |
| 1 dyn | $10^{-5}$ | 1 | $1.019716 \times 10^{-6}$ |
| 1 kp | **9.80665** | **9.80665** $\times 10^5$ | 1 |

| **Pressure** | Pa | bar | kp/m² | at | atm | Torr | lb/in² |
|---|---|---|---|---|---|---|---|
| 1 Pa (Pascal) = 1 N/m² | 1 | $10^{-5}$ | $1.019716 \times 10^{-1}$ | $1.019716 \times 10^{-5}$ | $0.986923 \times 10^{-5}$ | $0.750062 \times 10^{-2}$ | $145.0378 \times 10^{-6}$ |
| 1 bar = $10^6$ dyn/cm² | $10^5$ | 1 | $10.19716 \times 10^3$ | 1.019716 | 0.986923 | 750.062 | 14.50378 |
| 1 kp/m² = 1 mm $H_2O$ | **9.80665** | **0.980665** $\times 10^{-4}$ | 1 | $10^{-4}$ | $0.967841 \times 10^{-4}$ | $0.735559 \times 10^{-1}$ | $1.422335 \times 10^{-3}$ |
| 1 at = 1 kp/cm² | **0.980665** $\times 10^5$ | **0.980665** | $10^4$ | 1 | 0.967841 | 735.559 | 14.22335 |
| 1 atm = 76**0** Torr | 1.0132**5** $\times 10^5$ | 1.0132**5** | $1.033227 \times 10^4$ | 1.033227 | 1 | 760 | 14.69595 |
| 1 Torr = 1 mm Hg | 133.3224 | $1.333224 \times 10^{-3}$ | 13.59510 | $1.359510 \times 10^{-3}$ | $1.315789 \times 10^{-3}$ | 1 | $19.33678 \times 10^{-3}$ |
| 1 lb/in² = 1 psi | $6.89476 \times 10^3$ | $68.9476 \times 10^{-3}$ | 703.069 | $70.3069 \times 10^{-3}$ | $68.0460 \times 10^{-3}$ | 51.7149 | 1 |

| **Work, Energy, Heat** | J | kWh | kcal | Btu | MeV |
|---|---|---|---|---|---|
| 1 J (Joule) = 1 Ws = 1 Nm = $10^7$ erg | 1 | $2.778 \times 10^{-7}$ | $2.39006 \times 10^{-4}$ | $9.4781 \times 10^{-4}$ | $6.242 \times 10^{12}$ |
| 1 kWh | $3.\mathbf{6} \times 10^6$ | 1 | 860.4 | 3412.14 | $2.247 \times 10^{19}$ |
| 1 kcal | 4184.**0** | $1.1622 \times 10^{-3}$ | 1 | 3.96566 | $2.6117 \times 10^{16}$ |
| 1 Btu (British thermal unit) | 1055.06 | $2.93071 \times 10^{-4}$ | 0.25164 | 1 | $6.5858 \times 10^{15}$ |
| 1 MeV | $1.602 \times 10^{-13}$ | $4.450 \times 10^{-20}$ | $3.8289 \times 10^{-17}$ | $1.51840 \times 10^{-16}$ | 1 |

1 eV ≙ 23.0578 kcal/mol = 96.473 kJ/mol

| **Power** | kW | PS | kp·m/s | kcal/s |
|---|---|---|---|---|
| 1 kW = $10^{10}$ erg/s | 1 | 1.35962 | 101.972 | 0.239006 |
| 1 PS | 0.73550 | 1 | **75** | 0.17579 |
| 1 kp·m/s | $9.8066\mathbf{5} \times 10^{-3}$ | 0.01333 | 1 | $2.34384 \times 10^{-3}$ |
| 1 kcal/s | 4.184**0** | 5.6886 | 426.650 | 1 |

References:


[1] A. Sacklowski, Die neuen SI-Einheiten, Goldmann, München 1979. (Conversion tables in an appendix.)

[2] International Union of Pure and Applied Chemistry, Manual of Symbols and Terminology for Physicochemical Quantities and Units, Pergamon, London 1979; Pure Appl. Chem. **51** [1979] 1/41.

[3] The International System of Units (SI), National Bureau of Standards Spec. Publ. 330 [1972].

[4] H. Ebert, Physikalisches Taschenbuch, 5th Ed., Vieweg, Wiesbaden 1976.

[5] Kraftwerk Union Information, Technical and Economic Data on Power Engineering, Mülheim/Ruhr 1978.

[6] E. Padelt, H. Laporte, Einheiten und Größenarten der Naturwissenschaften, 3rd Ed., VEB Fachbuchverlag, Leipzig 1976.

[7] Landolt-Börnstein, 6th Ed., Vol. II, Pt. 1, 1971, pp. 1/14.

[8] ISO Standards Handbook 2, Units of Measurement, 2nd Ed., Geneva 1982.